Lightweight Electric/Hybrid Vehicle Design

Lightweight Electric/ Hybrid Vehicle Design

Ron Hodkinson and John Fenton

OXFORD AUCKLAND BOSTON JOHANNESBURG MELBOURNE NEW DELHI

Butterworth-Heinemann
Linacre House, Jordan Hill, Oxford OX2 8DP
225 Wildwood Avenue, Woburn, MA 01801-2041
A division of Reed Educational and Professional Publishing Ltd

A member of the Reed Elsevier plc group

First published 2001

British Library Cataloguing in Publication Data
A catalogue record for this book is available from the British Library

Library of Congress Cataloguing in Publication Data
A catalogue record for this book is available from the Library of Congress

ISBN 0 7506 5092 3

Printed and bound in Malta by Gutenberg Press Ltd

Contents

Preface

The stage is now reached when the transition from low-volume to high-volume manufacture of fuel cells is imminent and after an intense period of value engineering, suppliers are moving towards affordable stacks for automotive propulsion purposes. Since this book went to press, the automotive application of fuel cells for pilot-production vehicles has proceeded apace, with Daewoo, as an example, investing $5.9 million in a fuel-cell powered vehicle based on the Rezzo minivan, for which it is developing a methanol reforming system. Honda has also made an important advance with version 3 of its FCX fuel-cell vehicle, using a Ballard cell-stack and an ultracapacitor to boost acceleration. Its electric motor now weighs 25% less and develops 25% more power and start-up time has been reduced from 10 minutes to 10 seconds. Ballard have introduced the Mk900 fuel cell now developing 75 kW (50% up on the preceding model). Weight has decreased and power density increased, each by 30%, while size has dropped by 50%. The factory is to produce this stack in much higher volumes than its predecessor. While GM are following the environmentally-unfriendly route of reformed gasoline for obtaining hydrogen fuel, Daimler Chrysler are plumping for the methanol route, with the future option of fuel production from renewables; they are now heading for a market entry with this technology, according to press reports.

A recent DaimlerChrysler press release describes the latest NECAR, with new Ballard Stack, which is described in its earlier Phase 4 form in Chapter 5, pp. 139–140. NECAR 5 has now become a methanol-powered fuel cell vehicle suitable for normal practical use. The environmentally friendly vehicle reaches speeds of more than 150 kilometres per hour and the entire fuel cell drive system – including the methanol reformer – has been installed in the underbody of a Mercedes-Benz A-Class for the very first time. The vehicle therefore provides about as much space as a conventional A-Class. Since the NECAR 3 phase, in 1997, the engineers have succeeded in reducing the size of the system by half and fitting it within the sandwich floor. At the same time, they have managed to reduce the weight of the system, and therefore the weight of the car, by about 300 kg. While NECAR 3 required two fuel cell stacks to generate 50 kW of electric power, a single stack now delivers 75 kW in NECAR 5. And although the NECAR 5 experimental vehicle is heavier than a conventional car, it utilizes energy from its fuel over 25% more efficiently. The development engineers have also used more economical materials, to lower production cost.

Methanol 'fuel' could be sold through a network of filling stations similar to the ones we use today. The exhaust emissions from 'methanolized' hydrogen fuel cell vehicles are very much lower than from even the best internal combustion engines. The use of methanol-powered fuel-cell vehicles could reduce carbon-dioxide emissions by about a third and smog-causing emissions to nearly zero. Methanol can either be produced as a renewable energy source from biomass or from natural gas, which is often burned off as a waste product of petroleum production and is still available in many regions around the world. To quote D-C board members, 'there have already been two oil crises; we are obligated to prevent a third one,' says Jürgen E. Schrempp, Chairman of the Board Of Management of DaimlerChrysler. 'The fuel-cell offers a realistic opportunity to supplement the 'petroleum monoculture' over the long term.' The company will invest about DM 2 billion (over $ 1 billion) to develop the new drive system from the first prototype to the point of mass production. In the past six years the company has already equipped and presented 16 passenger cars, vans and buses with fuel cell drives–more than the total of all its competitors worldwide. Professor Klaus-Dieter Vöhringer, member of the Board of Management with responsibility for research and technology, predicts the fuel cell will be introduced into vehicles in several stages 'In 2002, the company will deliver the first city buses with fuel cells, followed in 2004 by the first passenger cars.'

The electric-drive vehicle has thus moved out of the 'back-room' of automotive research into a 'design for production' phase and already hybrid drive systems (IC engine plus electric drive) have entered series production from major Japanese manufacturers. In the USA, General Motors has also made very substantial investments with the same objective. There is also very considerable interest throughout the world by smaller high-technology companies who can use their knowledge base to successfully enter the automotive market with innovative and specialist-application solutions. This last group will have much benefit from this book, which covers automotive structure, and system design for ultra-light vehicles that can extend the range of electric propulsion, as well as electric-drive technology and EV layouts for its main-stream educational readership.

NECAR5 fuel-cell driven car.

About the authors

Electro-technology author **Ron Hodkinson** is very actively involved in the current value engineering of automotive fuel-cell drive systems through his company Fuel Cell Control Ltd and is particularly well placed to provide the basic electro-technology half of this work. He obtained his first degree in electrical engineering (power and telecommunications) from the Barking campus, of what is now the University of East London, on a four-year sandwich course with Plessey. At the end of the company's TSR2 programme he moved on to Brentford Electric in Sussex where he was seconded on contract to CERN in Switzerland to work on particle-accelerator magnetic power supplies of up to 9 MW. He returned to England in 1972 to take a master's degree at Sussex University, after which he became Head of R&D at Brentford Electric and began his long career in electric drive system design, being early into the development of transistorised inverter drives. In 1984 the company changed ownership and discontinued electronics developments, leading Ron to set up his own company, Motopak, also developing inverter drives for high performance machine tools used in aircraft construction. By 1989 his company was to be merged with Coercive Ltd who were active in EV drives and by 1993 Coercive had acquired Nelco, to become the largest UK producer of EV drives. In 1995 the company joined the Polaron Group and Ron became Group Technical Director. For the next four years he became involved in both machine tool drives and fuel cell controls. In 1999 the group discontinued fuel-cell system developments and Ron was able to acquire premises at Polaron's Watford operation to set up his own family company Fuel Cell Control Ltd, of which he is managing director. He has been an active member of ISATA (International Society for Automotive Technology and Automation) presenting numerous papers there and to the annual meetings of the EVS (Electric Vehicle Seminar). He is also active in the Power Electronics and Control committees of the Institution of Electrical Engineers. Some of his major EV projects include the Rover Metro hybrid concept vehicle; IAD electric and hybrid vehicles; the SAIC fuel-cell bus operating in California and Zetec taxicabs and vans.

Co-author **John Fenton** is a technology journalist who has plotted the recent course in EV design and layout, including hybrid-drive vehicles, in the second half of the book, which also includes his chapters on structure and systems design from his earlier industrial experience. He is an engineering graduate of the Manchester University Science Faculty and became a member of the first year's intake of Graduate Apprentices at General Motors' UK Vauxhall subsidiary. He later worked as a chassis-systems layout draughtsman with the company before moving to automotive consultants ERA as a chassis-systems development engineer, helping to develop the innovative mobile tyre and suspension test rig devised by David Hodkin, and working on running-gear systems for the Project 378 car design project for BMC. With ERA's subsequent specialization on engine systems, as a result of the Solex acquisition, he joined the Transport Division of Unilever, working with the Technical Manager on the development of monocoque sandwich-construction refrigerated container bodies and bulk carriers for ground-nut meal and shortening-fat. He was sponsored by the company on the first postgraduate automotive engineering degree course at Cranfield where lightweight sandwich-construction monocoque vehicle bodies was his thesis subject. He changed course to technology-journalism after graduating and joined the newly founded journal *Automotive Design Engineering* (*ADE*) as its first technical editor, and subsequently editor. A decade later he became a senior lecturer on the newly founded undergraduate Vehicle Engineering degree course at what is now Hertfordshire University and helped to set up the design teaching courses in body-structure and chassis-systems. He returned to industry for a short period, as a technology communicator, first Product Affairs Manager for Leyland Truck and Bus, then technical copywriter and sales engineer (special vehicle operations). With the merging of *ADE* with the Institution of Mechanical Engineers *JAE* journal he had the opportunity to move back to publishing and subsequently edited the combined journal *Automotive Engineer*, for fifteen years, prior to its recent transformation into an international auto-industry magazine.

Introduction

0.1 Preface

This book differs from other automotive engineering texts in that it covers a technology that is still very much in the emerging stages, and will be particularly valuable for design courses, and projects, within engineering degree studies. Whereas other works cover established automotive disciplines, this book focuses on the design stages, still in process for electric vehicles, and thus draws on a somewhat tentative source of references rather than a list of the known major works in the subject. The choice of design theory is also somewhat selective, coming from the considerable volume of works the disciplines of which are combining to make the production electric vehicle possible.

0.1.1 BIBLIOGRAPHIC SOURCES

Electrical propulsion systems date back virtually to the time of Faraday and a substantial body of literature exists in the library of the Institution of Electrical Engineers from which it is safe only to consider a small amount in relation to current road vehicle developments. Similarly a considerable quantity of works are available on aerospace structural design which can be found in the library of the Royal Aeronautical Society, and on automotive systems developments within the library of the Institution of Mechanical Engineers. With the massive recent step-changes in capital investment, first in the build-up to battery-electric vehicle development, then in the switch to hybrid drive engineering, and finally the move to fuel-cell development – it would be dangerous to predict an established EV technology at this stage.

A good deal of further reading has been added to the bibliographies of references at the ends of each chapter. This is intended to be a source of publications that might help readers look for wider background, while examining the changes of direction that EV designers are making at this formative stage of the industry. The final chapter also lists publications which seem to be likely sources of design calculations pertinent in designing for minimum weight and has a table of nomenclature for the principal parameters, with corresponding symbol notation used in the design calculations within the text of the chapters.

0.1.2 CONTEXT AND STRUCTURE

The current period of EV development could be seen as dating from a decade or so before the publication of Scott Cronk's pivotal work published by the Society of Automotive Engineers in 1995, *Building the E-motive Industry*. As well as pulling together the various strings of earlier EV development, the book takes a very broad-brush view of the many different factors likely to affect the industry as it emerges. Readers seeking to keep abreast of developing trends in EV technology could do little better than to follow the bound volumes of proceedings on the subject which have appeared annually following the SAE Congresses in February/March, as well as studying the proceedings of the annual worldwide FISITA and EVS conferences. One of these factors, put forward by Cronk, is the need for a combination of electromotive technology with those which went into the USA Supercar programme, aimed at unusually low fuel consumption born out of low-drag and lightweight construction. This is the philosophy that the authors of *Lightweight Electric/Hybrid Vehicle Design* are trying to follow in a work which looks into the technologies in greater depth. The book is in two parts, dealing with (a) electromotive technology and (b) EV design packages, lightweight design/construction and running-gear performance.

Ron Hodkinson draws on long experience in electric traction systems in industrial vehicles and more recently into hybrid-drive cars and control systems for fuel-celled vehicles. His Part One contains the first four chapters on electric propulsion and storage systems and includes, within his last chapter, a contributed section by Roger Booth, an expert in fuel-cell development, alongside his own account of EV development history which puts into context the review material of the following chapters. In Part Two, John Fenton, in his first two chapters, uses his recent experience as a technology writer to review past and present EV design package trends, and in his second two chapters on body construction and body-structural/running-gear design, uses his earlier industrial experience in body and running-gear design, to try and raise interest in light-weighting and structural/functional performance evaluation.

0.2 Design theory and practice

For the automotive engineer with background experience of IC-engine prime-moving power sources, the electrical aspects associated with engine ignition, starting and powering auxiliary lighting and occupant comfort/convenience devices have often been the province of resident electrical engineering specialists within the automotive design office. With the electric vehicle (EV), usually associated with an energy source that is portable and electrochemical in nature, and tractive effort only supplied by prime-moving electric motor, the historic distinctions between mechanical and electrical engineering become blurred. One day the division of engineering into professional institutions and academic faculties defined by these distinctions will no doubt also be questioned. Older generation auto-engineers have much to gain from an understanding of electrotechnology and a revision of conventional attitudes towards automotive systems such as transmission, braking and steering which are moving towards electromagnetic power and electronic control, like the prime-moving power unit.

In terms of reducing vehicle weight, to gain greatest benefit in terms of range from electromotive power, there also needs to be some rethinking of traditional approaches. The conventional design approach of automotive engineers seems to involve an instinctive prioritizing of minimizing production costs, which will have been instilled into them over generations of Fordist mass-production. There is something in this 'value-engineering' approach which might sacrifice light weight in the interests of simplicity of assembly, or the paring down of piece price to the barest minimum. Aerospace designers perhaps have a different instinctive approach and think of lightweight and performance-efficiency first. Both automotive and aerospace design engineers now have the benefit of sophisticated finite-element structural analysis packages to help them trade off performance efficiency with minimum weight. In earlier times the automotive engineer probably relied on substantial 'factors of safety' in structural calculations, if indeed they were performed at all on body structures, which were invariably supported by stout chassis frames. This is not to mention the long development periods of track and road proving before vehicles reached the customer, which may have led engineers to be less conscious of the weight/performance trade-off in detail design. Individual parts could well be specified on the basis of subjective judgement, without the sobering discipline of the above trade-off analysis.

Not so, of course, for the early aeronautical design engineers whose prototypes either 'flew or fell out of the sky'. Aircraft structural designers effectively pioneered techniques of thin-walled structural analysis to try to predict as far as possible the structural performance of parts 'before they left the drawing board', and in so doing usually economized on any surplus mass. These structural analysis techniques gave early warning of buckling collapse and provided a means of idealization that allowed load paths to be traced. In the dramatic weight reduction programmes called for by the 'supercar' design requirements, to be discussed in Chapters 4 and 6, these attitudes to design could again have great value.

Design calculations, using techniques for tracing loads and determining deflections and stresses in structures, many of which derive from pioneering aeronautical structural techniques, are also recommended for giving design engineers a 'feel' for the structures at the concept stage. The design engineer can thus make crucial styling and packaging decisions without the risk of weakening the structure or causing undue weight gain. While familiar to civil and aeronautical engineering graduates these 'theory of structures' techniques are usually absent from courses in mechanical and electrical engineering, which may be confined to the 'mechanics of solids' in their structures teaching. For students undertaking design courses, or projects, within their engineering degree studies, these days the norm rather than the exception, the timing of the book's publication is within the useful period of intense decision making throughout the EV industry. It is thus valuable in focusing on the very broad range of other factors–economic, ergonomic, aesthetic and even political–which have to be examined alongside the engineering science ones, during the conceptual period of engineering design.

0.2.1 FARTHER-REACHING FACTORS OF 'TOTAL DESIGN'

Since the electric vehicle has thus far, in marketing terms, been 'driven' by the state rather than the motoring public it behoves the stylist and product planner to shift the emphasis towards the consumer and show the potential owner the appeal of the vehicle. Some vehicle owners are also environmentalists, not because the two go together, but because car ownership is so wide that the non-driving 'idealist' is a rarity. The vast majority of people voting for local and national governments to enact antipollution regulation are vehicle owners and those who suffer urban traffic jams, either as pedestrians or motorists, and are swinging towards increased pollution control. The only publicized group who are against pollution control seem to be those industrialists who have tried to thwart the enactment of antipollution codes agreed at the international 1992 Earth Summit, fearful of their manufacturing costs rising and loss of international competitiveness. Several governments at the Summit agreed to hold 1990 levels of CO_2 emissions by the year 2000 and so might still have to reduce emission of that gas by 35% to stabilize output if car numbers and traffic density increase as predicted.

Electric vehicles have appeal in urban situations where governments are prepared to help cover the cost premium over conventional vehicles. EVs have an appeal in traffic jams, even, as their motors need not run while the vehicles are stationary, the occupant enjoying less noise pollution, as well as the freedom from choking on exhaust fumes. There is lower noise too during vehicle cruising and acceleration, which is becoming increasingly desired by motorists, as confirmed by the considerable sums of money being invested by makers of conventional vehicles to raise 'refinement' levels. In the 1960s, despite the public appeals made by Ralph Nader and his supporters, car safety would not sell. As traffic densities and potential maximum speed levels have increased over the years, safety protection has come home to people in a way which the appalling accident statistics did not, and safety devices are now a key part of media advertising for cars. Traffic densities are also now high enough to make the problems of pollution strike home.

The price premium necessary for electric-drive vehicles is not an intrinsic one, merely the price one has to pay for goods of relatively low volume manufacture. However, the torque characteristics of electric motors potentially allow for less complex vehicles to be built, probably without change-speed gearboxes and possibly even without differential gearing, drive-shafting, clutch and final-drive gears, pending the availability of cheaper materials with the appropriate electromagnetic properties. Complex ignition and fuel-injection systems disappear with the conventional IC engine, together with the balancing problems of converting reciprocating motion to rotary motion within the piston engine. The exhaust system, with its complex pollution controllers, also disappears along with the difficult mounting problems of a fire-hazardous petrol tank.

As well as offering potential low cost, as volumes build up, these absences also offer great aesthetic design freedom to stylists. Obviating the need for firewall bulkheads, and thick acoustic insulation, should also allow greater scope in the occupant space. The stylist thus has greater possibility to make interiors particularly attractive to potential buyers. The public has demonstrated its wish for wider choice of bodywork and the lightweight 'punt' type structure suggested in the final chapter gives the stylist almost as much freedom as had the traditional body-builders who constructed custom designs on the vehicle manufacturers' running chassis. The ability of the 'punt' structure, to hang its doors from the A- and C-posts without a centre pillar, provides considerable freedom of side access, and the ability to use seat rotation and possibly sliding to ease access promises a good sales point for a multi-stop urban vehicle. The resulting platform can also support a variety of body types, including open sports and sports utility, as no roof members need be involved in the overall structural integrity. Most important, though, is the freedom to mount almost any configuration of 'non-structural' plastic bodywork for maximum stylistic effect. Almost the only constraint on aesthetic design is the need for a floor level flush with the tops of the side sills and removable panels for battery access.

0.2.2 CHANGING PATTERNS OF PRODUCTION AND MARKETING

Some industry economists have argued that local body-builders might reappear in the market, even for 'conventional' cars as OEMs increasingly become platform system builders supplied by systems houses making power-unit and running-gear assemblies. Where monocoque structures are involved it has even been suggested that the systems houses could supply direct to the local body-builder who would become the specialist vehicle builder for his local market. The final chapter suggests the use of an alternative tubular monocoque for the sector of the market increasingly attracted by 'wagon' bodies on MPVs and minibuses. Here the stylist can use colour and texture variety to break up the plane surfaces of the tube and emphasize the integral structural glass. Although the suggested tubular shell would have a regular cross-section along the length of the passenger compartment, the stylist could do much to offer interior layout alternatives, along with a host of options for the passenger occupants, and for the driver too if 'hands-off' vehicle electronic guidance becomes the norm for certain stretches of motorway.

Somehow, too, the stylist and his marketing colleagues have to see that there is a realization among the public that only when a petrol engine runs at wide open-throttle at about 75% of its maximum rotational speed is it achieving its potential 25% efficiency, and this is of course only for relatively short durations in urban, or high density traffic, areas. It is suggested that a large engined car will average less that 3% efficiency over its life while a small engined car might reach 8%, one of the prices paid for using the IC engine as a variable speed and power source. This offsets the very high calorific value packed by a litre of petrol. An electric car has potential for very low cost per mile operation based on electrical recharge costs for the energy-storage batteries, and EVs are quite competitive even when the cost of battery replacement is included after the duration of charge/recharge cycles has been reached. It needs to be made apparent to the public that a change in batteries is akin to changing the cartridge in a photocopier–essentially the motive-force package is renewed while the remainder of the car platform (machine) has the much longer life associated with electric-driven than does the petrol-driven vehicle. In this sense batteries are amortizable capital items, to be related with the much longer replacement period for the vehicle platform which could well carry different style bodies during its overall lifetime.

The oversizing of petrol engines in conventional cars, referred to above, arises from several factors. Typical car masses, relative to the masses of the drivers they carry, mean that less than 2% of fuel energy is used in hauling the driver. Added to the specifying of engines that allow cars to travel at very large margins above the maximum speed limit is of course the conventional construction techniques and materials which make cars comparatively heavy. The weight itself grossly affects accelerative performance and gradient ability. Also some estimates consider six units of fuel are needed to deliver one unit of energy to the wheels: one-third wheel power being lost in acceleration (and heat in consequent braking), one-third in heating disturbed air as the vehicle pushes through the atmosphere and one-third in heating the tyre and road at the traction, braking and steering contact patch. This puts priorities on design for electric vehicles to cut tare weight, reduce aerodynamic drag and reduce tyre rolling resistance.

0.2.3 QUESTIONING THE INDUSTRY-STANDARD APPROACH

The design process in the main-line automotive industry is driven by the edicts of the car-makers' styling departments who ultimately draw their inspiration from the advertising gurus of Madison Avenue, whose influence has, of course, spread worldwide. The global motor industry has been predominately US dominated since Henry Ford's pioneering of systematic volume production and General Motors' remarkable ability to appeal to widely different market sectors with quite modestly varied versions of a standard basic vehicle. Thus far the electric, or hybrid drive, vehicle had to conform to historically developed design norms with the cautious conservatism of marketing management defining the basic scantlings. Conventional automotive design must conform to the requirements of Mr and Mrs Average, analysed by countless focus groups, while meeting the necessities of mass-production equipment developed during the first century of the motor vehicle.

When bold attempts have been made to achieve substantial reductions in weight below that of the standard industry product, the limitations of these major constraints have usually moderated the design objectives, Fig. 0.1. The overruling necessity to 'move metal' at the scale of ten million vehicles per year from each of the world's three main areas of motor manufacture makes radical design initiatives a scary business for 'corporate bosses'. Advertising professionals, with their colleagues in public relations, have skilfully built up customer expectations for the conventional automobile, from which it is difficult for the designer to digress in the interests of structural efficiency and light weight. Expectations are all about spacious interiors with deep soft seats and wide easy-access door openings; exterior shape is about pleasing fantasies of aggressiveness, speed and 'luxury' appearance. Performance expectations relate to accelerative ability rather than fuel economy, as Mr Average Company Representative strains to be 'first off the grid'.

Ecologists who seek the palliative effects of electric propulsion will need to face up to educating a market that will appreciate the technology as well as convincing motor industry management of the need for radical designs which will enable the best performance to be obtained from this propulsion technology. The massive sensitivity of the general public to unconventional vehicle

Fig. 0.1 Alcan's use of 5754 aluminium alloy substituted for steel in the Ford Taurus/Sable saved an impressive 318 kg. The client's constraint of minimal changes to the passenger compartment and use of existing production equipment must have constrained the possibilities for further weight reduction, however.

configurations was made abundantly clear from the reaction to the otherwise ingenious and low cost Sinclair C5 electric vehicle. While clearly launched as a motorized tricycle, with a price appropriate to that vehicle category, the C5 was nearly always referred to by its media critics as an 'electric car' when operationally it was more appropriate for use on reserved cycleways of which, of course, there are hardly enough in existence to create a market. While the Sunracer Challenge in Australia has shown the remarkable possibilities even for solar-batteried electric vehicles, it is doubtful whether the wider public appreciate the radical design of structure and running gear that make transcontinental journeys under solar power a reality, albeit an extremely expensive one for a single seater. Electric cars are perceived as 'coming to their own' in urban environments where high traffic densities reduce average speeds and short-distance average journeys are the norm. There is also long-term potential for battery-powered vehicles to derive additional 'long-distance' energy from the underground inductive power lines which might be built into the inside lanes of future motorways. It is not hard to envisage that telematics technology for vehicle guidance could be enhanced by such systems and make possible electronically spaced 'trains' of road vehicles operating over stretches of motorway between the major urban and/or rural recreational centres.

0.2.4 MARKET SEGMENTATION

At the time of writing some customer-appealing production hybrid and electric drive vehicles have already come onto the market. The Toyota Prius hybrid-drive car, described in Chapter 6, is already proving to be well received in the Japanese market where imaginative government operational incentives are in place. A variety of conversions have been made to series production compact cars which allow short-range urban operation where adequate battery recharging infrastructure is available. However, GM surprised the world with the technically advanced prototype Impact medium-range electric car, but the market has reportedly not responded well to its production successor and generally speaking there is not yet an unreservedly positive response.

Like the existing market for passenger cars, that for electric-drive cars will also be segmented, in time, with niches for sedan, convertible, dual-purpose, sports, utility, limousine and 'specialist' vehicles. The early decades of development, at least, may also be noted for the participation of both high and low volume builders. The low volume specialist is usually the builder prepared to investigate radical solutions and in the, thus far, 'difficult' market for electric cars it would seem a likely sector for those EVs which are more than drive-system conversions of existing vehicles.

With the high volume builders, already under pressure from overcapacity, their main attention is likely to be focused on retaining markets for current design vehicles, without the 'distraction' of radical redesigns. The ambitious, imaginative and high technology specialist has thus much to gain from an informed innovative approach and could benefit from a reported longer-term trend when drive systems will be manufactured by huge global producers and vehicle manufacturing will tend towards a regional basis of skilled body shops catering for local markets.

0.2.5 EV AS PART OF A WIDER TRANSPORTATION SYSTEM

The 'physical' design package for an electric vehicle will result from a much larger 'design package of affecting factors' which encompasses vehicle operational category, manufacturing systems/ techniques, marketing and distribution. Packages for industrial trucks and specialist delivery vehicles are already established but those for passenger-car variants much less so. It has been suggested that the first substantial sales of electric cars might well be to electricity generating companies in the public utilities sector, who would rent them to railway operators for end-use by rail travellers. Such people would purchase their hire with return travel tickets to destination stations at which EVs would be parked in forecourts for the use of travellers. Other potential customers might be city-centre car hire fleets, taxicab operators in fossil-fuel exhaust-free zones or local authorities setting up city-centre car pools.

One of the most imaginative EV applications is the lightweight mini-tram, Fig. 0.2, as exhibited at the Birmingham ElectriCity event in 1993. This is a vehicle that runs on low cost tracks which can be laid on an ordinary road surface without further foundation. The vehicle can travel up to 50 km/h and is a flywheel-assist hybrid machine having its batteries recharged via low voltage conductor rails positioned at intervals around the track. Each car weighs just over 3 tonnes unladen and can carry 14 seated and 11 standing passengers. A 5 km route, including rails, can be constructed, to include five trams, ten stops and four charge points, at a cost of just £1 million. It seems an ideal solution to the problem of congested cities that have roadways that date back to pre-automobile days, with the mini-trams able to transport both passengers and goods in potential 'pedestrian precincts' that would be spoilt by the operation of conventional omnibuses and tramcars. The proposal serves well to illustrate the opportunities for electric vehicles, given some imaginative lateral thinking.

Since launch, larger vehicles have been produced and entered service. The one seen at Bristol Docks (*Fig. 0.2, right*) has a steel frame with GRP body panels and weighs 13 tonnes, compared with the smallest railcar which weighs 48 tonnes. There are four production variants on offer, carrying 30, 35 or 50 passengers, and a twin-car variant of the latter. Use of continuously variable transmission now ensures the flywheels run at constant speed; a third rail at stations is used for taking in electricity for 'charging up' the flywheel. A 2-minute recharge would be required for the

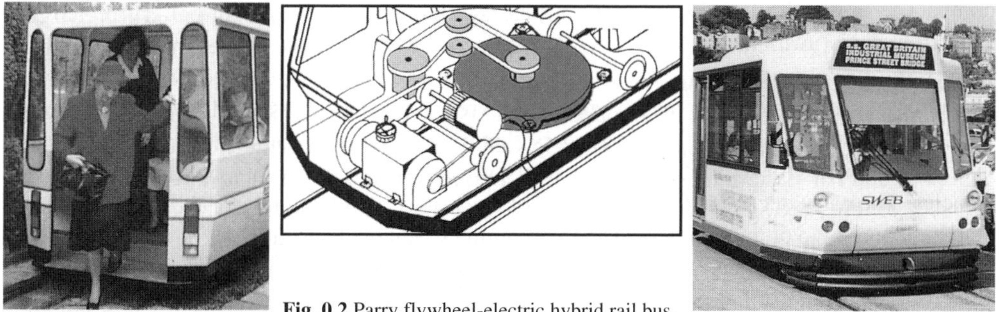

Fig. 0.2 Parry flywheel-electric hybrid rail bus.

flywheel to propel the vehicle its maximum distance of two miles; so more frequent stops are recommended to reduce recharge time, 0.5 km being the optimum. A hybrid version with additional LPG power was due for launch in Stourbridge, UK, as a railcar in early 2001.

Some of the above projects are all based on the proposition that the more conservative motor manufacturers may not follow the lead set by Toyota and Honda in offering hybrid-electric drive cars through conventional dealer networks. In the mid 1990s the US 'big-three' auto-makers were crying that there was little sales interest from their traditional customers for electric cars, after the disappointing performance of early low volume contenders from specialist builders. The major motor corporations are considered to operate on slender profit-margins after the dealers have taken their cut, but a change to supermarket selling might weaken the imperative from high volume products which could favour specialist EVs from the OEM's SVO departments. That the corporations have also jibbed violently against California's mandate for a fixed percentage of overall sales being EVs, and wanted to respond to market-led rather than government-led forces, suggests a present resistance to EVs.

A number of industrial players outside the conventional automotive industry are drawing comparisons between the computer industry and the possible future electric vehicle industry, saying that the high-tech nature of the product, and the rapid development of the technologies associated with it, might require the collaboration of companies in a variety of technical disciplines, together with banks and global trading companies, to share the risk of EV development and capitalize on quick-to-market strategies aimed at exploiting the continually improving technology, as has already been the case in personal computers. They even suggest that the conventional auto-industry is not adapting to post-Fordist economic and social conditions and is locking itself into the increasing high investment required of construction based on steel stampings, and ever more expensive emission control systems to make the IC engine meet future targets for noxious emissions. The automotive industry reacts with the view that its huge investment in existing manufacturing techniques gives them a impregnable defence against incomers and that its customers will not want to switch propulsion systems on the cars they purchase in future.

It may be that the US domestic market is more resistant to electric vehicles than the rest of the world because the cultural tradition of wide open spaces inaccessible to public transport, and the early history of local oilfields, must die hard in the North American market where petrol prices are maintained by government at the world's lowest level, for the world's richest consumers. Freedom of the automobile must not be far behind 'gun law' in the psyche of the American people. In Europe and the Far East where city-states have had a longer history, a mature urban population has existed for many centuries and the aversion to public transport is not so strong. Local authorities have long traditions of social provision and it may well be that the electric vehicle might well find a larger market outside America as an appendage to the various publicly provided rapid transit systems including the metro and pre-metro. And, according to a CARB contributor to Scott Cronk's remarkable study of the potential EV industry[1], with the control equipment in the most up-to-date power stations 'urban emissions which result from charging an electric vehicle will be 50–100 times less than the tail-pipe emissions from (even) ... ULEV' vehicles, a very different story to that put out by IC-engined auto-makers' PR departments.

It is also argued within Cronk's collection of essays that fuel savings from ultra-lightweight vehicles might predate the impact of electric vehicles, on public acceptance, particularly within European and Far-Eastern markets where petrol prices are at a premium and usually bear heavy social taxes. Fuel savings by such a course could be very substantial and the customer might, as a second stage, be more ready to take the smaller step to a zero-emissions vehicle. This is when he/she realizes that the cost of overnight battery charge, at off-peak rates from the utilities, could prove an irresistible economic incentive. The vehicles would be produced in a lean-production culture which would also help to pare the substantial overhead costs that are passed onto the customer in traditional auto-manufacture.

0.2.6 DIFFERENT CULTURE FROM THE PRIVATE I.C.E. CAR

The different performance package offered to the public by the EV involves disadvantages, such as comparatively low range and carrying capacity, which need to be offset in the customer's mind by advantages such as low maintenance, noise and vibration, creating the need for a different form of marketing and distribution from that of the conventional private car. The lower volume production rates also involve a quite different set of component and system suppliers, for servicing a specialist manufacture of this nature. The need for a charging infrastructure different from petrol stations also serves to distinguish EVs as a separate culture. Purchase price will be higher and resale price probably lower due to obsolescence in the face of advancing technology. The notion of periodically billing the customer for an ongoing personal mobility is likely to be preferable to just selling a car. The customer is thus spared the hassle of bargaining with dealers, obtaining finance, insurance and registration as well as the bother of refuelling and making arrangements for periodic servicing. Periodic servicing is likely to be extended to 50 000 mile intervals for EVs, and systems for refurbishing high mileage vehicles with updated technology systems might well be 'on the cards'. The interlinking of mobility providers by horizontal networks would obviously benefit the customer as he/she travels from one area to another, possibly using different transport modes. The provider might be a sort of cross between travel agent and customer liaison officer of a motoring organization, but principally the leaser of the EV, Fig. 0.3.

The need to perceive the EV as a function-specific addition to the family vehicle fleet is also important so that a town car for the school-run, shopping or commuting can complement the conventional car's use for weekend and holiday outings of longer distance. The local mobility provider will need PR skills to be regularly contactable by clients, but will not need the high cost service station premises of the conventional car dealer. In manufacturing the EV a different perception of OEM, from that of the conventional car assembler, is also apparent, because it is likely to be a company much smaller in size than that of its key specialist system suppliers who will probably serve many other industries as well. The OEM would become systems integrator for

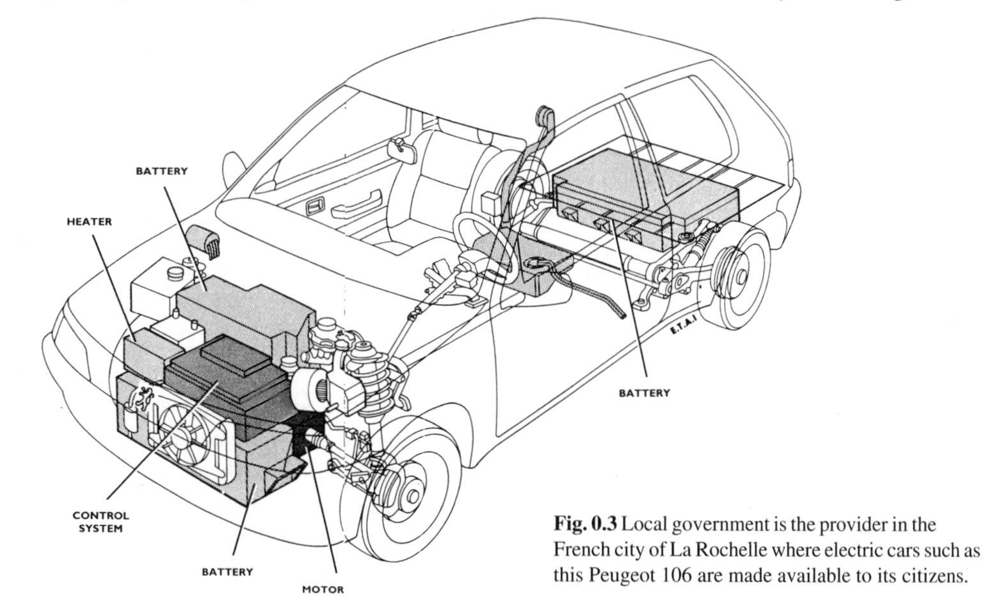

Fig. 0.3 Local government is the provider in the French city of La Rochelle where electric cars such as this Peugeot 106 are made available to its citizens.

a 'partnership' chain of long-term suppliers and appoint a project leader to coordinate design, development and production, leading a cross-company team. Such leadership would carry the authority for detailed cost investigations in any of the member firms. EV leasers would need to network with manufacturing project leaders and provide carefully researched hire schedules of potential lessees upon which series production could be planned. This is without need for large parks of finished vehicles which conventional OEMs use as a buffer between supply and demand, as well as their need to maintain excess idle production capacity in slack periods. Organizational innovation thus shares similar importance with technological innovation in EV production.

0.2.7 PUBLIC AUTHORITY INITIATIVES

National government programmes, such as the ARPA EV programme in the USA, can be used to unite heavy defence spending with value to civilian producers. As combat vehicles have very high auxiliary power demand they become almost hybrid in the sense of their power sources, albeit only one of them being conventionally the prime mover. Coupled with the need to operate tanks in silent mode during critical battle conditions, this makes the study of hybrid drive a reality for military as well as civilian operators. The idea of helping sustain civilian product development must be almost impossible to contemplate by British military hierarchies but if ever a cultural transformation could be brought about, the technological rewards might considerably improve on the efforts made by the military to sell technology to British industry. The USA has the tremendous built-in advantage of their military supremos caring deeply about maintaining the country's industrial base not normally part of the culture of UK military commanders!

Regional government initiatives can also be valuable in kick-starting cooperative ventures between companies from different industries. Again the US example, in California, is noteworthy where aerospace supplying companies have been encouraged to support pilot EV programmes. Valuable inputs to EV construction have therefore been made by companies skilled in structural design, computer simulation, lightweight materials, aerodynamics, fibre-optic instrumentation, head-up displays and advanced joining/fabrication. Of course, regional governments inevitably help EVs in the execution of environmental policies and already city authorities in many countries around the world have banned many vehicle categories from their central areas. National governments are also contemplating the huge sums of money spent in defending their oil supplies and probably noting the decreases in oil usage by industries such as building, manufacturing and power generation while transport oil usage continues to rise. The burgeoning use of computer and other electronics systems is also demanding more reliable electricity generation, that can accommodate heavy peak loads. Power generators will be increasingly pleased to step up utilization of the expanded facilities in off-peak periods by overnight charging of EVs. In the longer term, governments might even appreciate the reskilling of the workforce that could follow the return to specialization in the post-Fordist economic era and see that helping to generate new technological enterprises, as EV development and build could help recivilize a society condemned for generations to the mindlessness of mass production and the severe and dehumanizing work routines which accompany it.

0.2.8 REFINING THE CONVENTIONAL CAR PACKAGE FOR THE EV

The American 'supercar' programme, discussed in Chapters 4 and 7, has been an invaluable indicator as to how lightweight construction can dramatically improve the efficiency of automotive propulsion. As only 4% of a conventional car's engine is needed for city driving conditions, the oversizing of engines in multi-functional cars makes the reduction of exhaust pollution a particularly difficult task on IC-engined vehicles. Expert analysts maintain that half the engine efficiency gains made in the decade 1985–1995 were lost by making engines powerful enough, in the US, to drive at twice the speed limit on the open road. Obviously the situation is worsened if conventional heavyweight steel construction is used and the tare weight of cars rises with the increasing proliferation of on-board gadgetry. While 'supercar' construction has shown how structure weight can be reduced, advanced technology could also be used to reduce the 10% of engine power used in powering 'accessories' such as power steering, heating, lighting and in-car entertainment.

The imperative for power steering is removed by the ultra-light construction of the 'supercar', provided steering and handling dynamics are properly designed. In EV supercars, wheel motors might provide for ABS and ASR without further weight penalty. High intensity headlamp technology can considerable reduce power demand as can the use of fibre-optic systems which provide multiple illumination from a single light source. Light-emitting diode marker lamps can also save energy and experts believe that the energy consumption of air-conditioning systems could be reduced by 90%, if properly designed, and used in cars with sandwich panel roofs, heat-reflecting windows and solar-powered ventilation fans. But none of this compares with the savings made by high strength composite construction which has the potential to bring down average car weight from 3000 to 1000 lb. It is reported that many of the 2000 or so lightweight EVs operating in Switzerland already weigh only 575 lb without batteries.

The ability to achieve net shape and finish colour from the mould in polymer composite construction is important in offsetting the higher cost of high strength composites over steel. But also the cost of steel is only 15% of the conventional structure cost, the remainder being taken up in forming, fabrication and finishing. Around half the cost is taken up by painting. The cheaper tooling required for polymer composites is also important in making small-scale production a feasible proposition, alongside direct sales from the factory of 'made-to-order' cars. A number of these factors would help to remove the high mark-up to the customer of the factory price which is typical of conventional car sales and distribution.

0.3 Lean production, enterprise structures and networking

Lean production has grown out of post-Fordist 'flexible specialization' which has led to growing specialization of products, with a new emphasis on style and/or quality. The differentiated products require shorter production runs and more flexible production units, according to Clarke[2]. The flexibility is made possible by new technologies, the emerging economic structure being based on computerization and other microchip hardware. Rapid gains in productivity are made through full automation and computerized stock control within a system that allows more efficient small batch production. Automatic machine tools can be reprogrammed very quickly to produce small quantities of much more specialized products for particular market niches. Economies are set to be no longer dominated by competition between hierarchically organized corporations and open to those dominated by cooperation between networks of small and interrelated companies.

Lean enterprises are seen as groups of individuals, functions, and legally separate but operationally synchronized companies that create, sell, and service a family of products, according to Womack *et al*[3]. This is similar to the Japanese 'keiretsu' concept of large, loose groupings of companies with shareholding connections. They cooperate both technically and in sharing market information and the result is an array of business units competing in vertically and horizontally links with other companies within a single project. A trading company with well-developed worldwide networks is usually at the centre of the operation and can feed back vital market trends to the production companies. Of almost equal importance is the involvement of international banking corporations who can provide a source of industrial finance. Changes in legislation are required by European countries to make a similar system of common shareholdings plus private ownership acceptable to company law.

Lean production is the approach pioneered by Toyota in which the elimination of unnecessary steps and aligning all steps in a continuous flow, involves recombining the labour force into cross-functional teams dedicated to a particular activity, such as reducing the weight of an EV platform. The system is also defined by the objective of continually seeking improvement so that companies can develop, produce and distribute products with halved human effort, space, tools, time, and, vital to the customer, at overall halved expense.

Enterprise structures aim to exploit business opportunities in globally emerging products and markets; to unite diverse skills and reapply them in long-term cooperative relationships; to allocate leadership to the member best positioned to serve the activity involved regardless of the size of company to which he/she belongs; and finally to integrate the internal creation of products with the external consequences of the product. In EVs this would involve ensuring an adequate operational infrastructure be provided by an electricity generating company, in combination with local authorities. The products involved are those, such as the electric vehicle, that no one member company on its own could design, manufacture and market. Partners in an EV enterprise might also lead it into additional businesses such as power electronics, lift motors, low cost boat-hull structures and energy storage systems for power station load levelling, for example. Internally the use of combined resources in computer software technology could be used to develop simulation packages that would allow EVs to be virtual tested against worldwide crashworthiness standards. Managing of product external consequences could be facilitated by forming partnerships with electricity generators, material recyclers and urban planners, finance, repair and auto-rental service suppliers as well as government agencies and consumer groups.

0.3.1 COOPERATIVE NETWORKS

Unlike the Japanese networks of vertically integrated companies, such as the supply chains serving Toyota, an interesting Italian experience is one of horizontal networking between practitioners in specialist industries. Groups of small companies around Florence, in such areas as food processing, furniture making, shoe manufacturing, have been unusually successful and, in the case of tile manufacture, have managed to win an astonishing 50% of the world market. Export associations have been formed on behalf of these small companies and at Modena even a finance network has been formed between companies in which the participants guarantee one another's bank loans. The normal default rate of 7% for bank loans in this region has become just 0.15% for this industrial network, demonstrating the considerable pride built up by companies in meeting their repayment obligations. Commentators liken the degree of trust between participants as being akin to that between different branches of traditional farming families. Like the grandfathers of the farming families the 'elders' of the industrial networks offer their services for such tasks as teaching apprentices in local colleges. The secret, some say, is that these areas around Florence escaped the era of Fordism which affected northern Italy and many other industrial centres of Europe.

The approach to setting up such a network is to build on elements of consensus and commonalilty so as to create mutual facilities of benefit to groups of small companies wishing to compete successfully against the international giants. Generally a network has a coordinating structure of interlinked elements which are individuals, objects or events. The links can be in the form of friendship, dependence, subordination or communication. In a dense network everyone knows everyone else while some networks may, for example, comprise clusters of dense elements with ties between clusters perhaps only involving one individual in each. The specific definition of a network is the set of relations making up an interconnected chain for a given set of elements formed into a coordinating structure.

Analysts usually consider solidarity, altruism, reciprocity and trust when examining networks in general. Solidarity is largely brought about by sharing of common experience; so social class and economic position layers are sometimes seen as having solidarity as do family and ethnic groupings. With altruism, of course, people help each other without thought of gain. Because it is rare in most societies, rewards and penalties for actions tend to exist in its absence. Repeat commitment to a network is expressed as loyalty and individuals often react to disturbance either by 'exit', 'voice' (try and change things for better) or 'loyalty'. The latter may be expressed as 'symbolic relations' in which an individual is prepared to do his duty and meet his obligations. 'Voice' is important in the organization of networks as it involves argument, debate and persuasion, which is often fundamental to the direction taken by small to medium sized groups. Another stabilizing coordination is the reciprocity with which symmetry is maintained between giving and receiving. Of all the attributes, trust plays a central organizing role; essential if not all members behave absolutely honestly. Individuals bet against the opportunistic behaviour of others according to their reputations. Networks are often 'flat' organizations in the sense of having equality of membership. There is an underlying tendency for individuals to become involved with cooperative solidarity, if only because of the higher cost of not cooperating. Generally trust is built up over a period of recognizing and evaluating signals from other actors and having opportunities to test interpretations, over a rule-learning period, which leads to eventual solidification of mutual interest.

A study of French subcontracting companies to the engineering sector in the Lyons area, between 1975 and 1985, has shown that network coordination has improved performance relative to larger firms during that period, often becoming dynamic investors in flexible CNC machine tools. Essentially small firms benefited from large forms farming out some of their activities because they could not run flexible machines long enough to amortize the capital cost. But this was only the trigger and the firms later found the network of cooperation brought them trading advantages way beyond those available in a classic market. Recent economics approaches have dealt with transaction costs as a means of examining social ties between traders and such analysis involves the organizational implications of the transaction cost. Trust can lubricate the friction behind such costs. In the French study the small subcontractors were mainly supplying large engineering companies in the capital goods sector involved in large, complex, customized and expensive products for which client firms were unable to forecast requirements beyond a period of six months. Employees of the subcontracting firms undergo periods of training in the assembly shops of the client and the client firm becomes an expert in the engineering processes of the subcontractor so that mutual understanding can be built. Each subcontractor takes orders from one client of not more than 10–15% of total sales and the clients put themselves in the position of the subcontractors in determining optimal level of orders. The relatively low percentage figure allows the client a degree of flexibility without undermining the viability of the subcontractor. A 'partnership' exists in that in exchange for improved performance on quality and delivery the client firm guarantees a level of work for the subcontractor. Any defection of a subcontractor is made known to the whole community of suppliers and the full penalty has to be made for non-delivery, so that trustworthiness is not just judged by reputation; the long-term message from the experience was that 'trust is expedient'.

Other examples show that large companies often tend to divest themselves of activities to the extent that they become essentially 'systems integrators' among a specialized consortia of companies in the particular manufacturing environment. Quoted examples are Fiat, BMW and Volkswagen. This breaking up of vertical integration may involve affiliated organizations or separate suppliers, with many aspects of R&D and design being divested to systems suppliers. Relationships between sub-units are too delicate to be left to market-type arrangements in this 'associationalist' way of working.

0.4 Electric-drive fundamentals

While battery-electric vehicles were almost as common as IC-engined ones, at the beginnings of the commercialization of the powered road vehicle, it was not until the interwar years that serious studies were taken into operating efficiency of such systems, as a precursor to their introduction in industrial trucks and special purpose vehicles such as milk floats. Figure 0.4 illustrates some of the fundamental EV traction considerations as the technology developed. For the Mercedes Electromobile of the early 1920s, for example, seen at (a), more sophisticated wheel drives were introduced, with motors formed in the wheels to eliminate transmission gear losses. An energy diagram for this drive is seen at (b). The basic definitions and relationships of electromagnetism are helpful in the appreciation of the efficiency factors involved.

0.4.1 ELECTROMAGNETIC BASICS

While the familiar magnetic line-of-force gives the direction of magnetic force at any point, its field strength H is the force in dynes which would act on a unit pole when placed in the field. For magnetic material such as soft iron placed in the field, the strength of field, or magnetic intensity B, inside the iron is greater than H, such that $B = \mu H$, where μ is the permeability of the material (which is unity for non-metallics). When the cross-section of the object, at right angles to the magnetic field, is denoted by a, the magnetic flux ϕ is the product Ba in maxwells. Since it is taken that at unity field strength there is one line of force per square centimetre, then magnetic induction is measured in lines per cm^2 and flux is often spoken of as in 'lines'.

Faraday's law defined the induced EMF as rate of change of flux ($-d\phi/dt \times 10^{-8}$ volts) and Lenz's law defined the direction of the induced EMF as such that the current set up by it tends to stop the motion producing it. The field strength of windings having length l, with N turns, carrying current I is

$$H = 4\pi IN/10l \qquad \text{which can be rearranged as} \qquad \phi(l/ma) = 4\pi IN/10$$

where the flux corresponds to the current in an electrical circuit and the resistance in the magnetic circuit becomes the reluctance, the term on the right of the equation being the magneto-motive force. However, while in an electric circuit energy is expended as long as the current flows, in a magnetic circuit energy is expended only in creating the flux, not maintaining it. And while electrical resistance is independent of current strength, magnetic permeability is not independent of total flux. If H is increased from zero to a high value, and B plotted against H for a magnetic material, the relationship is initially linear but then falls off so there is very little increase in B for a large increase in H. Here the material is said to be saturated. When H is reduced from its high value a new BH curve lies above the original curve and when H is zero again the value of B is termed the retentivity. Likewise when H is increased in the negative direction, its value when B is zero again is the coercive force and as the procedure is repeated, (c), the familiar hysteresis loop is obtained.

In generating current electromagnetically, coils are rotated between the poles of a magnet, (d), and the current depends on both the strength of the magnetic field and the rate at which the coils rotate. Either AC or DC is obtained from the armature rotor on which the coils are mounted, depending on the arrangement of the slip-ring commutator. A greater number of coils, wound around an iron core, reduces DC current fluctuation. The magnetic field is produced by a number of poles projecting inwards from the circular yoke of the electromagnet. Laminated armature cores are used to prevent loss of energy by induced eddy currents. Armature coils may be lap-wound, with their ends connected to adjacent commutator segments, or wave-wound (series) when their ends are connected to segments diametrically opposite one another. The total EMF produced

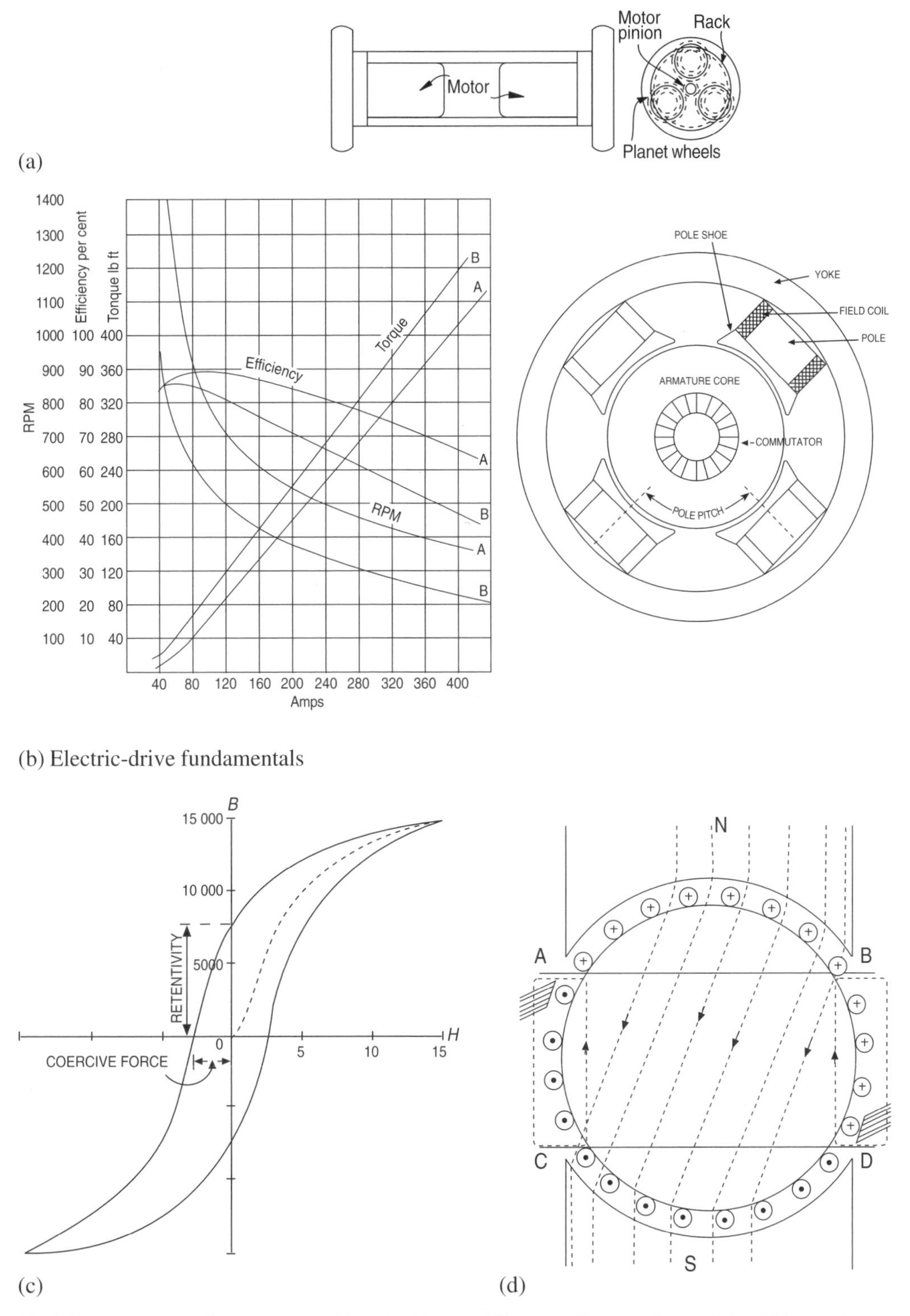

Fig. 0.4 Electric traction fundamentals: (a) Mercedes Electromobile motor; (b) motor characteristics; (c) hysteresis loop; (d) motor poles and their magnetic field.

is $(\phi nZ \times 10^{-8}/60)P/K$ where for lap-winding $K = P$ and for wave-winding $K = 2$. Z is the number of conductors in the armature and n is its rotational speed.

The armature-reaction effect is set up by the current in the armature windings affecting the magnetic field between the poles. In a simple 2 pole machine, armature current would produce transverse lines of force, and the resulting magnetic field would be as shown in the figure. Hence the brushes have to be moved forward so that they are in the neutral magnetic plane, at right angles to the resultant flux. Windings between AB and CD create a field opposed to that set up by the poles and are called demagnetizing turns while those above and below are called cross-magnetizing turns. Armature reaction can be reduced by using slotted pole pieces and by separate compensating field windings on the poles, in series with the armature. Also small subsidiary inter-poles, similarly wound, can be used.

When the machine runs as a motor, rather than generator, the armature rotates in the opposite direction and cuts field lines of force; an induced voltage known as a back-EMF is generated in the opposite direction to that of the supply and of the same value as that produced when the machine is generating. For current I, applied to the motor, and back-EMF E_b, the power developed is $E_b I$. By substituting the expression for E_b, the torque transmitted in lb ft is $(0.117I\phi ZP/K) \times 10^8$.

The field current can be separately excited (with no dependence on armature current) or can come from series-wound coils, so taking the same current from shunt-wound coils – connected in parallel with the armature and having relatively high resistance, so taking only a fraction of armature current. Compound wound machines involve a combination of series and shunt. In examining the different configurations, a motor would typically be run at a constant input voltage and the speed/torque curve (mechanical characteristic) examined. Since the torque of a motor is proportional to flux $\times$ armature current, and with a series wound machine flux itself varies with armature current, the torque is proportional to the square of current supplied. Starting torque is thus high and the machine attractive for traction purposes. Since the voltage applied to a motor in general remains constant, and back-EMF is proportional to ϕn which also remains constant, as the load increases, ϕ increases and therefore the speed decreases – an advantage for traction work since it prevents the motor from having to carry excessive loads.

The speed of a motor may be altered by varying either the brush voltage or the field flux. The first is altered by connecting a resistance in series with the armature, but power wastage is involved; the second, field control, is more economical – and, with a series motor, a shunt is placed across the field winding.

0.4.2 ELECTRIC TRANSMISSION

Electric transmission, Fig. 0.5, survived electric power sources in early vehicles and the engineers of the time established the parameters for optimizing the efficiency of the drive. In a 1920s paper by W. Burton[4], the author points out that for a given throttle opening and engine speed, the output in watts is fixed as the familiar product of voltage V and current I in the electrical generator. The ideal power characteristic thus becomes a rectangular hyperbola with equation VI = a constant. The simplest electrical connection between generator and electric transmission motor is as at (a). Generator and motor have to fulfil the function of clutch and gearbox, in a conventional transmission, and closure of the switch in the appropriate position provides for either forward or reverse motion 'clutching'. Below a nominal 300 rpm the generator provides insufficient power for vehicle motion and the engine idles in the normal way. The change speed function will depend on generator characteristic and a 'drooping' curve is required with generator voltage falling as load rises, to obtain near constant power – suggesting a shunt-wound machine. By adding a number of series turns the curve can be boosted to a near constant-power characteristic. These series windings also

Notation

H = Field strength
B = Magnetic intensity
μ = Permeability
ϕ = Magnetic flux
N = Number of field turns
Z = Number of armature turns
I = Current
V = Voltage
L = Length of windings
n = Rotational speed
E_b = Back-EMF

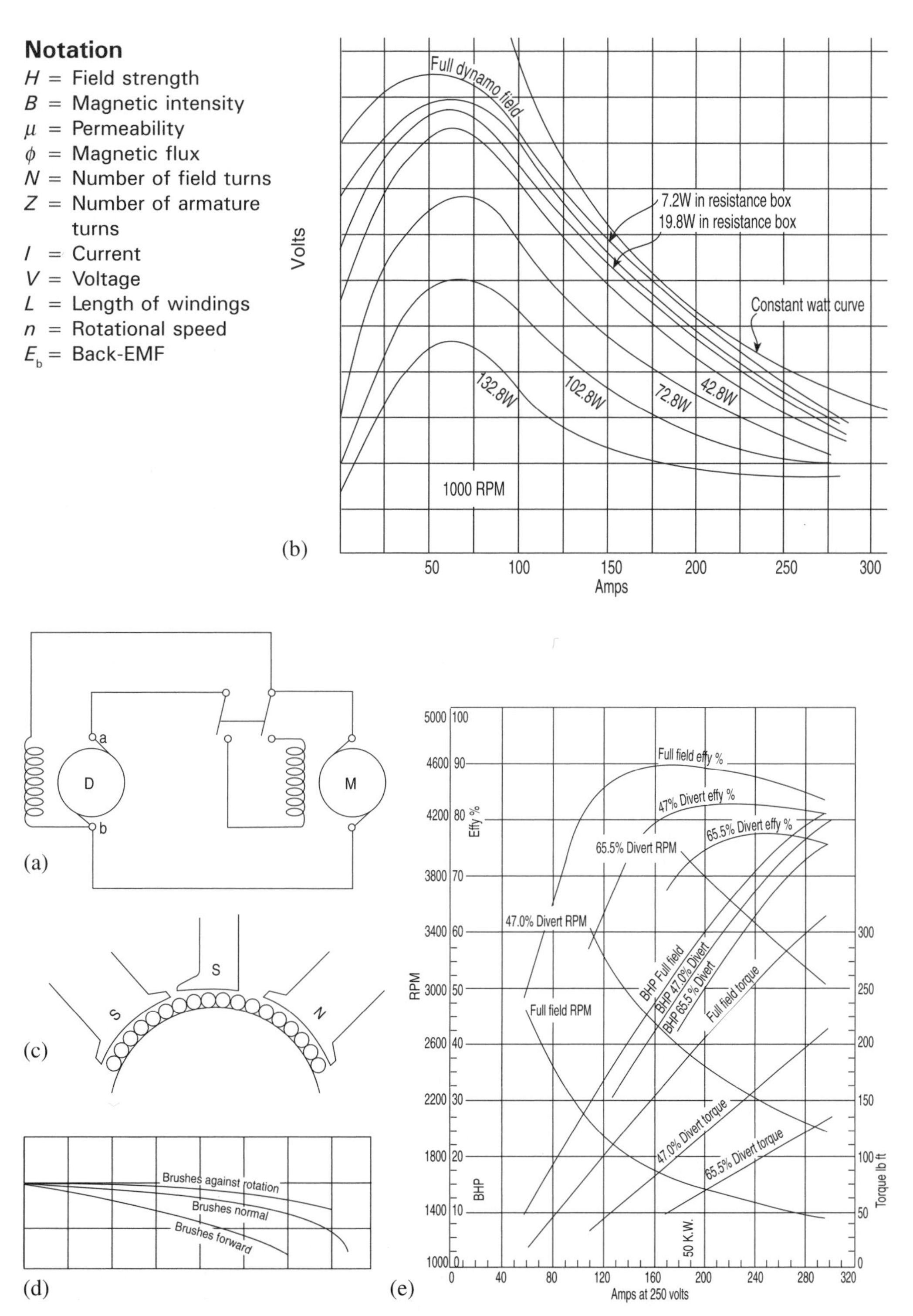

Fig. 0.5 Electric transmission basics: (a) 'clutching' of electric transmission; (b) high EMF at low loads; (c) horned interpoles; (d) brush movement effect; (e) motor characteristics.

help in rapid build-up of generator EMF. The resulting problem is heat build-up of these series windings under heavy vehicle-operating loads. Efforts to counteract this by reducing the length of the shunt coil creates the further difficulty of slow excitation after vehicle coasting. Since the brushes of the generator or motor short-circuit one or more sections of the armature winding, it is important that these sections are in the neutral zone between field magnets of opposite polarity at the moment they are shorted. To otherwise avoid destructive arcing under heavy load, the machine characteristic may be altered by moving the brushes either with or against the direction of armature rotation. This will provide more or less droop of the characteristic as shown at (b), but on interpole machines there is the added problem of the interpoles being prevented, under brush movement, of fulfilling their role of suppressing arcing.

Horned interpoles, (c), may be used to offset this effect. The shape of the horn is made such that the magnetic flux under the foot of the interpole is not altered but the additional shoe section is magnified sufficiently to act on a few turns of the armature, these turns providing sufficient induced EMF to give the required compounding effect for rapid excitation from standstill and under heavy loads. The view at (d) shows the performance characteristics by a machine of this type. While the curve for the full field (no series resistance) approximates to the constant power characteristic, its EMF rises at light loads. The effect of inserting resistance is also shown. However, for a given motor torque, speed is proportional to EMF applied so that if the engine speed is reduced, motor and thus vehicle speed will fall. To avoid this, the motor field windings have a diverter resistance connected in parallel to them, to weaken the motor field; the counter-EMF is reduced, and more current is taken from the generator, which increases motor speed again. Thus a wide speed ratio is provided. *In earlier times resistance was altered by handles on the steering column; with modern electronics, auto-control would, of course, be the norm.* Regenerative braking can be obtained by reversing the field coil connections of the motor which becomes a 'gravity-driven' series-wound generator, running on short-circuit through the generator armature. However, the currents involved would be too heavy and an alternative approach is required.

The theme is taken up by H.K. Whitehorne in a slightly later paper[5], who pays especial tribute to Burton's skewed horn interpole invention. He goes on to consider motor characteristics and favours the series-wound machine because its speed is approximately inversely proportional to the torque delivered, adjusting its current demand to the speed at which it runs and to the work it has to do. Characteristic curves of a motor running on a fixed voltage are shown at (e). Conditions are shown for full field, and for two stages of field diversion. Examination of the 50 kW line makes it apparent that the torque/amp curve is independent of voltage; speed is practically proportional to voltage and generally characteristics vary on the size of the motor, its windings and length of its core. However, on low voltage and heavy current, the efficiency falls rapidly which makes electric transmission a difficult option for steep gradients. There is considerable flexibility, though, as engine and generator running at 1500 rpm deliver 50 kW at 250 V, 200 A, the electric motor for this output being designed to run at 3800 rpm giving torque of 70 lb ft, for overdrive cruising, yet at 800 rpm giving 315 lb ft for gradients.

0.5 EV classification

EVs in common current use include handling trucks, golf carts, delivery vans/floats and airport people movers/baggage handlers. The more challenging on-road application is the subject of most of what follows in this book, where the categories include motor scooter, passenger car, passenger service vehicle, taxi and goods vehicle.

The smallest road-going EVs are probably the electric bicycles such as the Sinclair Zike and the Citibike product. Both these companies also produce bolt-on pedal assist systems for conventional bicycles. Electric motorcycles are less common than electric scooters, the BMW C1 being an example. Recent electric cars have divided between conversions of standard production models and a small number of purpose built vehicles. Japan's flourishing microcar market of smaller and lighter cars is an important target group for electric conversion, for which acceleration and efficient stop-start driving is more important than range. Such city cars are distinct from longer-range inter-urban cars and the latter market currently attracts hybrid drive cars of either gasoline or diesel auxiliary engines, with series or parallel drive configurations. Fuel-cell cars for the inter-urban market are still mostly in the development stage of value engineering for volume production.

Commercial and passenger service vehicle applications, that section of the market where downtime has to be kept to a minimum, and where low maintenance costs are at a premium, are particularly attractive to EVs. Municipal vehicles operating in environmentally sensitive zones are other prime targets. In passenger service applications battery-electric minibuses are a common application in city centres and IC-electric hybrids are increasingly used for urban and suburban duties. Gas-turbine/electric hybrids have also been used in buses and fuel-cell powered drives.

Guided buses include kerb-guided and bus/tram hybrids, the former having the possibility for dual-mode operation as conventionally steered vehicles. Guided buses have been used in Essen since 1980. Trolleybus and tramway systems are also enjoying a comeback.

At this relatively early stage in development of new generation EVs tabular classification is difficult with probably the only major variant being traction battery technology. A useful comparison was provided in a *Financial Times* report[6] on 'The future of the electric vehicle' as follows:

Battery	Advantages	Disadvantages	Comments
Lead–acid	Established technology; low cost and fairly long life (1000 cycles).	Low energy and power density.	Horizon and other high performance batteries greatly improve the suitability for EVs but must be made cheaper.
Nickel–cadmium	Higher energy density and cycle life than lead–acid.	Cadmium very toxic.	Being used for second generation, purpose-built EVs.
Lithium	High energy and power densities. Safety concerns overcome.	Expensive.	Research into scaling up to EV size will probably provide a mid-term battery.
Sodium–sulphur	High efficiency and energy density.	Thermal enclosure and thermal management is expensive. Corrosive components.	Several technical issues to be resolved before this could become an option.
Sodium–nickel chloride	High energy and power densities. Long life (over 1000 cycles).	Thermal enclosure and thermal management are expensive.	Promising mid-term option but currently over twice the cost of the USABC target.
Nickel–metal hydride	High power density, Long cycle life (over 2000 cycles). Twice the energy storage of lead–acid.	Expensive.	Promising mid-term option but currently over twice the cost of the USABC target.

Battery	Advantages	Disadvantages	Comments
Zinc–air	High energy density. Rapid mechanical recharging (3 minutes).	Infrastructural needs.	Interesting longer-term option for rapid recharging.
Nickel–iron	High energy density. Long life (over 1000 deep charge/discharge cycles).	Hydrogen emitted –safety concerns. Periodic topping up with water needed.	Research to increase efficiency and overcome disadvantage could lead to a long-term EV battery.
Nickel–hydrogen	High energy density. Robust and reliable, no overcharge/over-discharge damage. Very long life.	Fairly expensive (due to hand assembly).	Already used in communications satellites. Cost competitive for high cycle operations.

References

1. Cronk, S., Building the E-motive industry, SAE paper, 1995
2. Clarke, S., The crisis of Fordism or the crisis of social democracy, *Telos*, spring, No. 83, pp. 71–98, 1990
3. Womack *et al.*, From lean production to lean enterprise, *Harvard Business Review*, March–April, 1994
4. Burton W., *Proceedings of the Institute of Automobile Engineers*, 1926–1927
5. Whithorne, H., *Proc. IAE*, 1929–1930
6. Harrop, G., *The future of the electric vehicle, a viable market*? Pearson Professional, 1995

J.F.

PART ONE

ELECTROMOTIVE TECHNOLGY

1

Current EV design approaches

1.1 Introduction

The environmental arguments for electric propulsion become more compelling when they can be supported by an economic case that will appeal to the vehicle buyer. Here the current technology of electric and hybrid drive is reviewed in a way that shows the technical imperatives alongside the economic ones. After an analytical study of drive system comparisons for different vehicle categories, 'clean-sheet-design' integrated vehicle electric-drive systems are reviewed for small and medium cars and a concluding section encapsulates a procedure for optimizing motor, drive and batteries in the form of a power-pack solution.

A section on electric-drive fundamentals, establishing basic terminology, appears in the Introduction. In the preface to the case study chapters (5 and 6), contained in the second half of Chapter 4, the whole macro-economics of electric vehicles is discussed, with the wider aspects of the fuel infrastructure, as is a full analysis of competing electric-drive and energy-storage systems, for EVs.

1.2 Case for electric vehicles

1.2.1 ENVIRONMENTAL IMPERATIVE

The current world population of motor vehicles stands at 700 million, of which over 600 million are owned in G7 economies[1]. This number is set to increase to around 1000 million in the next ten years. The bulk of this growth is expected to occur in Second World countries where per capita income is reaching levels where car ownership is known to commence. This has two serious implications (Fig. 1.1): a large increase in the usage of hydrocarbon fuels and an increase in

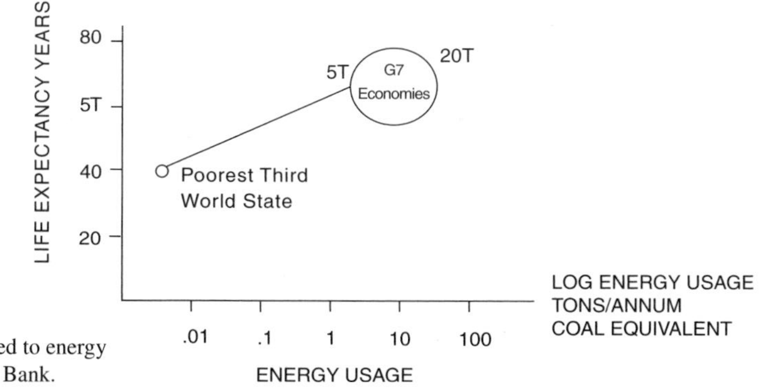

Fig. 1.1 Life expectancy related to energy usage, as seen by the World Bank.

pollution to globally unsustainable levels. Much has been heard of the so-called Greenhouse Effect. If carbon dioxide is on a scale of 1 as a greenhouse gas, methane is 25 and CFCs are 30 000–50 000. Clearly the release of hydrocarbons and CFCs by man must be curtailed as soon as possible; CO_2 is a different matter. If the quantity in the atmosphere was doubled from 20 to 40%, the temperature would increase by 5°C and the sea level would rise by 1 metre. However, the additional plant activity would eliminate famine for millions in Africa, the Middle East and Asia. In scientific circles, the 'jury is still out' on carbon dioxide.

The problem emissions are those of carbon monoxide, sulphur dioxide, nitrous oxide and lead, not to mention solid particles from the exhausts of diesels. In all of these, man is competing with nature. The problem is that man's emissions are now set to reach levels which history shows have had dramatic consequences in nature. For example, in 1815, a volcano emitted 200 million tons of sulphur dioxide into the atmosphere. In 1816 there was a cloud of sulphuric acid in the sky which blocked out the sun in the northern hemisphere for the whole of the summer. The temperature fell by 7° C and there were no crops. Every 2000 megawatt power station which runs on coal emits 150 000 tons of sulphur dioxide per annum. Acid rain destroys our forests and buildings in the northern hemisphere. Pollution on this scale in the southern hemisphere is unsustainable. Nitrous Oxide is emitted when nitrogen burns at 1500° C or above. This gas reaches high concentrations in cities and is converted by sunlight into photosynthesis smog, which is becoming a major health hazard worldwide. A change in the technology of motor transport could have the fastest impact on this problem as most vehicles are replaced every ten years.

1.2.2 ELECTRIC VEHICLES AS PRIMARY TRANSPORT

Consumers vote with their wallets! Electric vehicles will only have a healthy market based on a primary transport role using technology that achieves the performance of internal combustion engines. This means sources of energy other than batteries (Fig. 1.2). In reality we have a choice of IC engine, gas turbine and fuel cell, but how can we maintain performance whilst reducing pollution? The secret is to stop wasting the 72% of energy that currently goes out of the exhaust pipe or up from the radiator. The IC engine is currently operated with a fuel/air ratio of 14:1. This can be increased to 34:1 but the engine can no longer accelerate rapidly. Fortunately, this can be overcome by other means. The gas turbine is an efficient solution for large engines over 100 kW in commercial vehicles. Its performance is not as good as an IC engine's at lower powers, however, and fuel-cell electrics offer the best promise. Fuel cells are the technology of the future. There are many sorts but only one type of any immediate relevance to vehicles and this is the proton exchange membrane (PEM) cell. Using the Carnot cycle, this has a conversion efficiency limit of 83%. Scientists can achieve 58% now and are predicting 70% within ten years. Fuel cells have many excellent qualities. Small units are efficient – especially at light load. New construction techniques are reducing costs all the time and £200/kW was already achievable in 1992 using a hydrogen/air mixture. The real problem is providing the fuel.

1.2.3 THE FUEL INFRASTRUCTURE

Current engines obtain their energy by burning hydrocarbons such as propane, methane, petrol, diesel and so on. However, hydrogen is the fuel of the future. What powers a Saturn 5 Moon Rocket? Coincidence, or sheer necessity? Liquid hydrogen has an energy density of 55 000 BTUs per pound compared to 19 000 BTUs per pound for petrol and 17 000 BTUs per pound for propane. The problem is obtaining large amounts of hydrogen efficiently from hydrocarbon fuels. The percentage of hydrogen directly contained in these fuels is small in energy terms. For example, methane (CH_4) has 17.5% of its energy in carbon and 25% in hydrogen. However, there is now a solution to this problem, with a reforming process developed by Hydrogen Power Corporation/Engelhard called Thermal Catalytic Reforming. Put simply, it is the chemical process:

$$3Fe + 4H_2O = Fe_3O_4 + 4H_2 \qquad \text{and} \qquad Fe_3O_4 + 2C = 3Fe + 2CO_2$$

The first process takes place with a catalyst at 130° C. The hydrogen is stored in a hydride tank until required. The iron is returned to a central facility for reduction by the second process. The main points about this cycle are that a high proportion of hydrocarbon heat energy is converted into hydrogen and that 1 kg of iron provides enough hydrogen for a small car to travel 6 km on a fuel cell.

IC engines and gas turbines run well on most hydrocarbons and hydrogen. Fuel cells need hydrogen. Hydrogen has to be used and stored safely. This could be achieved by reforming it on demand at fuel stations – the waste heat would be used to generate electricity to be pumped back into the national grid. The primary fuel could be any hydrocarbon such as petrol, diesel, methanol, propane or methane. The only constraint is that the fuel source must have low sulphur content so as not to poison the catalyst. In the UK, we have a head start called the Natural Gas Grid. This is likely to become of critical importance for energy distribution, removing the need to distribute petrol and diesel by road. To satisfy future transport needs, we retain our 'fuel' stations as the means of distribution. This brings us to the problem of on-board hydrogen storage.

---

1. Petrol car: A journey of 68 miles each day consumes 2.5 gallons of fuel and takes 2 hours.
   Amount of energy in fuel = 5.14×10^8 joules
   Thermal power = 71.3 kW
   Mechanical power = 20 kW average
   Efficiency = 28%

2. Battery electric car as secondary transport.
   Power station efficiency 40%
   Electric car efficiency 80%
   OVERALL 32%
   CONCLUSION: Pollution is moved from car to power station. There is only an environmental return if the car's performance is sacrificed or the power station is non-thermal and range/performance is limited.

3. Hybrid car as primary transport.
   Hydrocarbon to electricity
   Via lean burn petrol engine 45%
   Electricity to mechanical power 90%
   OVERALL 40.5%
   CONCLUSION: Pollution reduced by 55% and fuel consumption is 70% of petrol vehicle with performance/range as the petrol vehicle.

4. Fuel-cell electric car as primary transport.
   Hydrocarbon to hydrogen conversion 80%
   Fuel-cell hydrogen to electricity 60%
   Electricity to mechanical power 90%
   OVERALL 43% (potential for 48% in 10 years)
   CONCLUSION: Pollution reduced by 90%; fuel consumption is 66% of petrol vehicle and performance/range is as petrol vehicle.

Fig. 1. 2 Some crude comparisons for fuel related to pollution.

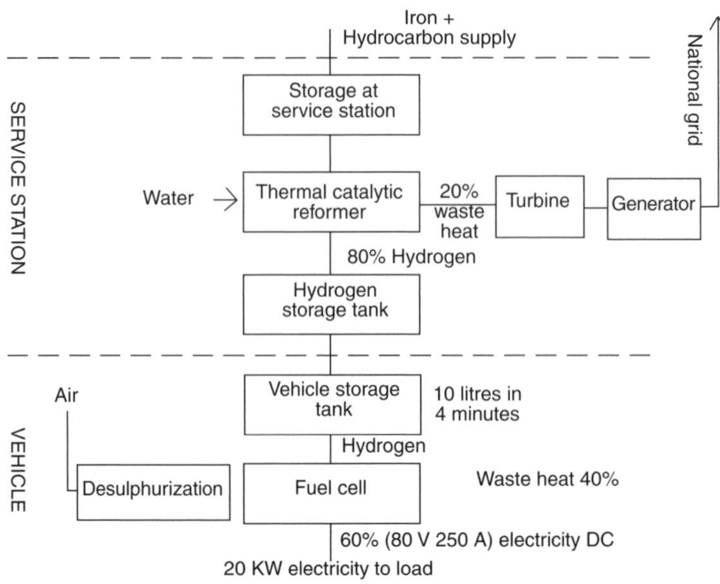

Fig. 1.3 Hydrogen distribution system.

Iron titanium hydride has long been known as a storage medium but one would need 500 kg to store 10 litres of hydrogen, at a cost of £3000 in 1992. The gas is stored in a standard propane tank filled with this material. If the tank is ruptured, the gas is given off slowly because of its absorption in the hydride. In the USA experiments are also taking place with cryogenic storage which is potentially cheaper and lighter. The overall distribution scheme is illustrated in Fig. 1.3. To summarize, the benefits of a change to hybrid/fuel-cell electric vehicles are: (i) engineering is practical; (ii) performance is acceptable to the consumer; (iii) it reduces fuel consumption; (iv) it reduces pollution, especially Nitrous oxide; (v) it reduces dependence on imported oil; (vi) it can be achieved quickly; (vii) it can be achieved at sensible cost; (viii) it prevents increased demand for oil; (ix) it fits in with the existing fuel infrastructure and (x) it solves the pollution problem in relation to projected pollution levels, not existing ones – the prime cause of the catalytic converter being ineffective.

1.2.4 FUEL-CELL ELECTRIC VEHICLE

This vehicle category, Fig. 1.4, will use a fuel cell to provide the motive power for the average power requirement and utilize a booster battery to provide the peak power for acceleration. Hydrogen would be stored in a tank full of metal hydride powder, or cryogenically. This system provides enough waste heat for cabin heating purposes. The fuel cell can recharge the battery when the vehicle is not in use. If the vehicle has an AC drive, it is possible for it to generate electricity for supply to portable tools, a house, or injection into the national grid. Fuel cells should reduce emission levels by a factor of 10, compared with IC engines on 14:1 air:fuel mixture.

1.2.5 CHARACTERISTICS OF FUEL CELLS

What is a fuel cell? It is an electrochemical cell which converts fuel gas and oxidant into electricity and water plus waste heat (see Chapter 4). The PEM cell has graphite electrodes with a layer of membrane sandwiched in between, plus gas-tight seals. Each cell is about 6 mm thick and produces 1 V off-load and 0.7 V on-load, at a current of around 250 amps. Consequently a fuel cell for a 15

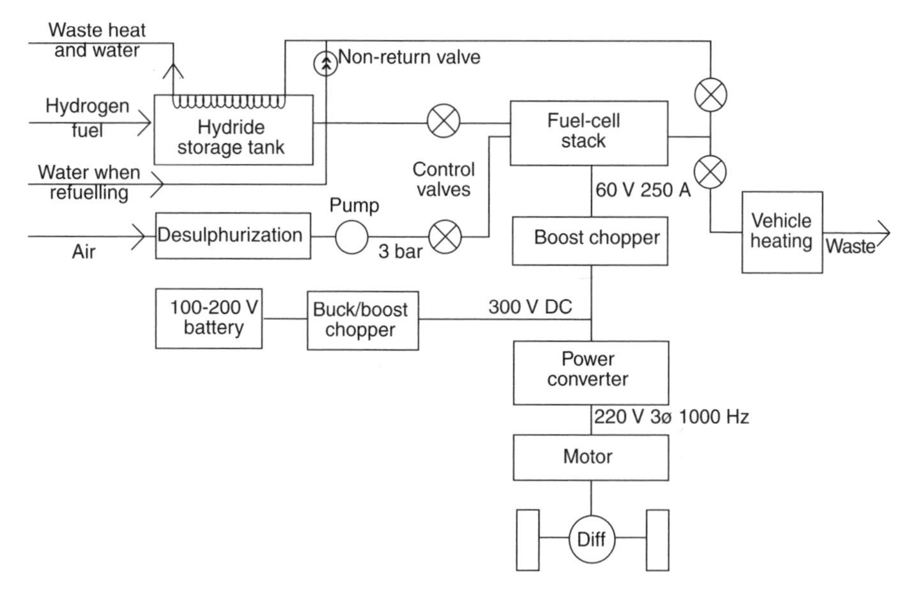

Fig. 1.4 Fuel-cell electric vehicle.

kW average power would produce about 60–70 V DC at 250 amps. In size it would be about 200 mm square and about 600 mm long. The cell operates at a temperature of 80°C. When cold, it can give 50% power instantly and full power after about 3 minutes. The units exhibit very long life. The problem until recently has been seal life when operated on air as opposed to oxygen. New materials have solved this problem. Output doubles when pure oxygen is used. Fuel cells do not like pollutants such as carbon monoxide in the source gases. Gas is normally injected at 0.66 atmospheres into the stack. The main challenge now is to refine the design so as to optimize the cost relative to performance. This will take time because the effort deployed at this time is small in relation to the effort put into batteries or other fuel-cell types. There is a very real case for a major multinational effort to train scientists and engineers in this technology in the short term, and to reduce the time to introduction on a large scale.

1.2.6 THE ROLE OF BATTERIES

Batteries have been with us for at least 150 years and have two main problems: they are heavy and they do not like repeated deep discharge. Batteries which are deep cycled, irrespective of the technology, deteriorate in performance with age. So the question must be asked 'what can batteries do well?'. The answer is to provide limited performance in deep discharge, or alternatively, much better performance as a provider of peak power for hybrid and fuel-cell vehicles.

Much work is under way on high temperature cells. These are unlikely to meet cost or weight constraints of primary transport applications. The best high temperature batteries can offer 100 Wh/kg. Overall, fuel cells already give 300 Wh/kg and this can be improved with development. What is needed is a battery with different capabilities to normal car starter batteries, namely very low internal resistance, long life, excellent gas recombination, room temperature operation, totally sealed, compact construction, reasonable deep discharge life as well as being physically robust.

The battery which satisfies the above criteria is the lead–acid foil battery, as manufactured by Hawker Siddeley. This type of construction has replaced nickel–cadmium pocket batteries on many aircraft. In particular the lead–acid foil battery retains far more charge from regeneration than conventional designs and can be charged and discharged rapidly. However, there is a trick to achieving this. Most batteries are made up of 'rectangular' arrays of cells so it is no wonder that the temperature of the cells varies with position in the stack. To charge a battery quickly it is vital to keep the cells at an even temperature. Consequently it is necessary to liquid cool the cells so as to obtain best performance and long life. Other points worthy of note are that batteries work best when hot; 40°C is ideal for lead–acid. The battery electrolyte is just the place to dump waste heat from the motor/engine/fuel cell.

Nickel–cadmium batteries offer better performance than lead–acid but are double the cost per Wh of storage at present and sealed versions are limited to 10 Ah but larger units are under development. The best nickel–cadmium units available at present are the SAFT STM/STH series. Sealed lead–acid and aqueous nickel–cadmium cells have peak power in W/kg of 90 and 180, with Wh/kg values being 35 and 55 respectively.

In terms of safety, long series strings of aqueous batteries are not a good idea. The leakage from tracking is high and they are very dangerous to work on. Consequently batteries should be of sealed construction with no more than 110 V in a single string. Ideally, the maximum voltage should be 220 V DC, that is +/– 110 V to ground arranged as two separate strings with a centre tap, so that no more than 110 V appears on a connector, with respect to ground, Fig. 1.5.

There is an opening in the market place for a low cost 2 pole, 220 V, 300 A remote-control circuit breaker to act as battery isolator with 5 kA short-circuit capacity. However, there is a problem with earthing the centre tap of the battery as one may need an isolating transformer in the battery charger. Consequently, in many of the new schemes proposed in the USA, a different route is implemented which is used in trolley buses – the all-insulated system. In most of these schemes, large capacity batteries are used (15–30 kWh) at a typical nominal voltage of 300 V. This will vary from 250 V fully discharged to 375 V at the end of charging. The electrical system is fully insulated from earth. During charging, the mains supply can be either centre tap ground or one-end ground. In the centre tap ground (typical USA situation,

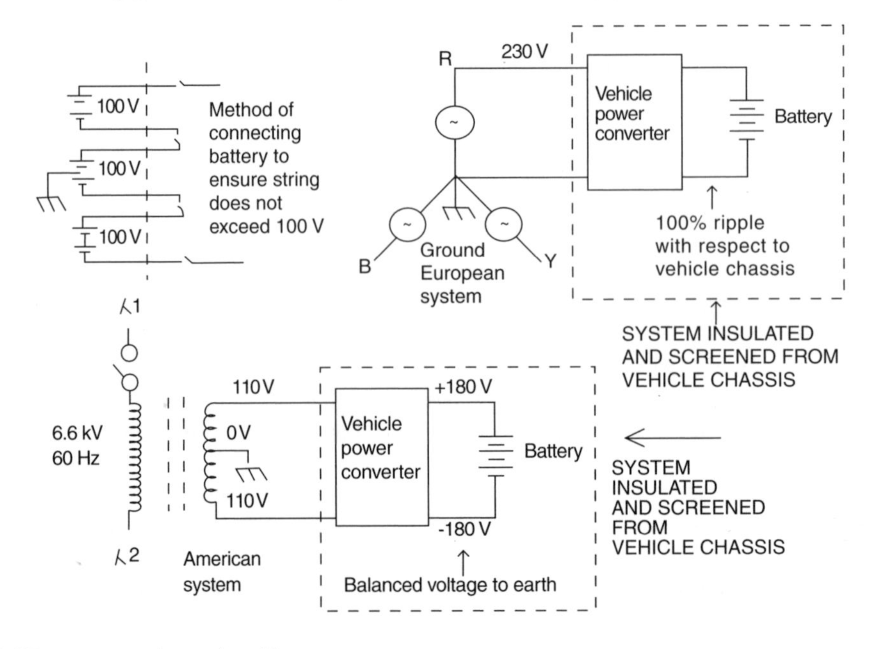

Fig. 1.5 Battery connections and earthing.

with 110/0/110) the potential of the vehicle electrics is balanced to earth. When one end is earth (typical European situation) the potential of the vehicle electrics will move up and down at the supply frequency with respect to ground and there is the prospect of earth leakage current through any capacitance to earth of the vehicle electrical system. However, this is very small, usually because the tyres isolate the vehicle. However, when charging it would be desirable to ground the vehicle body to prevent any shocks from people touching the vehicle and standing on a grounded surface.

1.2.7 ELECTRIC VEHICLE SPECIFICATIONS

From the previous considerations one can now start the task of specifying EV capability/ performance trade-offs. Polaron believe EVs will be partitioned as shown in Fig. 1.6. This does not pretend to be an exhaustive list but to show the range and scale of requirements to be provided for. The most interesting observation is that in the mass market, 30–150 kW, a solution is possible using just two sizes of drive, 45 and 75 kW. To complement the drives, motors are required of two speed ratings for each size, say 5000 rpm where compatibility with a prime mover is required, and 12 000 rpm for the direct drive series hybrid/pure electric case.

1.2.8 HYBRID VEHICLE EXAMPLES

It is now proposed to have a look at two cases (a) 45 kW parallel hybrid vehicle; (b) 90 kW series hybrid vehicle, as in (Fig. 1.7). The 45 kW parallel hybrid vehicle consists of, typically, a small engine driving through a motor directly into the differential gear and hence to the road wheels. Minimization of weight is the key issue on such a design along with low rolling resistance and low drag. At 60 mph a good design can expect to draw 8 kW to keep going on a flat level road. The vehicle would be fitted with an engine rated to supply about one-third of the peak requirement, that is 15 kW plus an allowance for air conditioning if relevant. The motor has to deliver up to 45 kW using energy stored in batteries. This can be done either by a constant torque motor operating via a gearbox or a constant power motor operating with only two gears

Power Rating	GVW	Engine	Motor type	Motor rating	Turbo alternator	Application
Below 40 kW	Less than 2 tons	None	Brush DC	Up to 40 KW	None	Straight battery electric van or car
40 kW –150 kW	2 ton	IC	Brushless DC	1 x 45 kW 5000 rpm	None	Parallel hybrid family car
	2 ton	GT	Brushless DC	1 x 75 kW 12 000 rpm	1 x 100 kW 60 000 rpm	Parallel hybrid performance saloon
	3 ton	IC	Brushless DC	1 x 75 kW 5000 rpm	None	Parallel hybrid 1 ton truck
	5 ton	GT	Brushless DC	2 x 45 kW 12 000 rpm	1 x 100 kW 60 000 rpm	Series hybrid 2 ton truck
	7 ton	GT	Brushless DC	2 x 75 kW 12 000 rpm	1 x 150 KW 50 000 rpm	Series hybrid single deck bus
150 kW to 1 MW	10 ton ↓ 40 tons	GT	Switched reluctance motor	1 x rating 5000 rpm	1 x rating 50 000 rpm at 150 kW 25 000 rpm at 1 MW	Heavy traction and road haulage Series hybrid configuration

Fig. 1.6 Short-term battery electric and hybrid vehicles.

or without a gearbox. The latter is rapidly becoming the standard for EVs using front wheel drive.

The vehicle uses the battery to provide peak acceleration power for overtaking, hill climbing and so on. On the flat a 0–60 mph acceleration time of around 12 seconds would be typical for this class of vehicle and a top speed of perhaps 80 mph, where permitted; the engine is started when the road speed exceeds 20 mph and then clutched into the motor. The engine then charges the batteries as well as satisfying the average demand of the car. During acceleration the electric drive and the engine work together to provide peak acceleration. It is in the cruise condition that optimum efficiency is required. Consequently more sophisticated designs use 3 way clutch units so that the motor can be mechanically disconnected when the battery is fully charged and only switched back in for acceleration. In this condition attention must also be paid to the minimization of rolling resistance and windage losses (Figs 1.6 and 1.8(a)).

The series hybrid vehicle corresponds to a high performance sports saloon. A 0–60 mph time of 7 seconds and a top speed of 120 mph could be expected (Figs 1.7 and 1.8(b)). The main power source would be a gas turbine which would operate through a PWM inverter stage to feed 300–500 V DC into the main bus. There are two separate drives, each driving a rear wheel of the vehicle. To reduce weight, the motors would be designed for 12 000 rpm and gearboxes employed to reduce the speed to the road wheels – about 1800 rpm at 120 mph. The gas turbine may operate over a 2:1 speed range to give good efficiency. Specific fuel consumption is doubled at 15 kW compared to 100 kW. However, overall consumption would still be that of a 'Mini', with emissions to match. The peak power for acceleration would come from batteries – probably nickel–cadmium in this case, where cost pressures are not so demanding.

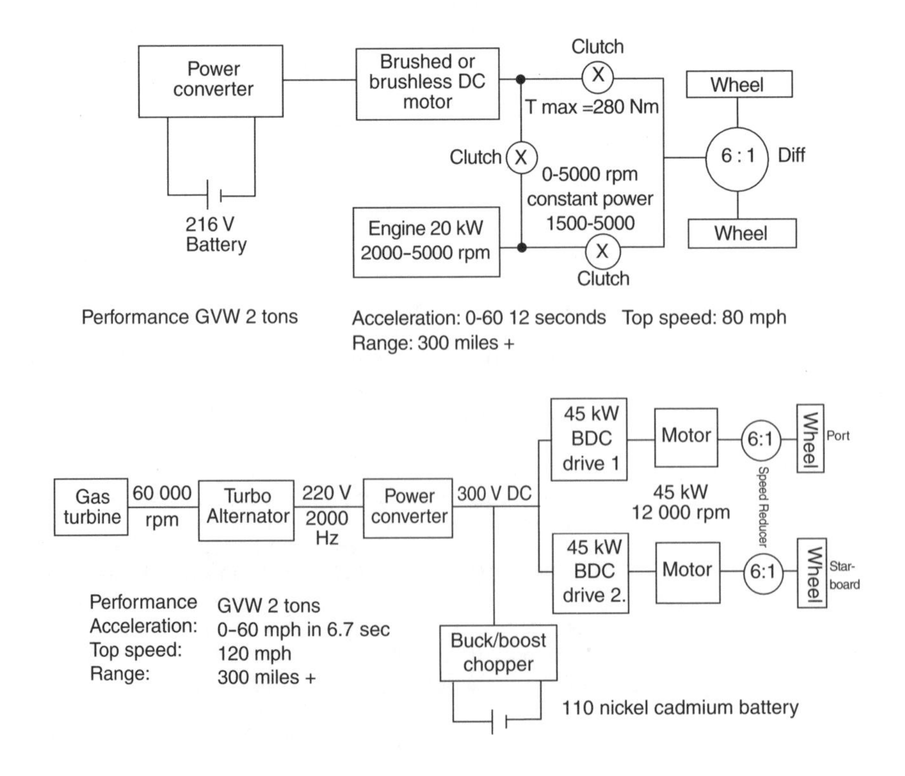

Fig. 1.7 A 45 kW parallel hybrid and 90 kW series hybrid.

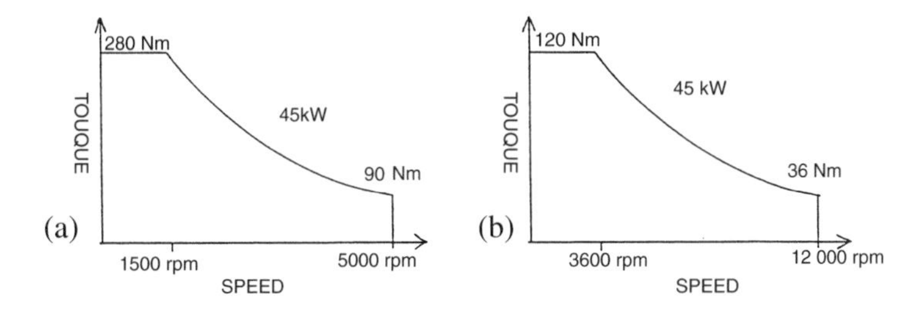

Fig. 1.8 Torque–speed curves for 45 kW vehicle (a) and each motor (b).

1.2.9 ELECTRICAL SYSTEM DESIGN CHALLENGE

What are the design problems for the electrical system? The first one is cost. Unless the final product is attractive to the consumer, we do not have a market. Where are we now? For 1000 off systems at 45 kW, a brushless DC motor would cost £1000, a controller £2000, and a battery £2000 (lead–acid). These 1992 prices will reduce with mass production. The second design challenge is one of methodology. Electric vehicles have been traditionally built by placing motor and batteries then spreading the electrical system over the vehicle. This needs to change. Polaron would like to suggest a modular approach to the problem whereby sealed batteries and controller power electronics are in one unit and the motor is in fact the second. The third design challenge is one of compatibility. Low performance vehicles can be built with 110 V electrical systems. However, as the power increases this is not practical. But both fuel cells and batteries are low voltage heavy current devices – how can this conflict be addressed?

The solution is to use power conversion. In Fig. 1.9 a 100 stage fuel cell is integrated with a 216 V battery to give a stabilized 300 V DC rail. The motor and controller are then built at 300 V where the currents are significantly reduced on the 100 V system. As the power level rises, voltages up to 500 V DC can be anticipated. However, when the power conversion is switched off the highest voltage will be the battery voltage. This additional power conversion will be needed for another reason. If vehicles are equipped with small booster batteries for acceleration, the DC link voltage will change significantly according to load conditions. The power conversion provides a means of stabilizing for this variation.

1.2.10 MOTOR TYPES AND LOCATIONS (FIG. 1.10)

Which is the best type of motor? Answer – the cheapest. Which is the cheapest motor? Answer – the lightest. Which is the lightest motor? Answer – the most efficient. On this criteria, there is no doubt that a permanent magnet brushless DC motor would sweep the board. However, our

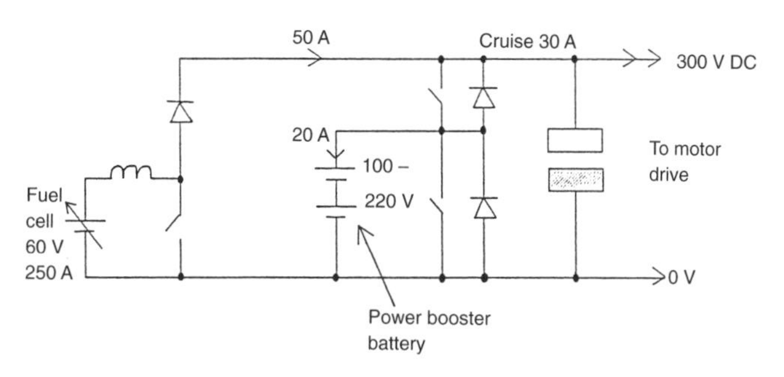

Fig. 1.9 Fuel-cell power conversion.

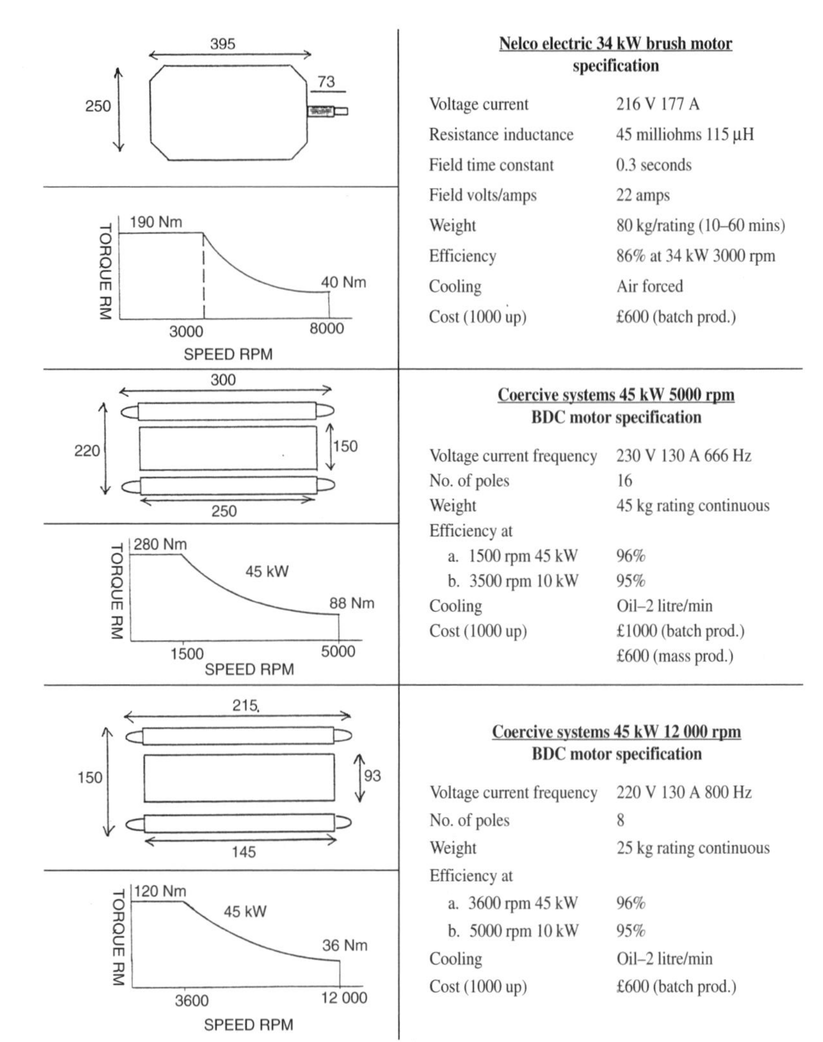

Fig. 1.10 Motor specifications.

enthusiasm must be tempered by two other considerations, cost of materials and controller costs. The factors affecting selection are covered in Section 1.3.

1.2.11 CHOPPER CONTROLLER FOR A 45 kW MOTOR

Figure 1.11 illustrates a typical pure battery electric vehicle scheme which could also be used in hybrid mode with an engine if required. The motor is a shunt field unit such as the Nelco Nexus 2 unit used in many industrial EVs. This machine is a 4 pole motor with interpoles and operates at a maximum voltage of 200 V DC. The field supply is typically 30 amps for maximum torque. The controller consists of a 2 quadrant chopper with a switch capacity of 400 amps. An electromechanical contactor shorts out the positive chopper switch in cruise mode for maximum efficiency. The chopper is fitted with input RF filtering and precharge to extend contactor life. The chopper switches at 16 kHz and the output contains a small L/C filter to remove the d*v*/d*t* from the machine armature. A Hall effect DCCT measures the armature current for the control system.

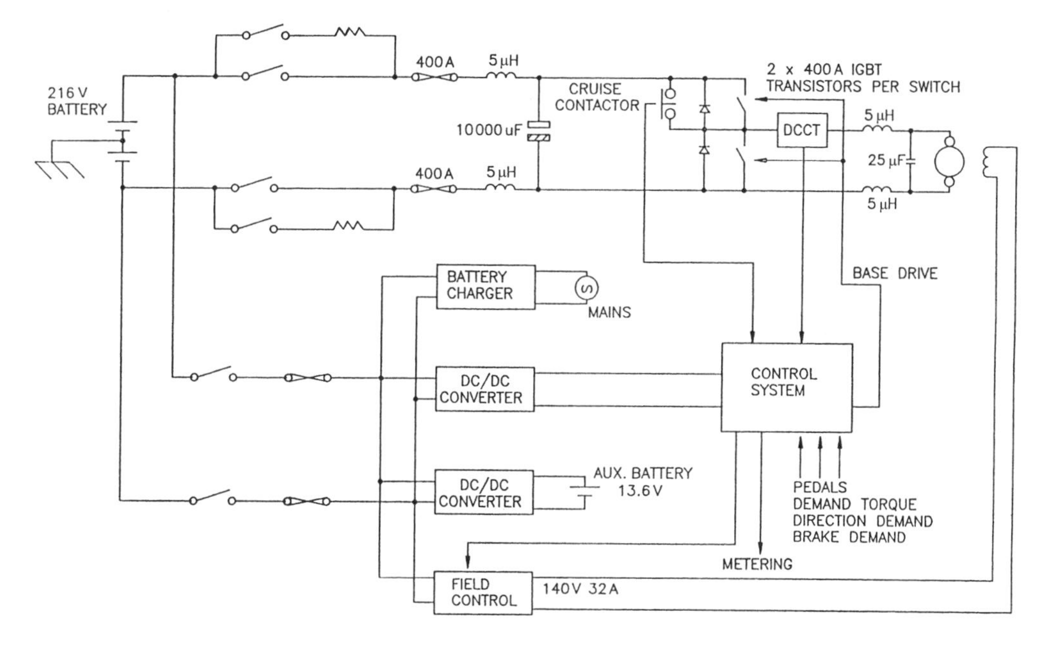

Fig. 1.11 Controller for 34 kW shunt field DC motor.

In the power supply area, there are four components: first is the battery charger, in this case a CUK converter, or a boost/buck chopper is also a possibility to make the mains current look like a sine wave for ensuring IEC555 compliance. Control of battery charging conditions is one of the most important considerations in extending battery life in deep discharge. For lead–acid batteries the level of float voltage is critical as well as maintaining cell temperature. The battery charger could incorporate a 20 kHz isolating transformer if costs permit. Experiments are under way with inductive power transfer which isolates the car and makes it necessary to plug in for charging. Another possibility is an automatic self-aligning connector which the car drives into when parking. The next consideration is the auxiliary 13.6 V battery supply. The vehicle seems likely to retain a separate 12 V battery for lighting and control functions. A 300 W DC/DC converter will satisfy this requirement. The third consideration is the control system power. This is a small (20 W) DC/DC converter which provides the control power for the chopper. It is likely to be incorporated with the main control PCB and could also be supplied from the 13.6 V battery. The final factor is the field controller. This is a 4 quadrant chopper which provides the motor field supply. It has to be able to reverse the current so that the motor can reverse without contactors in the armature circuit. If the motor has a tachometer fitted, this may be used for braking control and blending with electromechanical brakes. The important issue with this controller is that the power switching is contained in a single unit so that all the DC components are kept in one place. This is important for another reason to meet IEC555 RF interference legislation. Therefore all insulated systems will require an isolated conductive casing which can be connected to vehicle chassis.

1.2.12 CONTROLLER FOR A 45 kW AC MOTOR (BRUSHLESS DC OR INDUCTION)

This is illustrated in Fig. 1.12. The drive consists of a 3 phase PWM Drive which feeds the 3 phase motor. The beauty of this arrangement is that the motor can be disconnected and the mains fed to the inverter arms to give a high power battery charger, by phase locking the PWM to the mains.

An alternative to this arrangement is for the inverter to put power back into the mains. In case of fault, three alternistors provide current limit protection. In the brushless DC case, the motor permanent magnets provide 50% of the flux and the remainder comes from a 50 amp circulating current *Id* at right angles to the torque producing component *Iq*.

The inverter is constructed using 300 amp IBGT phase leg packages which minimize the inductance between transistors and associated bypass diodes. The inverter output is filtered by 6 x 10 μH capacitors plus 3 x 5 μH inductors. This reduces the 18 kHz carrier ripple current in the motor to about 20 AP/P. There is a real time digital signal processor (DSP) which performs vector control using state space techniques and this includes 3rd harmonic injection to maximize the inverter output voltage. Comprehensive overload protection is fitted. The inverter demand is a torque signal and a speed feedback is provided for the vehicle builder to close the speed loop. Both signals are PWM format (10–90%) on a 400 Hz carrier. The drive can be adapted for induction motor control but this is not so efficient, as explained in the motor section below.

1.2.13 TURBO ALTERNATOR SYSTEM FOR GAS TURBINES

Figure 1.13 illustrates a turbo alternator scheme for gas turbines. This scheme has two purposes: it starts the turbine, and provides a stabilized DC link voltage for a 2:1 change in turbine speed and changes in DC link current from no-load to full-load. The alternator itself is the result of many years' development in high speed gas compressors. It is a 4 pole unit which allows iron losses to be kept low and in particular the tooth tip temperature reasonable whilst still using silicon steel laminations (2 pole permanent magnet alternators are potential fireballs!). The magnet material is samarium cobalt with a carbon fibre or Kevlar sleeve. At these speeds, one needs every bit of strength possible. The magnets are capable of operation at 150° C. The use of metallic magnets is not a problem here because the weight is small. Hall sensors are fitted for machine timing during starting and voltage control purposes. A small L/C filter limits the amplitude of the carrier ripple on the alternator windings.

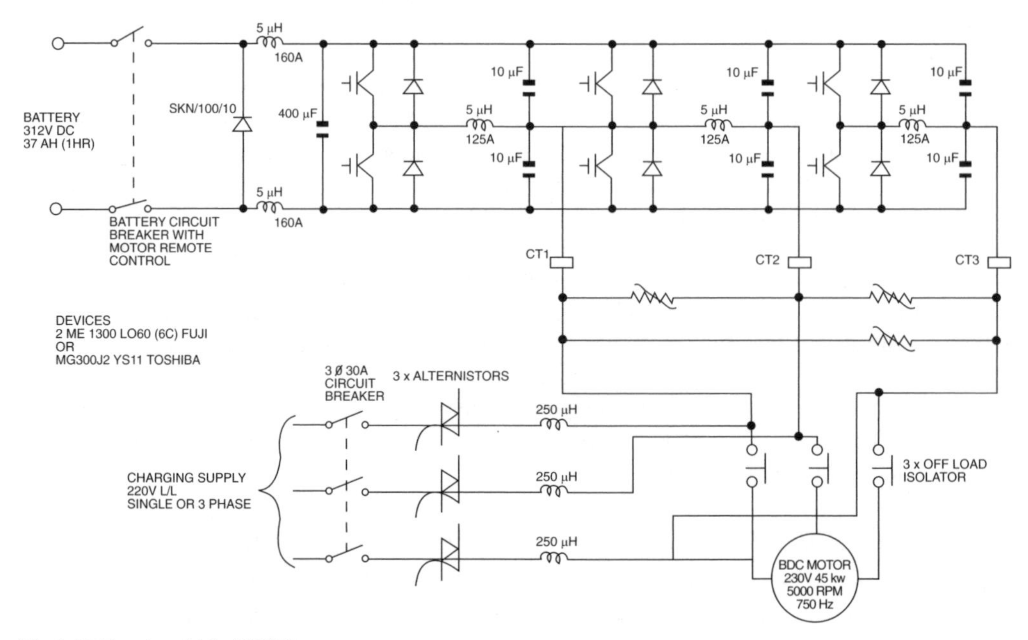

Fig. 1.12 Electric vehicle 45 kW inverter.

1.2.14 MODULAR SYSTEMS

From the foregoing considerations, it will be apparent that the motor car of the future needs power electronics to be viable. Fortunately, we now have the technology to satisfy the most demanding applications. There may be some rivalry between different types of power switches but cost will be the final judge. A manufacturer who constructs the power electronics as an all-insulated system in a single module permits module exchange as the first means of maintenance. Liquid cooling also makes sense. It can cool the motor, warm/cool the sealed batteries and provide power steering at the same time. This concept will make it possible to convert existing chassis as well as develop new ones, thus enabling product to be brought to market quickly. Standard electronics packages are the only way to achieve the unit costs necessary for product acceptance in the market. Interchangeable batteries will make it possible for maximum vehicle utilization in intensive duty applications, such as taxis and delivery vehicles. This method of construction also opens the door to new methods of financing EVs; for example, the user buys vehicle then rents battery/power electronics.

1.3 Selecting EV motor type for particular vehicle application

1.3.1 INTRODUCTION

Motor and drive characteristics are selected here for three different applications: an electric scooter; a two-seater electric car and a heavy goods vehicle, from four motor technologies: brushed DC motor, induction motor, permanent-magnet brushless DC and switched reluctance motor[2]. Any of the four machines could satisfy any application. This is not a battle of 'being able to do it', it is a battle to do it in the most cost-effective manner. There are two schools of thought regarding EVs – group A believe they should create protected subsidized markets for environmental reasons and are not too concerned with cost. Group B realize that until this technology can compete with

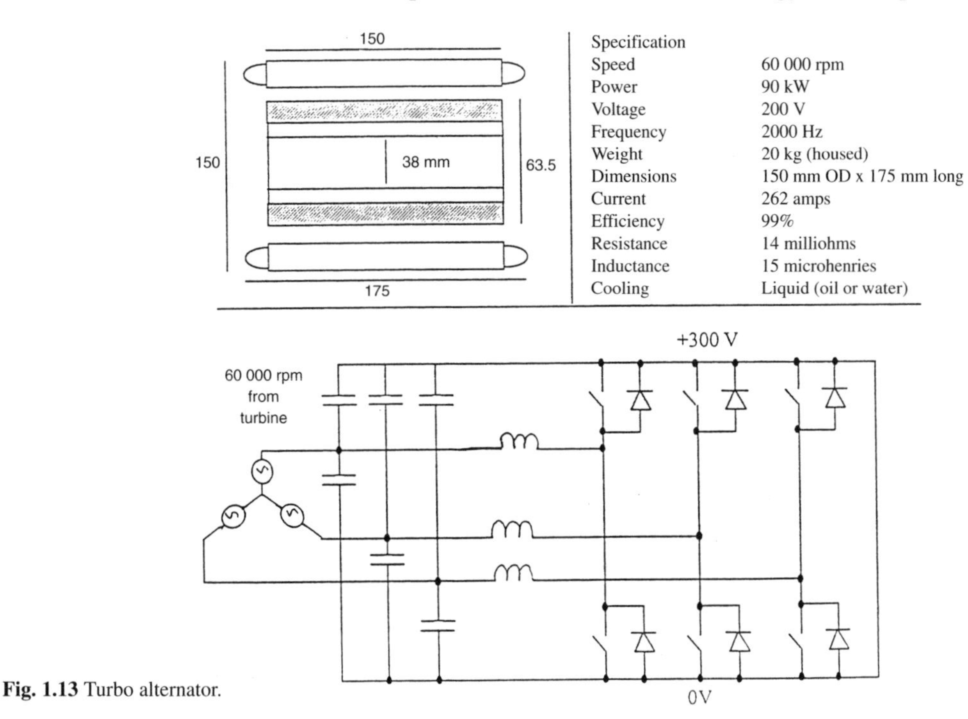

Specification	
Speed	60 000 rpm
Power	90 kW
Voltage	200 V
Frequency	2000 Hz
Weight	20 kg (housed)
Dimensions	150 mm OD x 175 mm long
Current	262 amps
Efficiency	99%
Resistance	14 milliohms
Inductance	15 microhenries
Cooling	Liquid (oil or water)

Fig. 1.13 Turbo alternator.

piston engines in terms of performance and cost there will be no significant competition, hence no major market share. Polaron are putting their money on group B. What is clear is that the economics will come right at lower powers first, then work upwards. Another fact is that a market needs to be established before custom designs can be justified and the most immediate need is for conversion technology for existing vehicle platforms.

1.3.2 BRUSHED DC MOTOR

This consists of a stationary field system and rotating armature/brushgear commutation system. The field can be series or shunt wound depending on the required characteristics. The technology is well established with more than a century and half of development. The main problem is one of weight compared with alternative technologies, consequently Polaron believe DC is best at lower powers overall, due to the built-in commutation scheme. As the power level rises many problems become significant: commutation limited to 200 Hz for high speed operation; problems with commutator contamination; significant levels of RF interference; brush life limitations and cooling/insulation life limitations. Polaron's Nelco division has made these machines for many years and

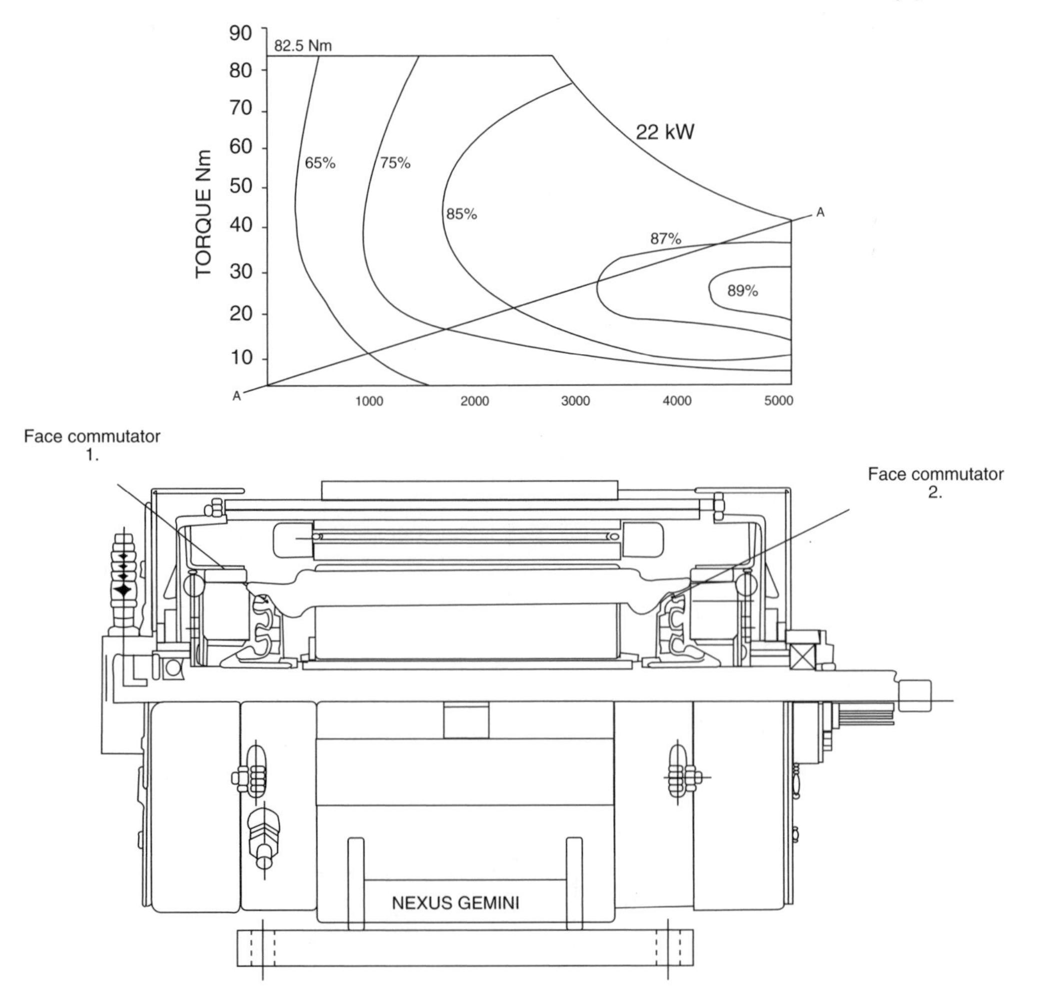

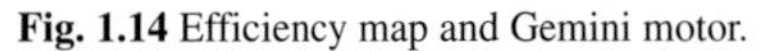

Fig. 1.14 Efficiency map and Gemini motor.

has introduced a new design to help overcome some of the problems. The so-called Gemini series consists of an armature with a face commutator at both ends of the armature. This permits two independent windings which may be connected in series or parallel. Improvements in the torque speed curve are seen in Fig. 1.14, while Fig. 1.15 shows a recently developed controller. While existing controllers have single quadrant choppers with contactors for reversing and braking, and field control is effected by a separate chopper unit, Polaron feel such a design gives limited overall performance and is better replaced by the arrangement shown. Brushed DC motors have a role in applications below 45 kW but, if power rises above this figure, mechanical considerations such as the removal of heat from the rotor become more important. There are also factors to take into account in terms of efficiency when partially loaded. In many of these respects, the use of brushless DC motors could provide a better alternative. These have a number of features acting in their favour, including high efficiency in the cruise mode and a readily adjustable field, plus the practical benefits of a more easily made rotor.

1.3.3 BRUSHLESS DC MOTOR

The term 'brushless DC motor', however, is a misnomer. More accurately it should be described as an AC synchronous motor with rotor position feedback providing the characteristics of a DC shunt motor when looking at the DC bus. It is mechanically different from the brushed DC motor in that there is no commutator and the rotor is made up of laminations with a series of discrete permanent magnets inserted into the periphery. In this type of machine, the field system is provided by the combined effects of the permanent magnets and armature reaction from vector control. Similar in principle to the synchronous motor, the rotor of this machine is fitted with permanent magnets which lock on to a rotating magnetic field produced by the stator. The rotating field has to be generated by an alternating current and in order to vary the speed, the frequency of the supply must be changed. This means that more complex controllers based on inverter technology have to be used.

Induction motors are used by many US battery-electric cars. The rotors are cooled with internal oil sprays which also lubricate the speed reducer. Operation at 12 000 rpm is common to minimize the torque and some designs operate under vacuum to reduce the noise. The one good point is that these motors are reasonably efficient under average cruise conditions (8000 rpm, 1/3 FLT). Polaron's view is their use will be short lived. Induction motors always have lagging power factors which cause significant switching losses in the inverter, and vector control is complex.

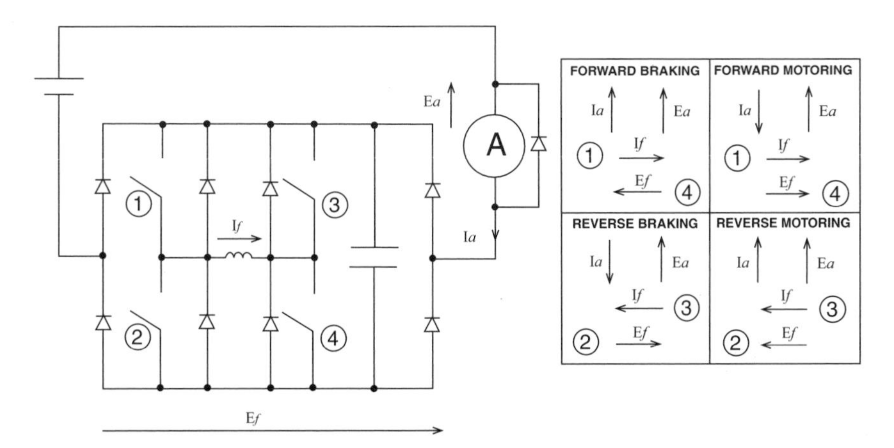

Fig. 1.15 Integral 4-quadrant chopper.

1.3.4 SWITCHED RELUCTANCE MOTORS

SRMs, Fig. 1.16, use controlled magnetic attraction in the 6/4 arrangement to produce torque. Existing SR drives are unipolar, in that the voltages applied to windings are of only one polarity. This was done to avoid shoot through problems in the power devices of the inverter. The 6/4 machine has a torque/speed curve similar to a DC series motor with a 4:1 constant power operating region. Torque ripple can be serious at low speed (20%).

In an attempt to improve the SR drive, two groups have made significant contributions: SR drives have worked with ERA Drives Club in developing the 8/12 SR motor, with much smoother operation; a University of Newcastle upon Tyne company, Mecrow, have postulated a bipolar switched reluctance machine using wave windings. This doubles copper utilization and increases output torque. It also uses a standard 3 phase bridge converter. Existing SR motors are both heavier and less efficient than PM BDC machines, for example a 45 kW unit (3.5:1 constant power/5000 rpm) would weigh 65 kg and have an efficiency of 94%. The new bipolar design should give a motor which is close to PM BDC in terms of weight (45 kg). However, in terms of efficiency, the BDC has the edge, both in the machine and the inverter, because it operates with a leading power factor under constant power conditions. However, SR motors are excellent for use in hostile environments and it is Polaron's expectation that they will be successful in heavy traction, where magnet cost may preclude brushless DC.

1.3.5 ELECTRIC MOTORCYCLE

An electric motorcycle is an interesting problem for electric drives. The ubiquitous 'Honda 50', an industry standard, is typical of personal transport in countries with large populations. The petrol machine weighs 70 kg and has an engine capable of about 5.5 bhp. Honda have developed an electric version where the engine is exchanged for an electric motor and lead–acid batteries. Honda's solution weighs 110 kg and has a range of 60 km; it is offered in prototype quantities at £2500 ($3500), 1996 prices. Some elementary modelling shows that the key problem is battery weight – especially using lead–acid. To minimize this requires good efficiency for both motor and driveline. The standard driveline from engine to wheel is about 65% efficient. A better solution is to use a low speed motor with direct chain drive onto the rear wheel. This solution offers a driveline efficiency of 90%. However, we need a machine to give constant power from 700 to 1500 rpm. Cruising power equates to 1.5 bhp at 40 km/h and 5 bhp at 60 km/h. Vital in achieving good rolling resistance figures is to use large diameter tyres of, say, 24 inches.

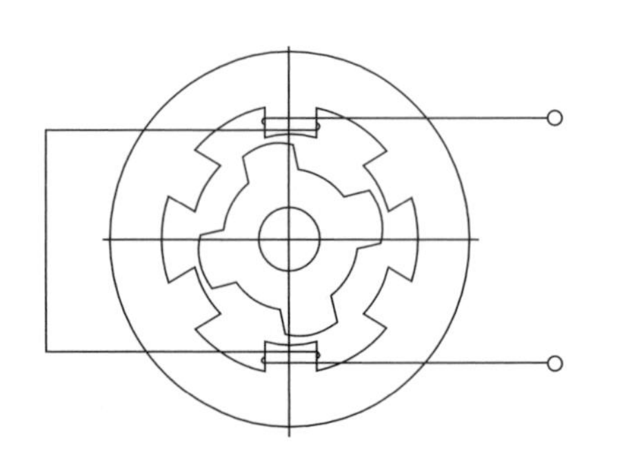

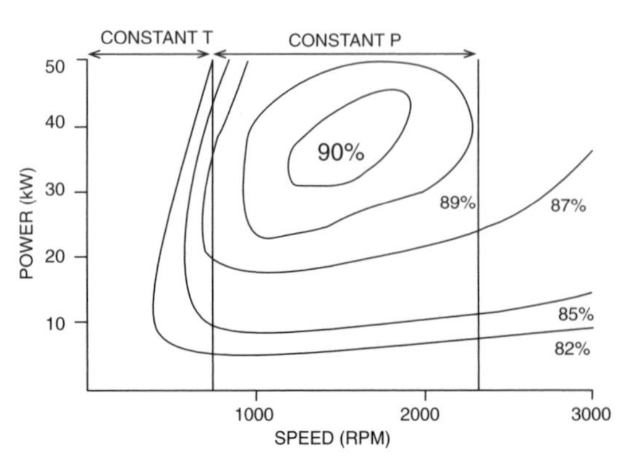

Fig. 1.16 Switched reluctance motor.

It is assumed that sealed batteries are to be used and consequently a battery voltage of 96 V was chosen to optimize the efficiency of motor and controller and particularly with an eye to controller cost. 200 V MOSFETS are near optimal at 100 V DC. A battery of 15 Ah 96 V weighs 40 kg (for comparison 24 V 60 Ah weighs 35 kg). In lead–acid 36 Wh/kg is achieved, while for comparison nickel hydride cells could offer 80 cells x 1.2 V x 25 Ah in a weight of 30 kg. The motor has to deliver a torque of about 40 Nm maximum and consequently a pancake-type design was chosen. Induction motors were rejected due to low efficiency and large mass for this duty. The four practical contenders are: permanent magnet brushless DC; permanent magnet DC brush pancake motor; DC series motor or switched reluctance motor. A tabulated comparison at Fig. 1.17(a) compares results. As can be seen, the permanent magnet brushless DC motor is the optimum performer at the two key cruise conditions. It has been estimated that with regenerative braking and flat terrain, a range of 70 km could be achieved with a 96 V 15 Ah lead–acid battery. The 25 Ah nickel hydride pack could give 120 km. However, 70 km is quite adequate for average daily use.

1.3.6 SMALL CAR

The small electric car is in the Mini or Fiat 500 class. Such a vehicle would weigh 750 kg and accelerate from 0 to 50 mph (80 km/h) in 12 seconds and have a range of 80 km with lead–acid batteries. The motor power would be 20 kW peak. As originally there were only aqueous batteries available, battery voltage was limited to 120 V DC by the tracking that took place across the terminals of the batteries due to electrolyte leakage. Two battery technologies were available: lead–acid and nickel–cadmium and vehicles were designed with efficiency = 25%, that is 188 kg of batteries if efficiency is expressed as battery mass/gross vehicle mass (for lead–acid 60 Ah 120 V 7.2 kWh and for nickel–cadmium 85 Ah 120 V 9.9 kWh).

Single quadrant MOSFET choppers were developed by Curtis and others to supply DC brushed series motors. The main advantage of this system was low cost (for example, lead–acid battery £900 in 1996; quadrant chopper £500; motor DC series £750). However, the apparent cheapness of this system is deceptive because: (a) fitting regeneration can raise the battery voltage to 150 V – an unsustainable level for some choppers – consequently friction braking was often used; (b) a separate battery charger was required. More recently sealed battery systems have become available and batteries of around 200 V are possible in two technologies, lead–acid foil and nickel hydride. These batteries are used with 600 V IGBT transistors which can operate at voltages up to 350 V DC. Battery capacity becomes limited if other services such as cabin temperature control/lighting/ battery thermal management are taken into consideration. A small engine driven generator transforms this problem and it is perhaps worth noting Honda have achieved full CARB approval for their small lean burn carburettor engines with the discovery that needle jet alignment is critical to emissions control and negates the need for catalytic converters.

All motor technologies are viable at 196 V; however, the practical consideration is that inverters are more costly than choppers which accounts for the popularity of DC brushed motors/choppers. To counteract the inverter cost premium, the electronically commutated machines have been designed for 12 000 rpm, to reduce the motor torque (DC brush machine 20 kW at 5000 rpm; other types 20 kW at 12 000 rpm). Another benefit of the higher transistor voltage capability is that the inverters/choppers can function as battery chargers direct off 220/240 V without additional equipment. High rate charging is possible where the supply permits. All electronically commutated machines provide regeneration. The motor comparison is tabulated at Fig. 1.17(b). All the machines deliver constant power (20 kW) over a 4:1 speed range, making gear changing unnecessary. The induction/brushless motors are assumed to use vector control.

(a)

	PM BDC	Brushed PM pancake	DC series motor	Switched reluctance
Size (mm)	200 × 100	200 × 100	200 × 175	200 × 150
Weight (kg)	10	10	18	14
Rating	3 @ 750	3 @750	3 @750	3 @1500
(kW@rpm@V)	40	40	60	70
	3 @1500	3 @1500	3 @1500	
	70	80	80	
Efficiency	0.3/750 80%	3/750 75%	3/750 70%	3/750 80%
(motor	0.75/750 94%	750/750 80%	750/750 70%	750/750 85%
only)	3/1500 93%	3/1500 85%	3/1500 80%	3/1500 85%

(b)

	Brushless DC PM motor	Induction motor	Switched reluctance	Brushed DC motor
Speed (rpm)	3000	3000	3000	1250
Torque (Nm)	64		64	154
rising to:				
Speed (rpm)	12 000		12 000	5000
Torque (Nm)	16		16	38.5
Voltage (V)	150 AC	150	75–150	192
Current (A)	126–81 AC	164–106 AC	180–90 DC	122 DC
Power (kW)	20	20	20	20
Frequency (Hz)	800	400	800	(equiv. 125 Hz)
Weight (kg)	12	25	20	50
Efficiency				
% @ 3000	95	90	92	(1250) 80
% @ 12 000	97	92	94	(5000) 85
Cooling	oil	oil	oil	air

(c)

	PM (DC) brushless	Induction motor	Switched reluctance	DC brush
Speed (rpm)	1000	1000	1000	1000
Torque (Nm)	2866	2866	2866	2866
at speed (rpm)	4000	4000	4000	4000
Torque (Nm)	716	716	716	716
Voltage (V)	380	380	190/380	500
Current (A)	753–486	980–630	1000/500	520
Power (kW)	300	300	300	300
Frequency (Hz)	1056	133	266	(133 equiv.)
Weight (kg)	300	600	500	1000
Efficiency				
% @ 1000	95	93	94	85
% @ 4000	97	95	96	89
Cooling	oil	oil	oil	air

Fig. 1.17 Motor comparisons for three vehicle categories (the four motor types are also discussed in Chapter 4).

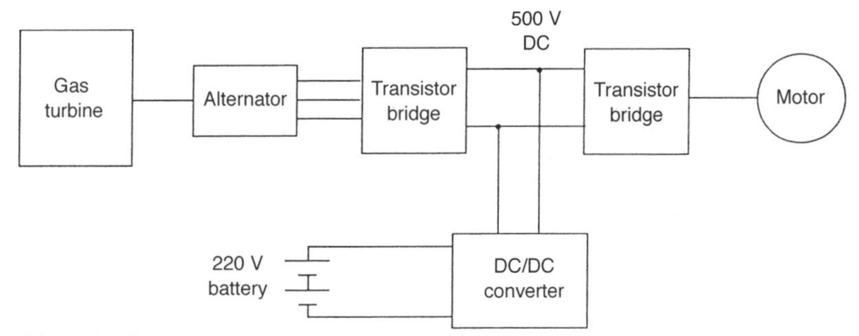

Fig. 1.18 Gas turbine technology.

1.3.7 HGV

The heavy goods vehicle is an articulated truck which weighs 40 tonnes. Often omitted from clean air schemes on the grounds of low numbers they travel intercontinental distances every year and are major emitters of NO_x and solid particles. Their presence is felt where there are congested urban motorways, and each one typically deposits a dustbin-full of carbon alone into the atmosphere every day, the industry declining to collect and dispose of this material! What is the solution? Use hybrid drivelines based on gas turbine technology; these vehicles would be series hybrids.

A gas turbine/alternator/transistor active rectifier, Fig. 1.18, provides a fixed DC link of 500 V. This is backed up by a battery plus DC/DC converter. A battery of 220 V (totally insulated) is used for safety. High quality thermal management would be vital to ensure long battery life; 2 tonnes of lead–acid units would be needed (144×6 V $\times$ 110 Ah) to be able to draw 400 bhp of peak power. It is likely that capital cost would be offset by fuel cost savings. Another benefit is that the gas turbine can be multifuel and operation from LNG could be especially beneficial. The drive wheels are typically 1 metre in diameter giving 683 rpm at 80 mph. Usually there are 3:1 hub reductions in the wheels and a 2:1 ratio in the rear axle, giving a motor top speed of 4000 rpm. Translated into torque speed this means 2866 Nm at 1000 rpm, falling to 716 Nm at 4000 rpm. All motors are viable at this power; however, two factors dominate: (a) low cost and (b) low maintenance. DC brushed motors with 3000 hour brush life are unlikely contenders! PM brushless DC is unlikely on cost grounds, requiring 36 kg of magnets for 2900 Nm of torque. Both induction motors and switched reluctance are viable contenders but switched reluctance wins on efficiency and weight. The contenders are tabulated at Fig. 1.17(c).

In the above review of four motor technologies for three vehicle categories, there is no clear winner under all situations but a range of technologies is evident which are optimal under specific conditions. Continuing development should improve the electronically commutated machines especially brushless DC and switched reluctance types. The relative success of these machines will be determined by improvements in magnet technology, especially plastic magnets, and cost reduction with volume of usage. On the device front, development is approaching a near ideal with 1/2 micron line width insulated gate bipolar transistors (40 kHz switching/l.5 V VCE saturated) but reduction in packaging cost must be the next major goal.

1.4 Inverter technology

Inverters are one area where progress is being made in just about every area[3]: silicon, packaging, control, processors and transducers. The task is to find a way down the learning curve as quickly as possible. Polaron believe the lowest cost will come from packaging motor and inverter as a single unit. The major development this year is that of reliable wire bond packaging for high

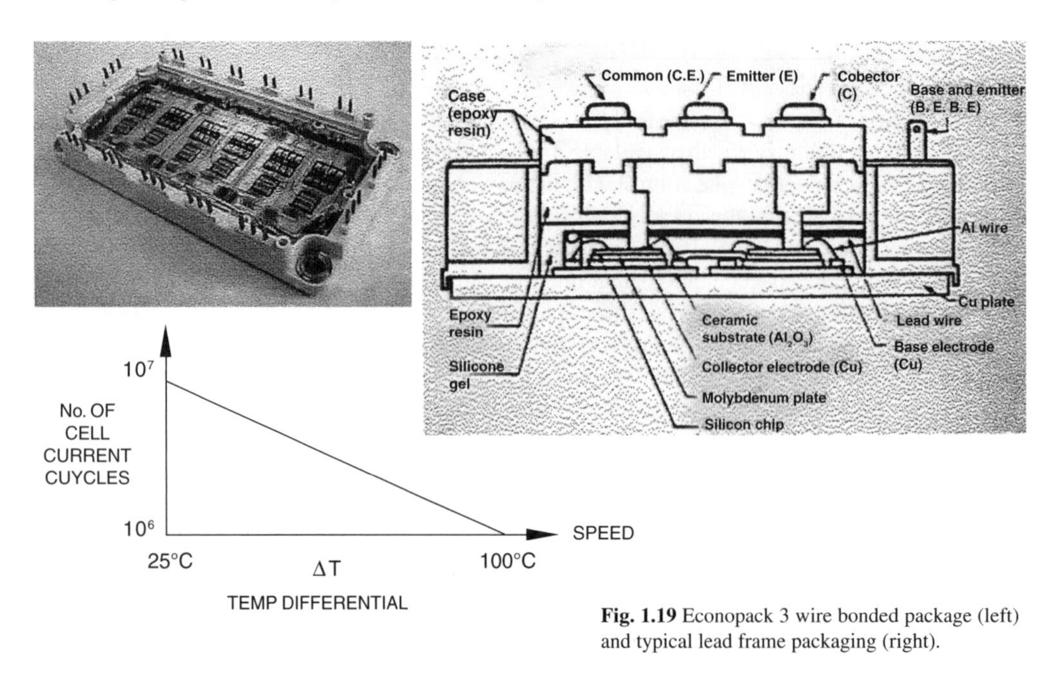

Fig. 1.19 Econopack 3 wire bonded package (left) and typical lead frame packaging (right).

power silicon. New wire bond materials can offer a fatigue life of up to 10 million full current cycles with a Delta T of 25°C across the wire bond. The shorter pins on the package coupled with liquid cooling give best results. In Fig. 1.19, note that the temperature differential is the temperature difference between the connection pins and the baseplate in °C.

Traditional insulated packaging uses lead frame construction with wire bonding to the chip to give fuse protection. This technique has a guaranteed life of 25 000 full current cycles but package cost is high. Connections are by bolted joints. Wire bonded packaging uses a plastic pin frame which is wire bonded to the die. This construction technique is standard for low power six packs (complete 3 phase bridge on a chip as used in air conditioners). What is new is the capability to offer this packaging in a high power device. In USA designers seem to prefer MOS gated thyristor MCTs. In Europe and the Far East insulated gate bipolar transistors (IGBTs) are popular. In fact both devices are converging on a common specification of: (a) maximum volt amp product per unit area of silicon; (b) saturation voltage of 1.5 V at *Ic* max; (c) high frequency forced commutation capability.

Currently MCTs have the better saturation but IGBTs have better commutation. In the coming years makers will see better saturation figures for IGBTs and even lower switching energies. This is the result of smaller line widths and thinner silicon device structures. Currently a 1200 V, 100 A six pack can switch 600 V DC at up to 16 kHz (V_{ce} Sat) 2.2 V/100 A, E_{on} 18 mJ E_{off} 14 mJ, cycle 32 mJ. The 600 V, 200 A six packs are now available as samples. Since the chips use non-punch

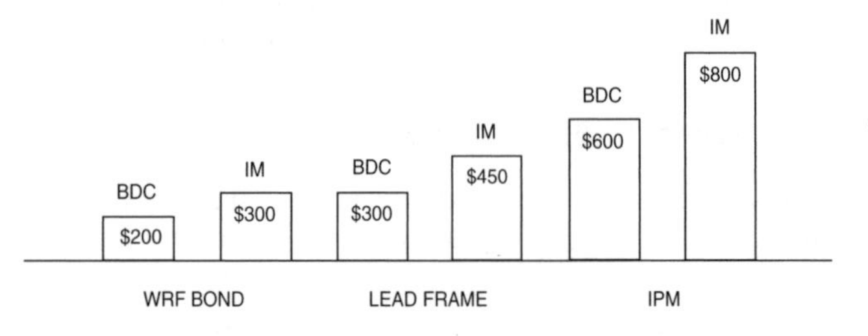

Fig. 1.20 Silicon cost for 70 kW drive.

through (NPT) technology, they may be connected in parallel without matching due to the inherent equalization characteristics of the die. Many vendors offer IGBTs in lead frame packaging, but this construction is not cost effective for electric vehicles. Devices of 1000 A, 1200 V are available. Intelligent power modules are also available, for example Semikron, SKIpacks Fuji, Toshiba and Mitsubishi. These integrate gate control with the power devices and have protection integral with the device. The cost of this approach is high at present; it is wire bonded packaging that offers the lowest device costs. A 1200 V, 100 A six pack is around 100 dollars (1997 price), Fig. 1.20.

Fundamental to the cost equation is that inverter cost is proportional to motor current. Electric and hybrid vehicles are tending to use drives of 70 kW because the vehicles weigh 1500 kg. What Fig. 1.21 illustrates is that the induction machine requires almost 1.8 times the current capacity of the brushless DC inverter for 3.5:1 constant power speed range. Typical circuit diagrams are illustrated in Figs 1.22a and b. The view in (a) is a typical induction motor drive with just six switches. This drive will need 3-off 600 V, 200 A six packs in parallel. Under US conditions, cars seldom require 70 kW for more than 10 seconds during overtaking. With current designs of battery peak power falls to 55 kW at minimum battery voltage limited by internal resistance (typically 1.75 V/cell for lead–acid). The view in (b) is a brushless DC drive using a double chopper circuit. Essentially a 300 V battery is increased to 600 V link with a 460 V motor. This inverter can be built with just two 1200 V, 100 A six packs. With oil at 40°C the package can operate at 140 A continuous. It will operate at 96 A RMS, 136 A peak on a 50% duty cycle for short periods. The brake resistor in the circuit prevents battery overcharging during regeneration. If the battery is overcharged its life may be reduced. In flat terrain the friction brakes may fulfil this role; however, in steep terrain the energy per 1000 metres height is 14.7 million joules or about 2400°C on average family car disc brakes. Electric cars do not have engine braking.

There are many benefits of using the high voltage circuit. First the motor current is 100 A or less. This makes the motor easier to wind and permits the use of printed circuit technology in the inverter. Second there is a major control benefit. An optimum control strategy is to use current-source PWM at low speed and voltage-source square wave at high speed. If a 300 V battery is used the DC link voltage is kept low until the motor voltage exceeds the DC link and then increased as the speed and voltage rise. This strategy reduces PWM carrier losses and permits better efficiency along the no-load-line of the vehicle. Use of printed circuit technology not only assists automatic assembly but also reduces EMC. EMC compliance is not too difficult in steel body cars but is much more of a challenge in composite structure vehicles. Having considered the inverter core some thoughts concerning the peripheral components are needed. Clearly this configuration requires an L/C filter for the chopper and an output filter for the motor to limit d*v*/d*t* on the motor windings. The dimensions of the L/C filter are determined by two factors: permitted inductor current ripple and permitted capacitor current ripple.

Polaron prefer to split the inductor to give good common mode rejection with respect to the battery. A value of 100 microhenries is suitable with a capacitance of 1250 microfarads. The inductors are made as air-core units with 10 mm microbore copper pipe. The turns may be close spaced by insulating the outside of the copper with epoxy powder coat paint. The spacing can be reduced further by using X extrusion copper which permits bending in two planes. The capacitors

	Induction motor	**Brushless DC motor**	**Brushless DC motor**
Motor speed	300 V	300 V	600 V
4000 rpm	120 V 361 A	220 V 197 A	460 V 94 A
13 500 rpm	220 V 197 A	220 V 223 A	460 V 96 A
Power	70 kW	70 kW	70 kW

Fig. 1.21 Base speed/max speed operating points for induction and brushless DC motors.

are ripple current dominated. With 100 A of motor current a capacitor that can handle 100 A peaks (30 A RMS) at temperatures of up to 50°C so oil immersion is the requirement. Polaron Group have chosen electrolytic units of 470 microfarads, 385 V, arranged five in parallel in series with five more in parallel. The cans are of the solder mount type choosing five more pins for mechanical strength in 40 mm × 50 mm cans.

The inductors for the d*v*/d*t* units are more challenging. An inductance of 10/20 microhenries is needed but it is advantageous if the inductance is more at low current. Consequently this application favours a cored inductor with low permeability iron powder and oil immersed litz wire winding. The core needs to have a moulded bobbin to provide inter-turn insulation for the litz wire and as a casting mould for the core material. A final point is that if one were prepared to hand wind the motor Polaron believe it would be possible to eliminate the d*v*/d*t* inductors by the use of an insulation extrusion to control the ground capacitance of the winding – the capacitance/inductance characteristic as a uniform transmission line.

In summary, for motors, Polaron believe brushless DC will prove to be the dominant technology especially for hybrid vehicles where efficiency at peak power matters. Machines for 12 000 rpm are well established. Successful operation of 70 kW machines at 20 000 rpm has been demonstrated and 150 kW machines are in development. Currently, higher speeds present a number of technical/cost obstacles (there are successful company designs operating to 150 000 rpm but not using low cost methods). Improvements in materials could radically change this in the next few years. In the inverter area, cost is proportional to current and the brushless DC motor requires 60% of the current of the induction motor to achieve a 3.5:1 constant power operating envelope. The double chopper circuit offers many benefits over the single bridge solution and is cheaper to construct. For 70 kW a 600 V DC link is best. The use of a controlled DC link becomes even more important in hybrid vehicles where smaller batteries lead to greater voltage variations between peak motoring and peak regeneration. The use of high voltage is not a safety hazard so long as the motor and inverter are contained in a single enclosure where the active components are not accessible. Oil immersed construction offers the lowest temperature rises and the best component reliability, especially for the silicon and filter capacitors. This method of construction permits complete subsystem testing before mounting in a vehicle.

1.5 Electric vehicle drives: optimum solutions for motors, drives and batteries

Optimum supply of voltage for the power electronics of EVs is around 300 V DC using the latest IGBT power transistors[4]. This also provides a sensible solution for the motor because in the power range of 30–150 kW the line currents are quite reasonable. A consequence of using a 300 V battery is that the rail voltage will vary from 250 to 400 V under different service conditions.

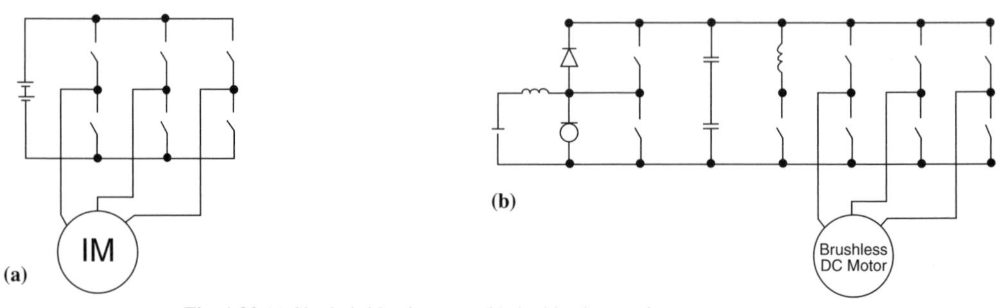

Fig. 1.22 (a) Single bridge inverter, (b) double chopper inverter.

1.5.1 BATTERY CONSIDERATIONS

A good commercial battery for deep discharge work is the Trojan 220 Ah 6 V golf cart unit. This gives 75 A for 75 mins and weighs 65 lb, consequently a 108 V stack weighs 1170 lb and cost $1080 in 1991. It also requires considerable maintenance and occupies a projected area of 1342 square inches and is 10 5/8 inches high.

In comparison, sister company Nelco have available a sealed lead–acid battery of 12 V, 60 Ah and arranged into 18 cells to give 108 V. It occupies 720 square inches of plan area and weighs 697 lb. This arrangement can also provide 75 A for 75 minutes. The problem area is cost. This battery cost $2700 in 1991. If the voltage was increased to 312 V, with the same stored energy, the cost rises by 20% at 45 kW. Such 300 V battery systems require great attention to safety; 100 V batteries may be feasible at 45 kW but this ceases to be true at 150 kW. In fact, one can draw the graph in Fig. 1.23(a) to define minimum voltage for a given output power. Other areas worthy of comment are maintenance and battery life. High voltage strings of aqueous batteries are dangerous and should be banned by legislation. This is not so of sealed lead–acid batteries as there is no need for maintenance access. However, no voltage greater that 110 V should be present in a single string or an individual connector. Long series strings present a potential maintenance problem with respect to cell equalization. The problem may only be resolved by keeping all cells at the same temperature. A final problem is fast charging; this is ternperature limited to 60°C max cell temperature. The newer cells may be fast charged so long as the temperature is contained and the individual cell voltage is below 2.1 V.

1.5.2 FUEL-CELL CONSIDERATIONS

There is no doubt that the long-term power supply for electric vehicles will be some form of hydrogen fuel cell, the leading current technology being the PEM membrane system as manufactured by Vickers/Ballard. This is a complete system measuring 30 × 18 × 12 inches which produces about 5 kW at 45% efficiency.

The unit consists of 36 plates of 250 A rating and the fuel gases operate at 3 bar and the exhaust temperature is around 80°C. This arrangement leads to the relation in Fig. 1.23(b). Hence for a vehicle with a storage battery approximately one-third maximum power +10 kW is the peak fuel-cell

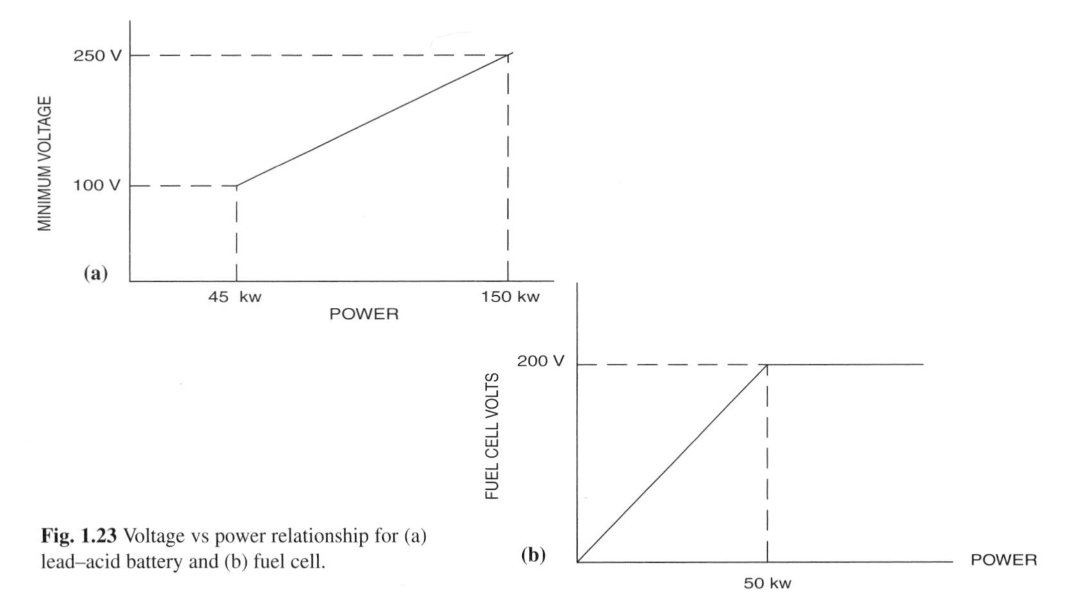

Fig. 1.23 Voltage vs power relationship for (a) lead–acid battery and (b) fuel cell.

load. Hence for 45 kW this amounts to five modules producing 100 V at 250 A. For a 150 kW system, vehicle builders will need ten modules giving 50 kW at 200 V. The voltage may not rise above 200 V due to problems relating to the hydrogen. Warm up takes about 5 minutes from cold with units producing 50% output at 20°C. Once hot, response is 1–2 seconds for load steps and endurance has been confirmed as greater than 20 000 hours. One of the more intriguing possibilities offered by fuel cells is to use the power converter to produce 50 Hz for powering lights and portable tools on site vehicles.

1.5.3 OVERALL SOLUTION

There is a basic incompatibility between the power source voltage and the motor voltage; so how can this problem be addressed?

The solution is to put a reversible chopper between the battery/fuel cell and the inverter (Fig. 1.24). This means that the supply to the inverter is stabilized under all conditions resulting in full performance during receding battery conditions and no overvoltage during battery charging mode. By using the inverter as the battery charger express charging can be performed, where mains supply permits, in approximately 3 hours.

1.5.4 MOTORPAK SAFETY CONSIDERATIONS

To charge and discharge the battery quickly whilst optimizing battery use requires perfect control of the battery temperature. Since the battery is sealed this is best achieved by immersing in silicon fluid. A circulating pump passes fluid to and from the motor. This keeps the batteries cool and at equal temperature during charging using the motor as a heatsink and, during discharge, the motor warms the batteries to give optimum performance. Hence the batteries are built into a tank and this prevents access by the operator.

The next concept is to make the battery module interchangeable. This permits refuelling either by recharging the battery or by exchanging the battery module.

1.5.5 MOTORPAK CONSTRUCTIONAL CONSIDERATIONS

If costs are to be optimized, it makes sense to locate the power controller close to the battery. In the above case, Nelco have taken the concept one stage further. The power controller is located in the base of the battery tank. We call this concept Motorpak, Fig. 1.25, and as can be seen the mechanical execution could not be made much simpler. The motor and PCU pack are mounted

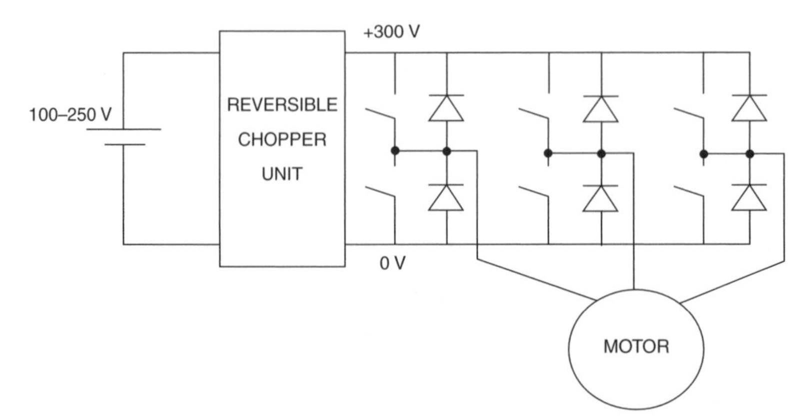

Fig. 1.24 Reversible chopper.

under the vehicle either in place of or in addition to the conventional power train. No gearbox is needed and the motor provides nearly 300 Nm of torque directly. The following specification applies for a 45 kW Motorpak:

Input	50–240 V AC, 40–65 Hz single or 3 phase up to 30 A; recharge time 3 hours
Output	0–220 V, 3 phase up to 750 Hz 60 kVA, 13.6 V DC 500 W
Batteries	18 off, 12 V, 60 Ah sealed lead–acid units, may be configured as 108 or 216 V unit
Weight	800 lb (362 kg)
Dimensions	30 in long, 27 in wide, 14 in high
Construction	Weatherproof
Controls	Function switch, accelerator pedal, voltmeter/ammeter/amp hour meter, 13.6 V for auxiliaries, 2 oil pipes to motor (4 litres/min)
Deep discharge performance	800 cycles to 80%
Stored energy	10 kWh
Cost in 1991	£3000 at 1000 off ex batteries (£5000 with batteries), price includes motor
Temp. range	–20°C to + 40°C

45 kW traction motor

Type	Brushless DC permanent magnet
Size	375 long × 250 diameter, weight 50 kg

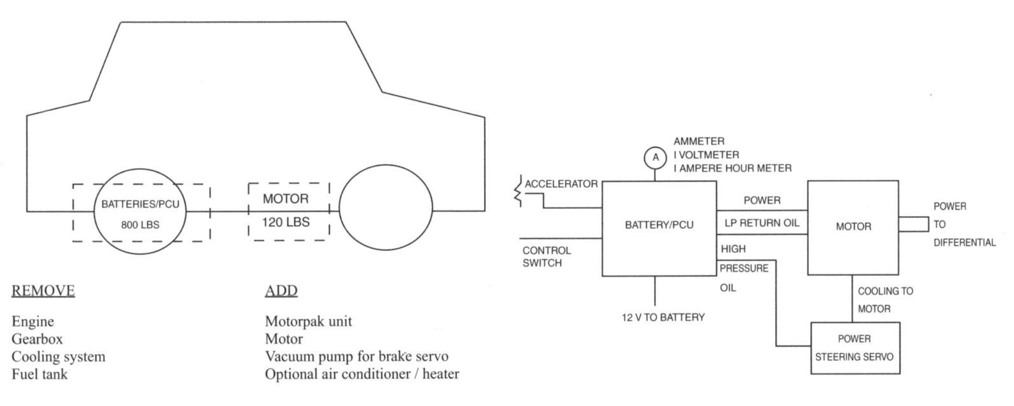

Fig. 1.25 Motorpak concept.

Torque	0–1500 rpm, 280 Nm falling to 70 Nm at 5000 rpm on 45 kW constant power curve
Construction	Flange mount with double ended shaft and integral encoder
Cooling	Silicon oil, 4 litres/min
Electrical rating	220 V, 130 A, 750 Hz

Power pack contains: batteries, power conversion unit, 12 V, 500 W supply for auxiliaries and hydraulic power steering supply/cooling for motor. This unit is interchangeable in seconds. A 45 kW unit weighs just 800 lb; the motor is oil cooled and weighs 130 lb.

1.5.6 ADVANTAGES OF THE SYSTEM DESCRIBED

If the conventional engine is replaced by a battery/motor the weight increases by approximately 300 lb for a 1 tonne vehicle. This means the system can be fitted to existing chassis designs or retrofitted to cars. The system can be used standalone or as a hybrid. The complete electrics pack is interchangeable for instant refuelling and the PCU works with any battery input supply for 100–250 V. Batteries are rated for 800 deep discharge cycles to 80% depth of discharge. On a 45 kW unit, the battery can supply 75 A for 75 minutes at 108 V DC or 37.5 A for 75 minutes at 216 V.

Total safety is ensured by all electrical parts except the motor which is contained in a single totally insulated module with no parts distributed over the vehicle. Batteries are sealed to give best resistance to crash situations. Electrics are protected against short-circuits with both fuses and circuit breaker. The oil cooling system can supply the power steering if required. Minimized technical risk is ensured by a total package solution and if technology improves only one module has to be changed. The module approach makes many finance packages feasible, facilitating user acceptance; for example, the user buys the vehicle and motor but hires the battery and PCU. The battery pack can be recharged in 3 hours where mains supply permits. The PCU functions as a battery charger and the drive system can supply up to 45 kW of mains electricity for short periods – longer if used with a fuel-cell prime mover. The PCU makes use of portable power appliances viable which is particularly useful for the building industry. Finally, the concept makes conversion of existing vehicles possible.

References

1. Hodkinson, R., 45 kW integrated vehicle drive, *EVS 11,* Florence, 1992
2. Hodkinson, R., Machine and drive characteristics for hybrid and electric vehicles, ISATA 29, Stuttgart, 1996
3. Hodkinson, R., Towards 4 dollars per kW, p. 4 *et seq., EVS 14,* Orlando, December 1997
4. Hodkinson and Scarlett, Electric Vehicle Drives, Coercive Ltd report, December 1991

Further reading

Electric vehicle technology, bound volume of SAE papers, 1990
Electric and hybrid vehicle technology, bound volume of SAE papers, 1992
Electric and hybrid vehicle design studies, bound volume of SAE papers, 1997
Technology for electric and hybrid vehicles, bound volume of SAE papers, 1998
Strategies in electric and hybrid vehicle design, bound volume of SAE papers, 1996
Electric vehicle design and development, bound volume of SAE papers, 1991
Breaking paradigms, the seamless electro-mechanical vehicle, Convergence 96, SAE, 1996

2
Viable energy storage systems

2.1 Electronic battery

Electric vehicles are at a historical turning point – the point where technology permits the performance of electric vehicles to exceed the performance of thermal engines[1]. Currently quality battery technology is expensive and heavy. This favours hybrids with small peaking batteries – typically 10 Ah at 300 V, with a capability of 70 kW for 2 minutes. New battery geometries have been developed for this application in the form of high performance D cells with 6 mm thick end caps and M6 terminals. A single string of cells handles 100 amps for 10 seconds. The heat is transferred to the end caps then removed by forced air cooling. It is vital to maintain an even cell state of charge in long strings. This solution has two drawbacks at present: (1) Cost – a 300 V 6.5 Ah stack costs more than $10 000 in January 98; (2) reliability – with only one string a single high resistance cell disables the battery.

Using the best available nickel–cadmium cells, it is possible to build a reliable peak power stack and use electronic control to maintain equal currents in strings. A later section will consider the next generation, an all electric hybrid with aluminium battery and alkaline fuel cell; this last important energy storage system is also discussed in depth in Chapter 4. A review of different battery types and performances is given in Chapter 5.

2.2 Battery performance: existing systems

All battery technologies can offer some solution to the peak power problem but there is only one parameter which ultimately matters, and this is the internal resistance of the cell. This is much more related to cell geometry than cell chemistry, as we shall discover. When AA, C, D and F cells were originally designed, nobody was thinking of discharging them at hundreds of amps, so it is hardly surprising that they are not ideal for the purpose. This problem will become even more extreme as power density improves:

D cell characteristics

	Lead Acid	NiCad	NiMh	Lithium	Aluminium
Amp Hr (20hr)	2.5	4.0	6.5	7.5	50
Cell voltage	2.0	1.2	1.2	3.6	3.0
Max C rate	40.0	25.0	11.0	40.0	Unknown
Ah/25°C	0.5	3.0	5.0	6.0	Unknown

The first volume hybrid electrics in the market came from Toyota (Prius) and Nissan, Fig. 2.1. Toyota uses a 288 V string of NiMh D cells to give a peak power of 21 kW. The battery is mounted in horizontal strings of 6 cells in a matrix weighing 75 kg and is made by Matsushita Panasonic EV Energy. Nissan has worked with Sony who have developed the lithium ion battery, of AEA Technology (UK), into a high performance D cell, of 3.6 V 18 Ah, with a peak power of 1700 W/kg. (This cell is 250 mm long × 32 mm OD and includes an electronic regulator.) Both solutions use advanced thermal management and both deliver the performance but at high cost. The task in hand is to reduce costs by 50%.

2.2.1 BATTERY PERFORMANCE: ALTERNATIVE APPROACH

Question: Which application gets the best peak power battery performance at present? Answers: Portable power tools and model aeroplanes; chemistry: nickel–cadmium; cell size: RR/C 1–2 ampere hour; peak discharge: 30° C; maximum current: 32 amps limited by connections (tags),

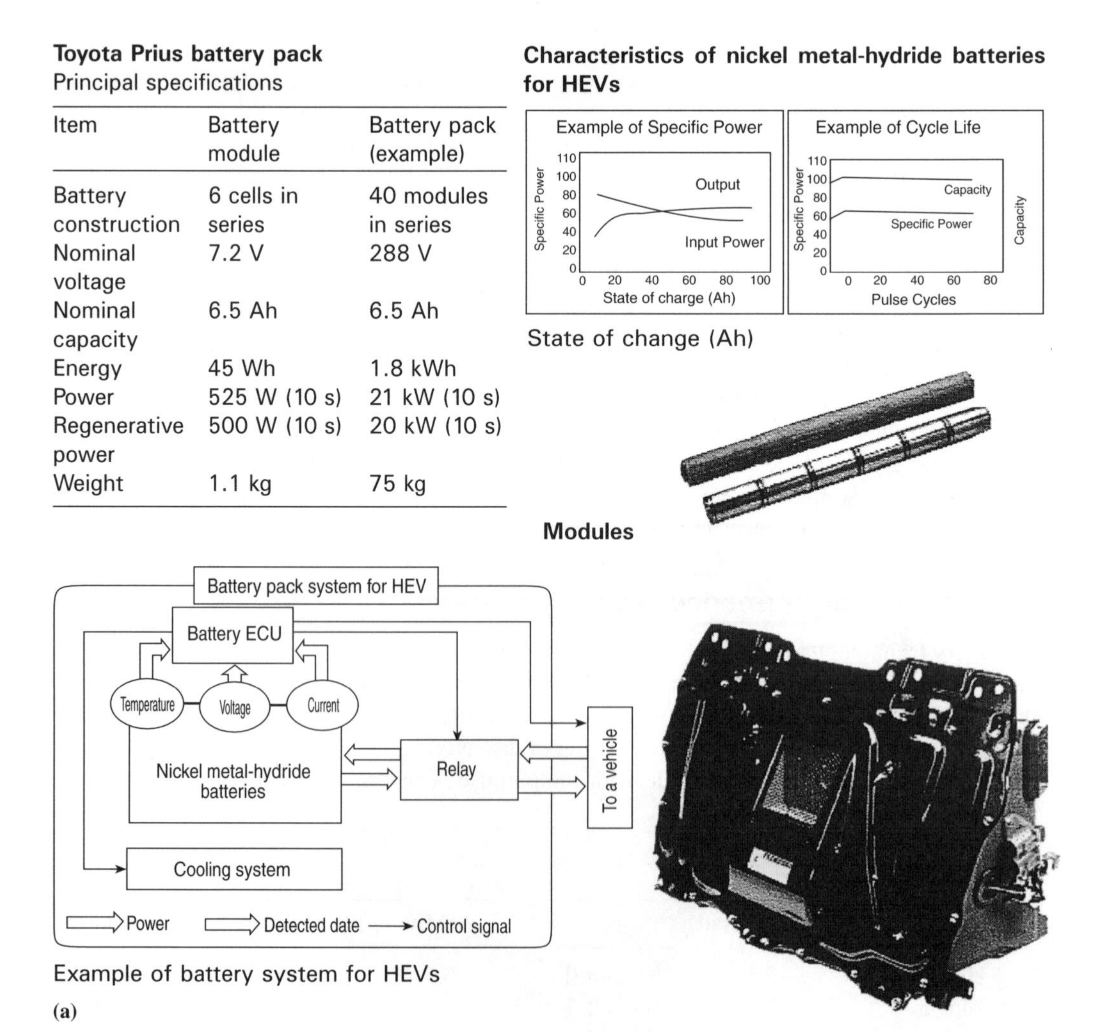

Toyota Prius battery pack
Principal specifications

Item	Battery module	Battery pack (example)
Battery construction	6 cells in series	40 modules in series
Nominal voltage	7.2 V	288 V
Nominal capacity	6.5 Ah	6.5 Ah
Energy	45 Wh	1.8 kWh
Power	525 W (10 s)	21 kW (10 s)
Regenerative power	500 W (10 s)	20 kW (10 s)
Weight	1.1 kg	75 kg

Fig. 2.1 Production hybrid battery technologies compared: (a) Toyota Prius; (b) Sony/Nissan.

Example of battery pack system for HEVs (Toyota spec.)

Fig. 2.2. The performance issues of these cells are complex. Amazingly there are 14 different types of sealed Ni–Cad cells in four main product groups: (a) standard, (b) high discharge current, (c) fast charge and (d) high temperature.

It is the double-sintered fast charge cells that are used in model aeroplanes and in the world championships the Sanyo Cadnica is specified using a 1.7 Ah cell. A Tamiya connector can deliver up to 32 amps. Battery packs of 7.2 V six packs are usually fitted. In the power tool area, Black and Decker have the Versapack, a three cell string. Two or four strings are used in series with larger cordless power tools. One problem is that sealed Ni–Cad cells cannot be connected in parallel. At the end of the charge cycle, the cell voltage falls, causing rapid charge/discharge between parallel cells. On the face of it, the packaging problem is daunting, with more than 2000 cells in a pack.

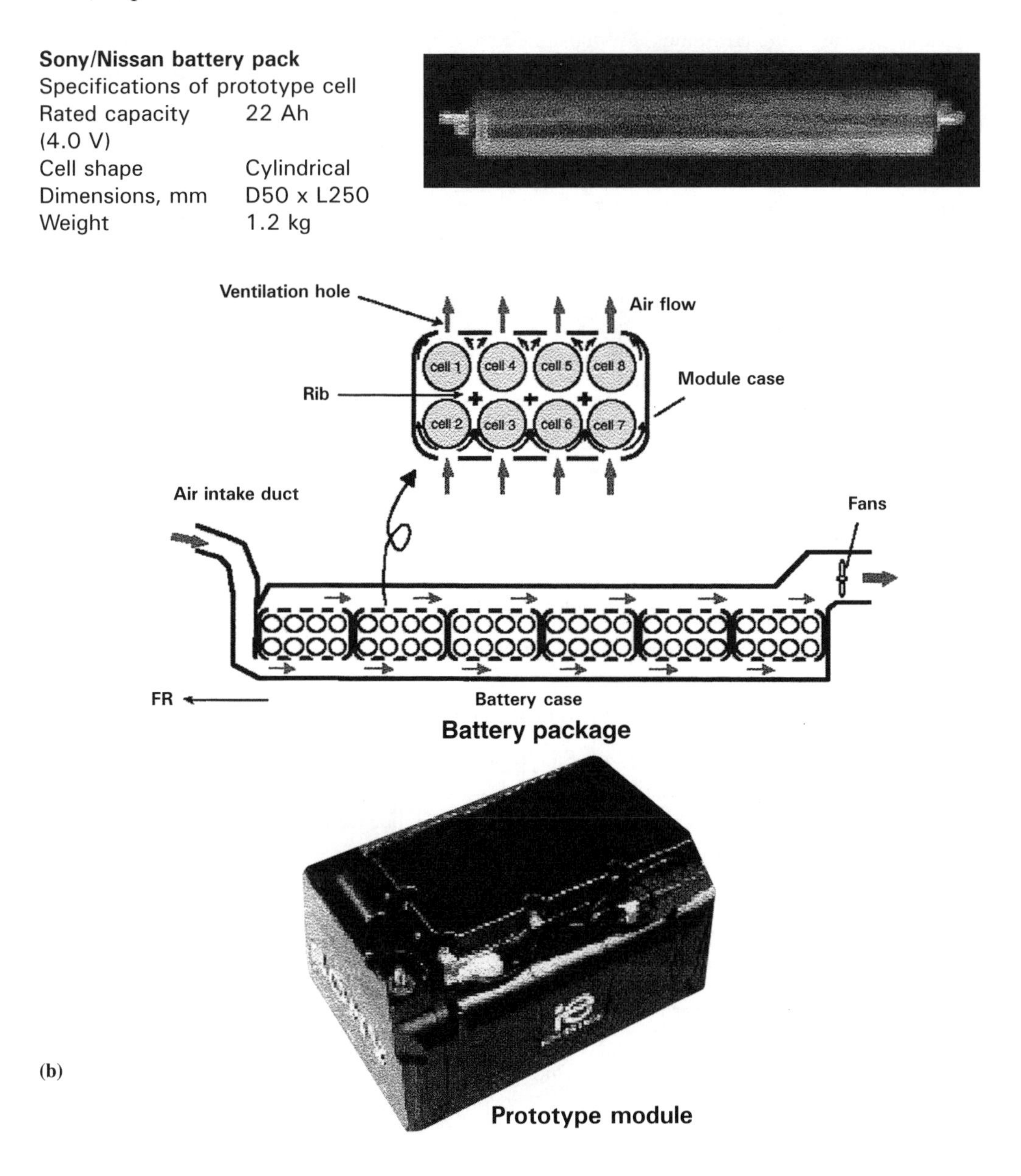

(b)

Ni–Cad cells

BatteryType	RR	C	Sub D	D
Cell capacity (Ah)	1.0	1.7	2.8	4.0
Int. resistance* (milliohms)	4.5	3.5	3.0	2.8
Cell dimensions, DxL (mm)	22x34	22x56	32x42	32x56
Weight (gm)	42	65	130	200
Int. Resistance of 4 Ah milliohms	1.1255	1.6	2.5	2.8

*Internal Resistance is at 20° C and 50% state of charge.

Fig. 2.2 Typical Ni–Cad packages and capacities.

2.2.2 INTERNAL RESISTANCE PERFORMANCE OF FAST-CHARGE NI–CAD CELLS

The results discussed here are based on Sanyo products but the same trends are seen in Varta, Panasonic and Saft cells. See Fig. 2.2.

Why does internal resistance matter so much? This is because at 25° C discharge rate, the voltage drop on a 1.2 V cell is as tabulated below:

Voltage drop on a 1.2 V cell at 25° C discharge rate

RR	C	Sub D	D
112.5 mV	160 mV	250 mV	280 mV

The 1.2 V cell is thus no longer a 1.2 V cell but closer to 1 V at 20° C. Why does the 1 Ah cell win? It is because it is short and fat – the others are long and thin. Cell geometry is the decisive factor for low internal resistance.

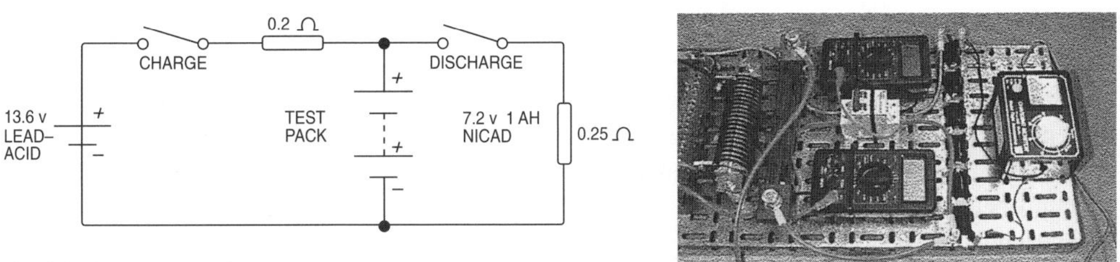

Fig. 2.3 Discharge test rig.

To test the performance of individual cells, Polaron built a string of six and charged/discharged at 24° C; that is 24 amps on 1 Ah cells. After five cycles, discharge time increased to 130 seconds, and temperature rise was about 10° C. It ran for several hundred cycles with virtually no change in characteristics. This is very severe compared to the true operating conditions, where the cells will have to supply 24 amps for perhaps 10 seconds under real world conditions, Fig. 2.3.

2.2.3 BUILDING A STRING OF CELLS

To achieve 70 kW for 2 minutes will require ten strings of 260 cells at 1 Ah. This is economical but not optimal. Three strings of 3.3 Ah would be optimal. This corresponds to using short D cells. The use of 1 Ah packages does have some benefits. Spot welded connections can handle the current without special packaging. The key problem is that of automatic assembly with so many cells. This is easily accomplished using welding robots. One technique is to use two parallel plates which each can hold two strings of 260 cells. The cells are sealed with O-rings so that the centre of each cell is oil cooled. The connections have to be in air, because the gas seal should not be immersed in oil as the seal may be damaged – and the oil contaminated with potassium hydroxide. At a link current of 25 A nickel-plated steel links seem to be quite adequate. A second interesting packaging concept would be to create a ten pack version of the Versapack concept. These could be mounted in horizontal strings and air cooled for cell equalization.

2.2.4 CELL CURRENT SHARING

Given ten parallel strings for 70 kW peak power, there is only one way to ensure equal currents in each string – active regulation. This seems a very expensive proposition but it fits in well with the structure of modern inverter drives. The present trend is to use a 300 V battery with a boost chopper and increase the battery voltage to give a DC bus of 600 V, as used in industrial drives. Normally one would parallel a number of transistors to give a current rating of 300 amps. Three times 100 amps would be optimal as this is a complete 3-phase pack of IGBTs. In this case it is necessary to use smaller packs of 30 amps where each leg has its own independent current regulator. This would not be an attractive proposition if it were not for the fact that at this current the required control circuitry is available economically. Ten such circuits may be connected to a common DC bus. This arrangement ensures excellent current sharing in both charge and discharge and prevents the situation where strings charge/discharge into one another. What we need now is some improvement in battery packaging. To operate three strings in parallel would be optimal from a cost versus reliability standpoint, as one would use the separate 100 amp phase legs in a six pack to control the current in the individual strings, Fig. 2.4.

2.2.5 BATTERY SYSTEM PROSPECTS

It has been shown that a matrix of small rechargeable cells can be made to give large peak powers on a repeated basis with excellent life performance. The new D cell designs in NiMh and lithium ion are very expensive and a cheaper alternative is to use l Ah Ni–Cad cells with ultra-low internal resistance. Using this technique it would be possible to buy the cells for a 21 kW/10 second pack for about $2000 in 1999. To this the packaging and control cost must be added. However, at present this is still significantly cheaper than the use of custom battery packages.

Cell geometry is the decisive factor in achieving low internal resistance and there is much room for improvement on existing cell packages. Short cells with minimum distance from foil to terminal give best results. The use of multiple strings of cells in parallel, with active current sharing, improves reliability and reduces cost since the currents in individual packages are modest compared to single strings. Temperature control of the strings helps to maintain even state of charge at high charge/discharge rates and keeps cells cool, while extending cell life.

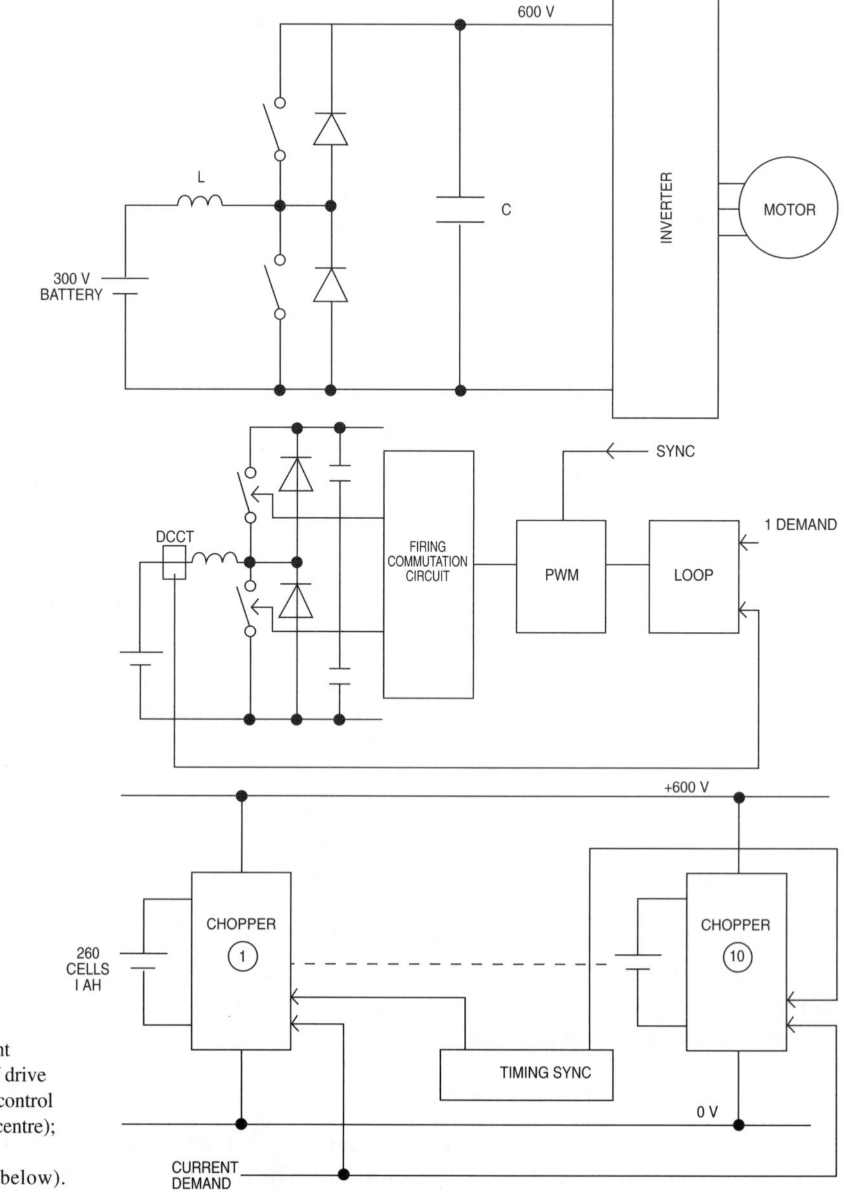

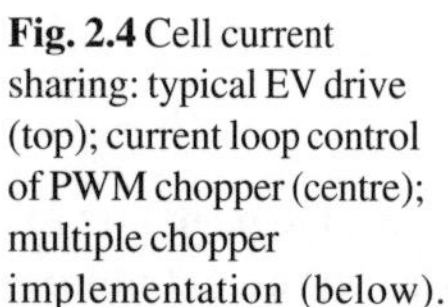
Fig. 2.4 Cell current sharing: typical EV drive (top); current loop control of PWM chopper (centre); multiple chopper implementation (below).

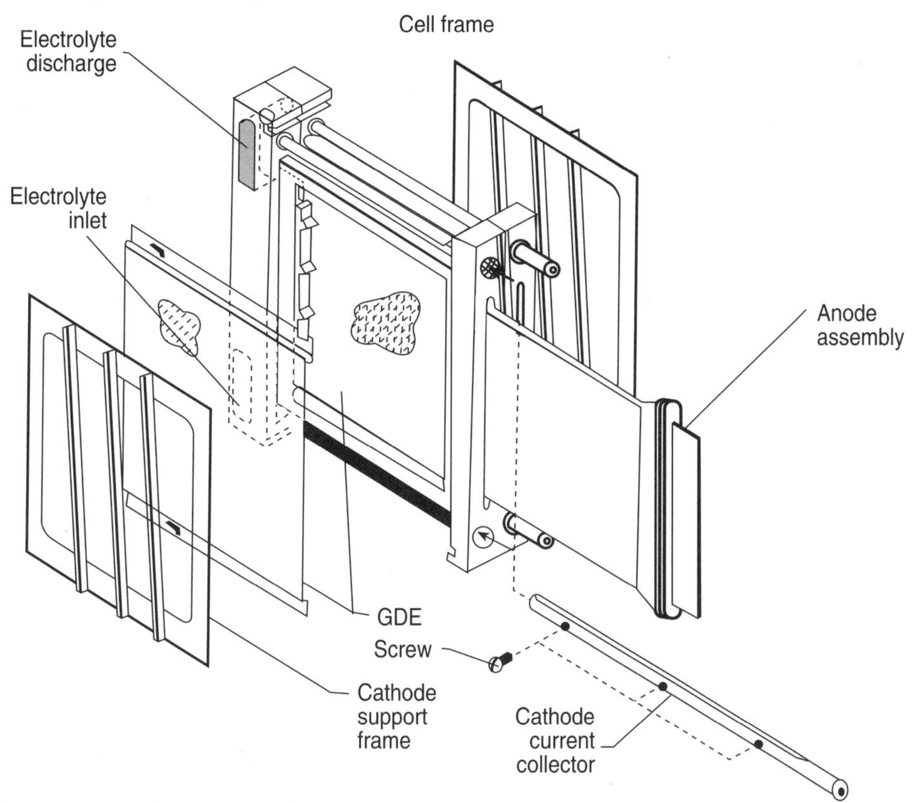

Fig. 2.5 Exploded view of aluminium/air bipolar battery (courtesy Eltech Systems).

The development of advanced battery chemistries with increased power and energy density will place even greater demands on cell packaging in the future and a new family of optimum proportions needs to be designed for the job.

2.3 Status of the aluminium battery

In 1997, patents were filed in Finland for a new aluminium secondary battery. The inventor was Rainer Partanen of Europositron Corporation who claims major improvements in power density and energy density for the new cell based on a 1.5 V EMF[2]. The author is interested in this problem because it represents one of the last major barriers to be overcome before the widespread introduction of electric and hybrid vehicles. In recent years, significant effort has been directed at improving secondary battery performance and this effort is beginning to bear fruit. We can now see advanced lead acid, nickel metal hydride and Lithium Ion products out in the market place with performance of up to 100 Wh/kg and 200 W/kg.

Market requirements fall into two distinct categories: (a) small peak power batteries of 500 Wh (2 kWh for hybrids) and (b) 30–100 kWh for pure electric vehicles. Each of the cell types has its own distinctive attributes but none has so far succeeded in making the breakthrough required for mass market EV implementation. The fundamental problem is one of weight. At the factory gate, vehicle cost is almost proportional to mass, as is vehicle accelerative and gradient performance. Consequently it will take at least 300 Wh/kg and 600 W/kg to achieve the performance/weight ratio for long range electrics we really desire. This would make one type of hybrid particularly attractive – the small fuel cell running continuously together with a large battery.

If we consider a low loss platform for a passenger car at 850 kg with $N = 0.3$, we have 250 kg for the battery. At 300 Wh/kg we obtain 75 kWh, which would give a range of more than 250 miles after allowing for auxiliary losses. A low loss platform would consume 5 kW at 60 mph + 5 kW for auxiliary losses – total 10 kW – this equates to 7.5 hours at 60 mph = 450 miles steady state. This level of performance is an order of magnitude better than lead–acid at the current time. Clearly a new approach to the problem is required.

2.3.1 WHY ALUMINIUM?

In simple terms the answer involves (a) abundancy, (b) low cost and (c) high energy storage. If we consider the recent developments in batteries they all seem to use materials like nickel which are highly dense and limited in supply. Likewise in fuel cells using platinum catalysts material scarcity is implicit, the annual production being about 80 tons worldwide. Any mass market battery needs to use materials available in abundance. In the bumper year of 1985, 77 million tons of bauxite were mined worldwide; aluminium is one of the most plentiful materials available on Earth. In terms of 1999 costs, aluminium is $2000 per ton so 250 kg would cost $500 – an acceptable sum.

In terms of energy storage, aluminium has one of the highest electrical charge storage per unit weight except for the alkali metals:

Aluminium	0.11 coulombs per gram	2.98 Ah per gram
Lithium	0.14 coulombs per gram	3.86 Ah per gram
Beryllium	0.22 coulombs per gram	5.94 Ah per gram
Zinc	0.03 coulombs per gram	0.82 Ah per gram

Lithium and beryllium are alkali metals and are not suitable for use with liquid electrolytes, due to rapid corrosion, so are normally used with solid electrolytes.

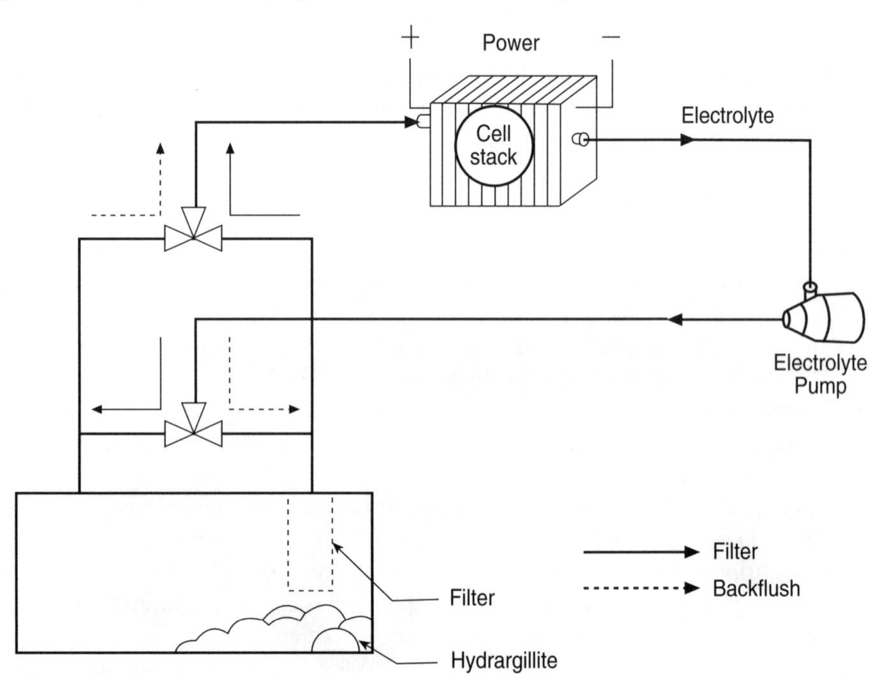

Fig. 2.6 Conceptual design of filter/precipitator system when integrated with aluminium/oxygen battery (courtesy Eltech Systems).

2.3.2 DEVELOPMENT HISTORY OF ALUMINIUM BATTERIES

The first serious attempt to build an aluminium battery was made in 1960 by Solomon Zaromb[3] working for the US Philco Company. In Zaromb's concept for an aluminium air cell, the anode was aluminium, partnered with potassium hydroxide, and air was the cathode. This battery could store 15 times the energy of lead–acid, achieving 500 Wh/kg and a plate current density of 1 A/sq. cm. The main drawback was corrosion in the off-condition, which resulted in the production of a jelly of aluminium hydroxide and the evolution of hydrogen gas. To overcome this problem Zaromb developed polycyclic/aromatic inhibitors and had a space below the cell for the aluminium hydroxide to collect. The chemical reaction is

$$Al + 3H_2O = Al(OH)_3 + 3/2H_2$$

In 1985 another attempt was made by DESPIC[4], using a saline electrolyte. Additions of small quantities of trace elements such as tin, titanium, indium or gallium move the corrosion potential in the negative direction. DESPIC built this cell with wedge-shaped anodes which permitted mechanical recharging, using sea water as the electrolyte in some cases. The battery was developed by ALUPOWER commercially. The battery had limited peak power capability because of conductivity limitations of the electrolyte, but provided substantial watt-hour capacity.

Other attempts have involved aluminium chloride (chloroaluminate) which is a molten salt at room temperature, with chlorine held in a graphite electrode. This attempt in 1988 by Gifford and Palmisano[5] gives limited capacity due to high ohmic resistance of the graphite. Equally significant is work by Gileadi and co-workers[6] who have succeeded in depositing aluminium from organic solvents though the mechanisms of the reactions are not well understood at this time.

Between 1990 and 1995 Dr E. J. Rudd[7] led a team at Eltech Research in Fairport Harbor, Ohio, USA, which built a mechanically recharged aluminium battery for the PNGV programme, Fig. 2.5. It had 280 cells and stored 190 kWh with a peak power of 55 kW, and weighed 195 kg. This battery used a pumped electrolyte system with a separate filter/precipitator to remove the aluminium hydroxide jelly, Fig. 2.6. Alupower[8] built a 6 kW aluminium–air range-extender system under the same programme, Fig. 2.7.

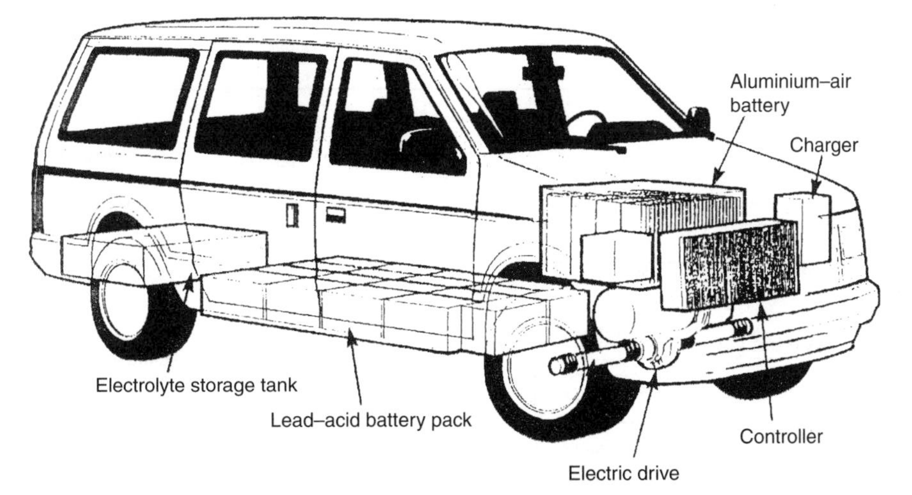

Fig. 2.7 A 6 kW range extender by Alupower, with lead–acid main battery.

2.3.3 NEW-CONCEPT ALUMINIUM BATTERY

The cell invented by Rainer Partanen, Fig. 2.8, is an attempt to defeat the disadvantages of the aluminium–air cell. It is a secondary battery which uses coated aluminium for the anode and pure aluminium for the cathode. The electrolyte is a mixture of two elements: (a) an anion/cation solution currently consisting in proportion of 68 g of 25% ammonia water mixed with 208 g of aluminium hydroxide, and made up with water to give 1 litre of solution; (b) a semi-organic additive consisting of metal amines.

The exact formulation of the additive is a commercial secret. The inventor claims that this electrolyte achieves a large increase in charge carrier mobility and this results in figures of up to 1246 Wh/kg and 2100 Wh/litre, which have been achieved in many prototype cells that have been constructed. The figures relate to active materials, without casing. It is suggested that the technology is suitable for the construction of plate (wet cell) and foil (sealed) cells, with no limitations on capacity. The test cells have achieved a life of up to 3000 cycles, the main degradation mechanism being corrosion of the coating on the anode during recharging. One remaining hurdle to be overcome is the identification of a better coating material to reduce the corrosion.

The battery has some unusual characteristics in that it operates over a very wide temperature range, –40 to +70° C. This is in stark contrast to most batteries whose low temperature/high temperature performance is poor. The cell voltage is a nominal 1.5 V. Some interesting consequences arise if one assumes that the claims are true. The most significant is packaging. If we take the D cell which is 32 mm diameter × 58 mm long, as used by Panasonic/Toyota in the PRIUS battery pack, a battery with 150 g active mass stores 6.8 Ah and has a peak discharge current of around 100 amps. If we build a D Cell at a value of 1246 Wh/kg, this leads to a figure of 150 Ah. Polaron understand that very high levels of discharge current are possible – the inventor claims up to 20 times more power than existing cells in the market – but finding methods of supporting these currents in such a small space is a major challenge to achieve low terminal resistance, lead-outs and sealing, Fig. 2.9. It is claimed that the new technology uses environmentally safe materials which are fully recyclable.

Other developments which lend support to this invention[9] are the emergence of ultracapacitors and electrolytic capacitors, both using aluminium electrodes with biological

	Leclanch	Alkaline	Lead–acid	Ni–Cad	Ni–Mh	Lithium–ion	Aluminium
Amp. hrs (20)	4.5	18	2.5	4.0	6.5	18	150 (75 Dem)
Cell voltage	1.5	1.5	2.0	1.2	1.2	3.6	1.5
Max C rate	2C	2C	40C	25C	11C	40C	3C
AH/25° C present	Not possible IR limited 2		0.5	3.0	5.0	14	Not possible at Package IR limits current to 500 A

Aluminium metal in anode/cathode	486 g	180 cm³
Anion/cation reactant solution	1199 g	820 cm³

Theoretical maximum energy and current capacity

2100 Wh/litre	1448 Ah/litre
1246 Wh/kg	859 Ah/kg

Practical cells in a package should achieve 70–80% of the above values when package mass is included.

Fig. 2.8 Characteristics of D cell (32 × 62 mm) against those of a 1 litre Partanen cell.

electrolytes. Very significantly, ultracapacitors operate well at low temperatures. In Russia 24 V modules, 150 mm diameter × 600 mm long store 20 000 joules and are used for starting diesel engines at –40° C.

2.3.4 PATENT PROTECTION

The technical background to the invention is the result of a remarkable discovery in the field of complex electrochemistry and is based on the composition of the solution for electrical analysis and catalysis, releasing the energy potential of aluminium. Patent protection is being applied in three areas:

(1) The first is a solution which, under discharge, generates a reaction on the cathode side causing the energy potential of the aluminium to be released, and by ionization changes the molecular structure from metal to solution. Patent application Fi 954902 PCT/EPO (published).

(2) The second is a solution which, under discharge, generates a decomposition reaction in the chemical reactant mass. This is in crystal form which dissolves into solution and produces electrical potential on the anode side. Patent application Fi 981229 PCT/EPO (registered).

(3) The third component are the electrodes which have a dual role. They are formed of materials which enable them to act concurrently as non-ionized anode and ionized cathode. These electrodes are used in a multicell configuration as in existing battery technology. Patent application Fi 981379 PCT/EPO (registered).

The composition of these solutions, and the reactant mass, have the capability of producing an electrical current from the non-ionized anode and aluminium cathode when conductivity (resistance) is placed between them. When discharging, power is generated by the energy of the released aluminium, which reduces to about 35% of its original molecular density. When recharging, the reactant solution returns to its original form in solution and crystal mass and the aluminium atoms are deposited back onto the electrodes.

2.3.5 ALUMINIUM PROSPECTS

An aluminium secondary battery looks to be a very promising candidate for the storage of substantial energy. Whether the inventor Rainer Partanen has found the correct technique remains to be demonstrated. Although the claims for peak power and energy density seem very high, Sony have demonstrated 1800 watts per kg in lithium–ion recently and aluminium–air cells achieved 500 Wh/kg in 1964. The author considers aluminium to be a worthy contender for advanced battery construction and clearly this is an area which merits much greater investigation in the future. One point is clear – by making the active aluminium electrode the cathode, the parasitic reaction that is the big drawback of the aluminium–air cell is avoided, because the 1.5 V potential across the cell suppresses the reaction. Two questions that remain to be answered concern the levels of conductivity and mobility that will need to be exceptional to justify the claims made for the Partanen cell, also whether cell packaging will be a significant problem, requiring a new range of packages to be developed.

2.4 Advanced fuel-cell control systems

This section considers the development of a fuel-cell controller and power converter for a vehicle weighing 2 tons, for operation in an urban environment[10]. The techniques employed can be used with either PEM membrane fuel cells or alkaline units. The main challenge is to re-engineer a high cost system into a volume-manufactured product but this is unlikely to be achieved'overnight'. What is required is a new generation of components which are plastic as opposed to metal based.

The power electronics are practical, but need integrated packaging to reduce costs. Equally important is improvement in the fuel-cell stack specifications. This section considers the requirements and performance of a low pressure scheme at the current state of the art and predicts the measures needed to achieve significant cost reduction.

Modern hybrid cars are demonstrating major improvements in fuel consumption (3 litres/100 km) and emissions (ULEV limits) compared to conventional thermal engines. These designs use small peaking batteries which weigh less than 100 kg, for a family sedan, and store perhaps 2 kWh.

A new aluminium battery chemistry has been identified whereby it should be possible to store 50 kWh in a weight of 150 kg in perhaps 3/5 years from now. Nickel–metal hydride needs 500 kg with current technology to achieve 50 kWh. This makes a new type of hybrid an interesting long-term contender – the electric hybrid with a small fuel cell. In this vehicle a 2–5 kW fuel cell would charge the battery continuously. The only time the battery would

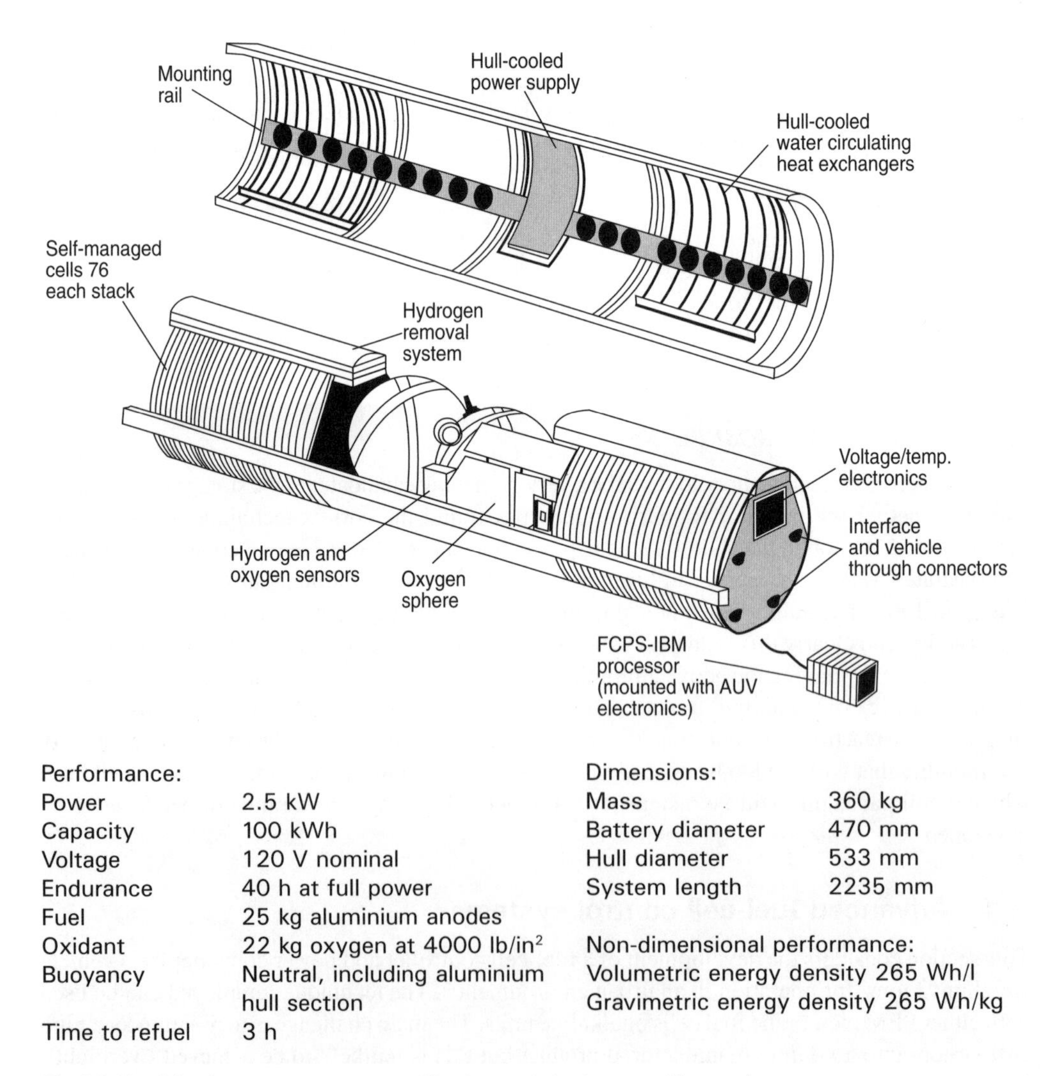

Performance:	
Power	2.5 kW
Capacity	100 kWh
Voltage	120 V nominal
Endurance	40 h at full power
Fuel	25 kg aluminium anodes
Oxidant	22 kg oxygen at 4000 lb/in^2
Buoyancy	Neutral, including aluminium hull section
Time to refuel	3 h

Dimensions:	
Mass	360 kg
Battery diameter	470 mm
Hull diameter	533 mm
System length	2235 mm

Non-dimensional performance:
Volumetric energy density 265 Wh/l
Gravimetric energy density 265 Wh/kg

Fig. 2.9 Aluminium/oxygen power system and its characteristics (courtesy Alupower).

become discharged would be if one travelled more than 400 km in one day. In this case the battery would be rapidly charged at a service station. Since the battery is light the cost is moderate and because it is not normally deep cycled a long life can be expected. Aluminium test cells have already demonstrated over 3000 deep discharge cycles and operation down to –80°C, as seen in the previous section.

At the present time we need to use larger fuel cells and smaller batteries similar to the hybrids with thermal engines. The vehicle which is going to be the development testbed is the new TX1 London taxi chassis made by LTI International, a division of Manganese Bronze in Coventry, shown in Fig. 2.10. This vehicle has been chosen because of growing air quality problems in London. The City of Westminster is now an Air Quality Improvement Area. This is mainly due to a large increase in diesel use which has resulted in unacceptable levels of PM10 emissions. Public Transport is a major contributor, with the concentration of large numbers of vehicles in the central zone.

Two types of fuel cell are attractive for use in vehicles – the PEM membrane and the alkaline types, as described in the following chapter. Both types have undergone a revolution in stack design in the last few years with the result that the stack (Fig. 2.11) is no longer the major cost item in small systems, it is the fuel-cell controller and the power converter. In this section we shall review the problems to be solved and offer some suggestions as to the likely course of development. As always the fundamental issue is to convert a high cost technology for mass production civilian use. Current (1998) fuel cells cost $1000 per kW and most of that cost lies in the control system and power conversion. Stacks will cost less than $100 per kW in mass production. The challenge is to reduce the control system cost. It is for this reason that most vehicle fuel-cell manufacturers are opting to supply the stacks, and leave the car industry to manufacture the controller, Fig. 2.12. This is an opportunity that Fuel Cell Control Ltd intends to take up by offering control systems commercially.

Fig. 2.10 TXI London taxi.

(a)

(b)

(c)

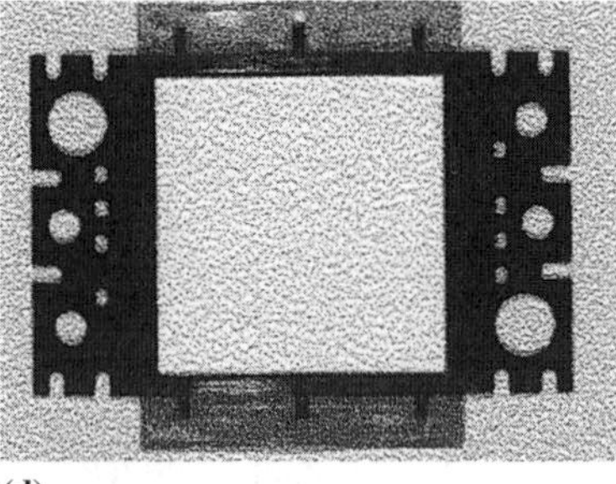

(d)

Fig. 2.11 Developed PEM fuel cell: (a) plate; (b) stack; (c) anode; (d) cathode.

2.4.1 WHAT IS IN A FUEL-CELL SYSTEM?

Here is a typical specification:

Power:	7.2 kW max	
Output voltage:	96 V DC no-load	
	64 V DC full-load	
Output current:	110 amps	
Operating temperature:	70°C	
Fuel:	Air	45 cubic metres per hour
	Pure hydrogen	5 cubic metres per hour
Hydrogen storage:	Cryogenic – 180°C	
	High pressure 200 bar	
DC/DC converter 1		
Input:	60–100 V DC	
Output:	0–396 V at 2.45 V per cell, lead–acid - 18 A	
	Current ripple less than 1 part in 10 000	
Fuel-cell controller		
Intelligence:	80 I/O programmable logic controller at 24 V DC	
Control:	Close loop: hydrogen 0–1/10 bar; air 0–45 cubic metres/hour proportional to demand	
Purge:	Dry nitrogen loop	

Pumps:	(1) Hydrogen 80 watts
	(2) Air 320 watts
	(3) KOH 85 watts
	(4) Water 10 watts
Valves:	10 off electropneumatic control
Preheat:	2 kW – 312 V DC
DC/DC converter 2	
Input:	200/400 V DC
Output:	27.6 V DC 600 W for control system

Fig. 2.12 Fuel-cell system specification.

Figure 2.13 shows the cell layout of the two fuel-cell types, alkaline and proton exchange membrane (PEM). In the alkaline type, the electrolyte is a liquid – potassium hydroxide or KOH. This is the same material used in alkaline batteries. The anode membrane is porous and has eight very small amounts of platinum catalyst (1 g would cover three football fields of plate area). The cathode side has a silver catalyst made by Hoechst. It is possible to use platinum but the cost is much greater.

In the PEM type the electrolyte is a solid, Nafion 115 sheet – a proprietary Du Pont product; there are competitors such as Dow Chemical and Ashai in Japan. The anode is similar to the alkaline type. The cathode membrane has a platinum catalyst and much research has been aimed at reducing the cathode loading which is why many PEM cells use the high pressure approach, since it helps to reduce the amount of catalyst for a given current density. Catalysts are the main cost in stack construction and optimizing their use is a major research area. Other differences between the two types are cooling and source gas purity.

In both systems about 40% of the fuel expended is given out as heat. In the PEM type, water cooling plates are used to remove this heat. In the alkaline type the electrolyte does this job and has the added advantage that it does not freeze at 0°C. Consequently both systems need a liquid cooling system.

In Europe, Esso is already committed to making hydrogen available at vehicle service stations. Hydrogen may also be used to power aircraft in the future. In America, petrol is effectively subsidized (see Chapter 4) which makes it very hard for other fuels to compete. One of the main interests there has been reforming petrol and methanol to produce hydrogen. If done in the vehicle this produces a hydrogen supply which contains a high concentration of carbon monoxide. PEM systems can be made to tolerate this impurity. To date alkaline stacks need pure hydrogen.

However, the whole business of on-board reforming is undesirable in terms of cost and complexity and is inefficient in terms of fuel consumption compared to using pure hydrogen made at a central

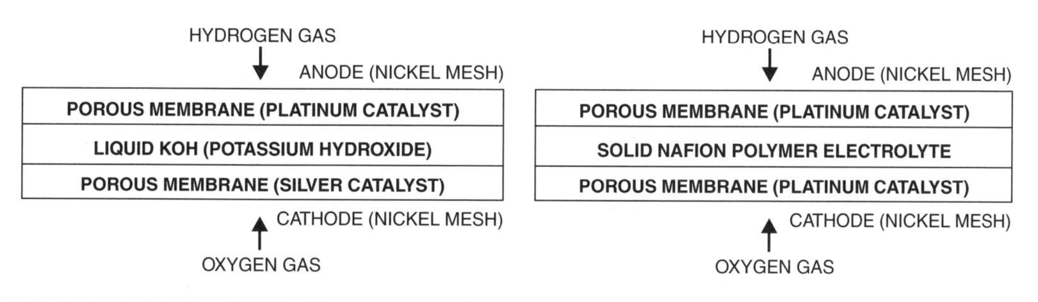

Fig. 2.13 Alkali (left) and PEM cell layouts compared.

facility. There are two main ways hydrogen can be stored: gas or liquid. As a gas it is usually compressed to 200 bar and stored in steel tanks with man-made fibre reinforcement and carbon additives to assist in the absorption. This technique works for large vehicles where bottles can be roof mounted – buses, for example. As a liquid the energy density is three times that of petrol, being 57 000 BTUs per lb compared with 19 000 BTUs per lb for gasoline. Two gallons would be needed to travel 500 miles in a 3 litre/100 km (80/100 mpg) PNGV specification vehicle. The gas liquefies at –180°C and 20 bar. In modern super-insulated vehicle tanks, hydrogen can be kept liquid for 2 weeks without refrigeration. A 20 watt Sterling cycle refrigerator can keep it liquid indefinitely. This system is suitable for application where space is limited, such as aeroplanes and cars. Many people believe the compression process uses too much energy. In fact it is the LIND refrigeration cycle, which is used to take hydrogen down to –269°C from –180°C where hydrogen is liquid at atmospheric pressure, that is the heavy consumer of compressor energy.

Polaron believe the technique permits the early use of hydrogen because tank exchange is possible until the investment in on-board refuelling is possible. The tank for a car would only be the size of an outboard-engined boat fuel tank. Cryogenic storage is already well established in the natural gas industry where liquid natural gas (methane) at –160°C is used to fuel 1000 bhp heavy duty trucks in Europe and Japan.

Considering again fuel-cell stacks, in either system an anode or cathode plate is 2.5 mm thick, so a pair of plates give a 5 mm build-up. Each pair of plates gives 1 V at no-load and typically 0.66 V at full load. This means that the stack length is around 500 mm, plus manifolds, for a 64 V, 7.2 kW continuous rating stack. It should also be pointed out that stack power doubles, at least, if pure oxygen is used instead of air. This is unlikely, however, as on-board enrichment to 40% oxygen is promised in the near future, as are some significant improvements in stack chemistry, especially in the catalyst area.

The fuel-cell stack is controlled by regulating the hydrogen pressure in the range 0–30 millibars. A recirculation loop permits water vapour to be added as PEM fuel cells work best with wet hydrogen. The air pressure is regulated by changing the blower speed in conjunction with the fuel-cell current demand. This takes 10 seconds to rise in a low pressure system, but may fall rapidly. The DC/DC converter determines the load applied to the fuel cell.

2.4.1 FUEL-CELL OPERATING STRATEGIES

As we can now see in Fig. 2.14, a fuel cell is a complex system and the key problems are that the feedstock must be kept pure and power consumption minimized in the auxiliaries. There are two main fuel-cell operating strategies: high pressure, 1–3.5 bar, and low pressure, at 1/20 bar The benefit of high pressure systems is fast hydrogen diffusion in the membrane which results in fast response – less than 1 second. Consequently it is possible for the fuel cell to follow the vehicle load profile and operate without a battery. This strategy is spoilt by warm-up issues. The stack must be at 70°C to deliver rated power. Warm-up can take 15 minutes. Another problem is that the power to supply the compressed services is significant – perhaps 25% of output at peak power.

Low pressure systems have modest auxiliary power needs, perhaps 10% of rated output at full power and proportionally less at low power. The main consumers are the air compressor and the KOH pump. The price is slower response. It typically takes 10 seconds for the fuel cell to ramp up to full power, consequently a peaking battery is needed to provide power during acceleration. This means that generally a smaller fuel cell may be used.

Fuel cells are the opposite of most electrical devices in that peak efficiency occurs at minimum load. In a high pressure system this profile is ideal for a motorway express coach where most time

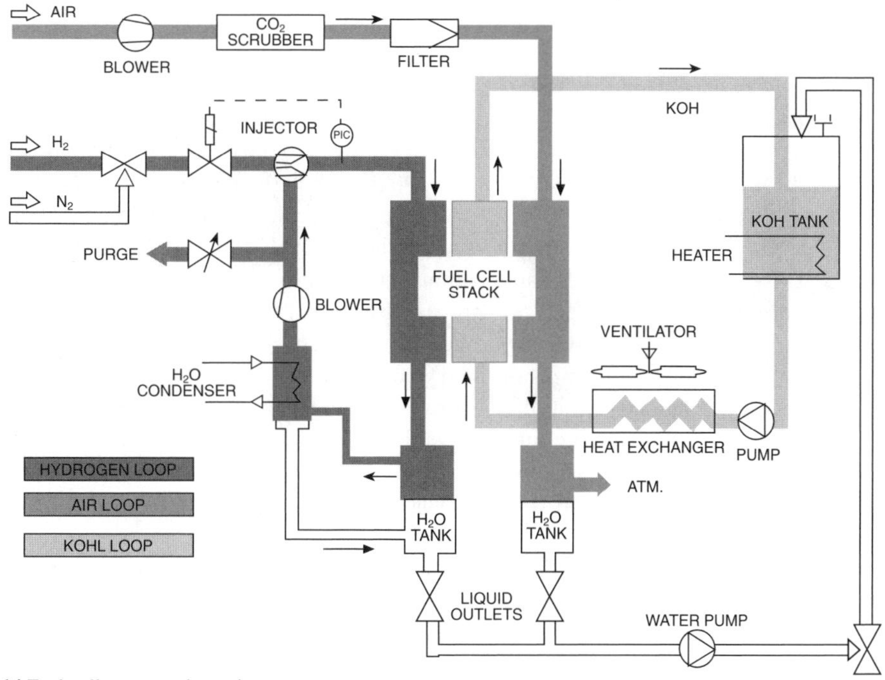

Fig. 2.14 Fuel-cell system schematic.

is spent cruising at 20/30% of maximum power. Hence efficiency is good in cruise and less so in continuous urban cycle duty. However, efficiency and emissions are always better than the equivalent thermal engine. Efficiency is 60% at no-load and 40% at full-load, at the current state of development – and the theoretical maximum is 83%.

2.4.2 *COST REDUCTION STRATEGIES FOR FUEL-CELL CONTROLS*

The cost of the fuel-cell controller (Fig. 2.15) is split up, at present, as follows: 20% each on pumps and compressors; valves and actuators; programmable logic controller; DC/DC converters; sensors and transducers. The challenge is to achieve a 5:1 cost reduction for mass-market viability.

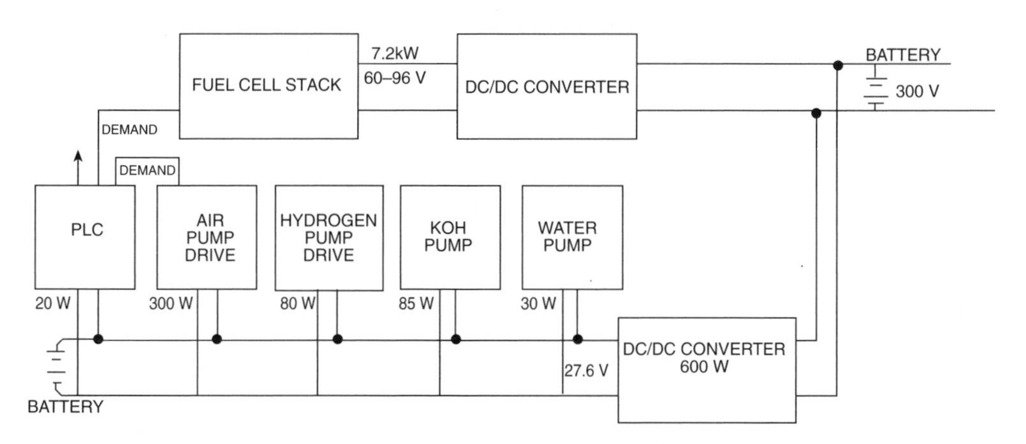

Fig. 2.15 Fuel-cell control system.

Fig. 2.16 Hydrogen/air blowers shown to left of drive electronics.

The hydrogen pump (shown left in Fig. 2.16) is a side channel blower and has to operate at 1/30 bar at 5 cubic metres/hour with wet hydrogen at 70°C, plus slight KOH contamination. The pressure criterion usually results in a choice of blower made by Gast and Rietschle. The standard unit is a 150 mm cube and weighs 5 kg. The operating point is 2800 rpm and power consumption by the pump is around 85 W, with an additional 36 W of copper loss in the motor. Fuel Cell Control Ltd rewound the 2 pole D56 induction motor as a 4 pole 20 V unit. Our second attempt will be an 8 pole design which should reduce the copper loss to about 8 watts. The low voltage is chosen for safety and the unit will be driven with a linear sine wave inverter (shown centre in Fig. 2.16). The dv/dt is kept low to avoid spontaneous ignition, in case hydrogen enters the motor chamber. The windings are potted, to avoid direct electrical contact and reduce the free volume in the motor chamber. All parts in the hydrogen contact area are nickel plated (zinc–copper–nickel). The blower is made of aluminium.

The air pump (shown right in Fig. 2.16) is about a 300 mm cube and weighs about 15 kg. The motor is a 1/2 hp D63 induction machine and has been rewound as 8 pole 20 V, 15 A, 325 W at 192 Hz (2900 rpm). The inverter is a switching unit to minimize losses, as the ignition risk is lower than with hydrogen. A 30 amp inverter provides good efficiency and the speed of this blower is adjusted with variations in power demand.

For the future, the company are working on a high speed channel blower, to operate at 10 000 rpm, using a brushless DC motor. In star-winding form, at 4000 rpm, it will satisfy 5 cubic metres per hour and at 10 000 rpm 45 cubic metres per hour. Thus a single design could do both jobs and it only weighs about l kg. However, silencing must be carefully considered.

The water and KOH pumps are standard 10 and 85 watt capacity permanent-magnet brushed DC driven pumps running at 3000 rpm on 28 V DC. The pumps are magnetically coupled with Talcum parts to resist aggressive fluids (such as potassium hydroxide).

2.4.3 FUEL-CELL CONTROL VALVES AND ACTUATION

Selecting suitable valves with such a diverse array of media and operating conditions has not been easy, Fig. 2.17. In fact the valves themselves are neither expensive nor heavy. The problem area is the actuators and it is intended to redesign these for the next version. Currently there is no safety legislation in place for hydrogen powered vehicles. The onus is on the supplier to demonstrate

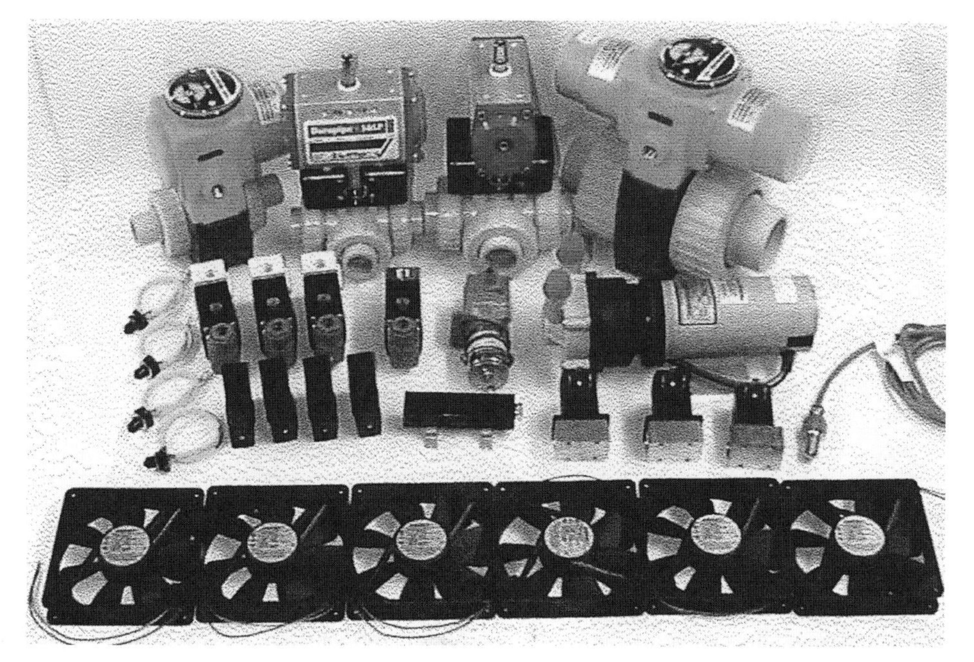

Fig. 2.17 Kit of valves.

fitness for purpose and that all reasonable precautions have been taken. It is felt this will change once meaningful experience has been achieved. Clearly, declaring a vehicle to be a class 1 safety area would destroy all economic viability. Consequently, as with petrol and propane, safe techniques need to be established and demonstrated before regulations are enforced. Some are obvious, such as no fuel or fuel processes to be contained in the passenger compartment. Others require experience such as fuel storage and distribution. Storage in a closed building needs careful consideration.

The valves can be neatly split into two groups, high and low pressure. The high pressure units are standard metal valves with electric solenoid actuators and spring return; they operate with 28 V coils. The larger units were chosen as 2 and 3 way plastic ball valves, using polypropylene bodies and EPDM seals for KOH compatibility plus high temperature operation (70–80°C).

Polaron had a major problem with the actuators. Fail-safe operation with low power consumption was needed. Solenoid valves in larger sizes use the controlled medium as a pilot fluid and consequently do not operate reliably with pressures as low as 1/30 bar. The solenoids are direct acting, with no economy measures or permanent magnet biasing, and thus consume significant power. In the end nitrogen was used as a pilot fluid, with 4 watt pilot valves to control the opening. This approach works well but the valves use up a lot of space, especially the actuators.

The intention for the future is to design a plate with spring-loaded clutches operating from a common motor drive. This should cut down on the volume and permit a much lower cost solution. Actuation accounted for 70% of the cost and 70% of the volume of the valves. It was found to be a niche sector market where nobody has a comprehensive system of interchangeable valves, seals and actuators suitable for onerous conditions.

2.4.4 *PROGRAMMABLE LOGIC CONTROLLER (PLC)*

An 80 I/O PLC with interface modules cost $1500 in 1998. Quantity build could halve this price – but still nowhere near the objectives. A Mitsubishi F Series was chosen for development, Fig.

2.18. Production units are destined to use a custom-engineered microprocessor unit based on a Siemens/Thompson C167 CAN bus processor, which is becoming a standard in the European and American car industry. This unit must represent one of the toughest design challenges. To convert low voltage, heavy current into higher voltage with galvanic isolation, ultra-low current ripple and high efficiency. Many solutions have been analysed, although this one offers the best combination of characteristics. Figure 2.19 shows electronic system circuits.

Let us consider a square wave phase-shift chopper: at minimum volts input, we need full reinforcement to achieve 396 V output. However, at low load we have a 96 V DC link and perhaps 90 degrees phase shift between A and B. This is not too bad, except that we only draw output current for 50% of the time: this means that the DC link contains 100% current ripple at 2F switching frequency. Since the pulse width should always be 50% plus, the solution is to have three such choppers with 120° phase shift between them. This has the effect of overlapping the converters if the correct measures are taken. Consequently, the supply now only contains 30% ripple worst case at 6F, but when the current is largest we have maximum overlap; see Fig. 2.19 (bottom).

The first attempt is to build this converter with each stage operating at 3 kHz, with torroidal 0.08 mm silicon steel cores. This is both silent and efficient. The edges are deliberately softened to reduce dv/dt (capacitive ripple). At low frequency this does not cost much in losses, with 10 microsecond edges, but reduces the spikes when the diodes reverse recover. The design is adaptable to different output voltages by rewinding the output transformers and chokes. It is believed 90% efficiency can be achieved with this design at 60 V input, 7.2 kW. A double L/C Filter attenuates the current to the fuel-cell stack to ensure compliance with 0.01% ripple current rating. The reason for this is to prevent poisoning of the fuel-cell catalysts.

The cost of this unit is a problem. The silicon for the main switches are LAPT transistors, at 15 A and 200 V, using 2SA1302 and 2SC3276; the six switches cost $150 in parts (1998 prices). It is intended to have these parts integrated into a high power package. The chokes cost $150 and capacitors $60. To improve the costs higher frequency is needed; time will tell if this can be achieved without sacrificing efficiency and current ripple.

2.4.5 SENSORS AND TRANSDUCERS

This is perhaps the hardest area in which to reduce the cost. The devices operate under hostile conditions and are currently made to order. One success has been to reduce the cost of the 2 kW preheater from $1000 to $200 by a complete redesign. In other areas there has been success in combining functions such as the flow meter and the temperature controller in the KOH loop. A complete new family of low cost sensors is needed before production cost targets can be met.

Fig. 2.18 Mitsubishi 80 I/O F-series PLC with DAC, ADC and RS232 modules.

2.4.6 FUEL-CELL FUTURE PROSPECTS

It is early days for vehicle fuel cells and the main challenge is better, lighter, cheaper, more convenient to use parts – preferably plastics. Insulated packaging of semiconductors is the main issue in the DC/DC converters at low voltage and heavy current. This in time will lead to control systems with lower parts count and greater reliability. As we get down the learning curve, and volumes increase, costs will fall. Major improvements in fuel storage, oxygen enrichment and stack materials should lead to continuing increases in current density and hence smaller stacks for the same power output.

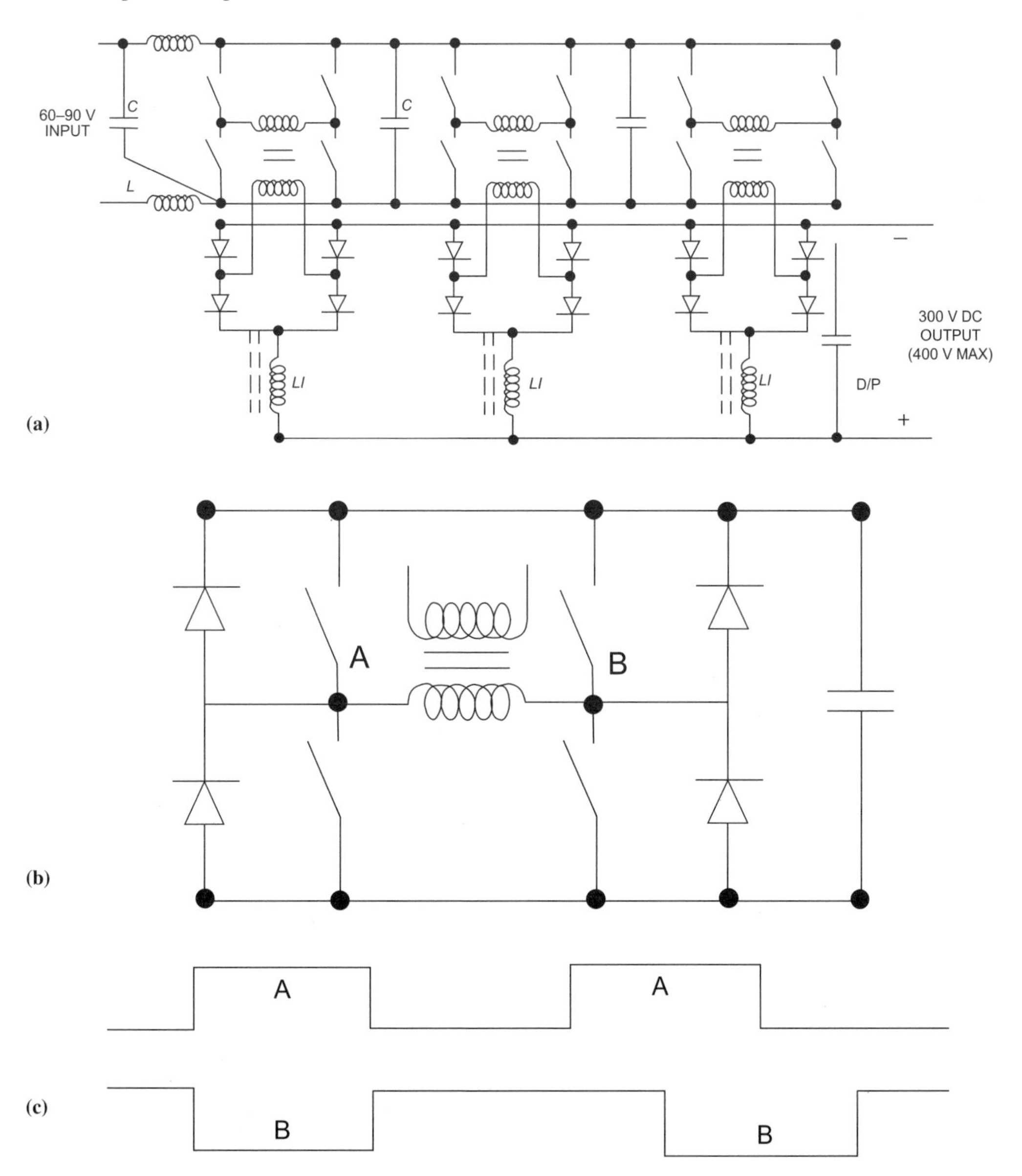

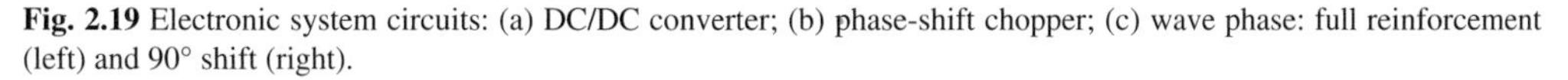

Fig. 2.19 Electronic system circuits: (a) DC/DC converter; (b) phase-shift chopper; (c) wave phase: full reinforcement (left) and 90° shift (right).

2.5 Waste heat recovery, key element in supercar efficiency

In the longer term vehicles will be electrically propelled using flywheel storage, hydrogen fuel cells or both. These systems potentially offer high cycle efficiencies, and low emissions vital to improving air quality in our big cities[11].

Bill Clinton's initiative for US family cars to achieve 100 miles per gallon by 2003 is a serious challenge to the engineering industry; see Chapter 4. Current solutions include lightweight structures, reduced running losses and small engines. However, the author believes that the target can be achieved using conventional vehicle platforms with low drag floorpans and instead attention is focused on high efficiency drivetrains. Small engines give poor acceleration so to achieve acceptable performance hybrid technology is required. The internal combustion engine car achieves 28% efficiency under motorway conditions and half that on an urban cycle. The key problem is how to convert more of that energy into useful work. The account below will investigate two schemes: (a) turbine recovery system and (b) thermoelectric recovery system.

In both schemes the energy produced is converted into electricity. This is because such an arrangement provides a plausible method for matching the power into the electrical drive. It is accepted that all mechanical solutions are also viable in a hybrid vehicle.

2.5.1 HYBRID ELECTRIC DRIVE

Figure 2.20 illustrates a parallel hybrid driveline using a Wankel engine and a brushless DC Motor. The package can produce 70 kW peak power and 20 kW average. This combination provides excellent acceleration using energy stored in a small flat plate lead-acid battery. Tests to date show this battery still delivers 100% peak power of 45 kW and 80% capacity after 22 000 cycles to 30% depth of discharge. Thermal management is vital to achieving these figures.

The engine operates on a two stroke cycle and produces high quality waste heat at the exhaust. The temperature of the gas is around 1000°C. This gas contains 72% of the energy in the fuel. If we can convert a third of this energy into electricity we can nearly double the NTG of the vehicle under motorway conditions.

Why is this important? Hybrid solutions are very effective at improving efficiency under urban cycle conditions but make no impact under motorway conditions. Here is a system that can begin to solve this problem. Supposing in the existing scheme we require 20 kW to operate a vehicle at

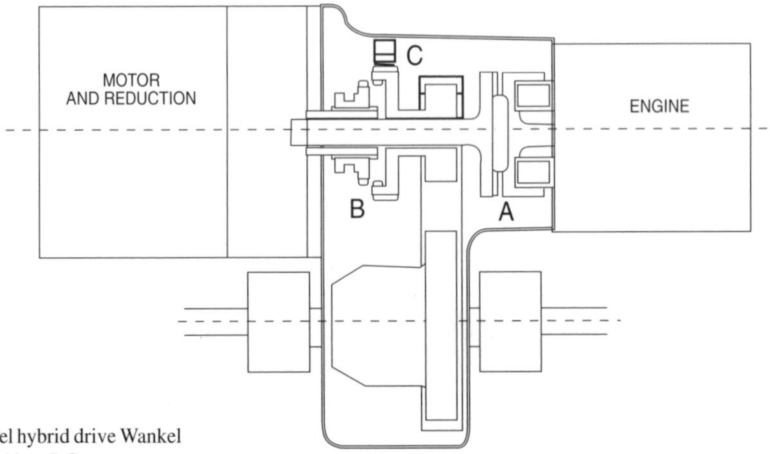

Fig. 2.20 Parallel hybrid drive Wankel engine and brushless DC motor.

70 mph on a motorway. The engine produces 55 kW of waste heat. If a third is converted to electricity 18.33 kW is made available and perhaps 15 kW of this is additional mechanical power. Consequently the engine only has to produce 60% of the mechanical power to give the same output of 20 kW and the electrical system could be reduced to 10 kW. To summarize, the 20 kW requirement could be met by using 12 kW of engine power and 8 kW of waste heat recovery power. Future schemes could possibly improve on these figures.

2.5.2 *THE TURBINE RECOVERY SCHEME*

In this scheme the concept is to use a small turbine system in association with an electric generator. Figure 2.21 illustrates the conceptual realization of the idea. Some modelling of the turbine shows that, for reasonable efficiency, speeds of 150 000 rpm are necessary. If such speeds can be achieved the sizes of the components will be tiny. For example, for 10 kW, the rotor of the generator will be 25 mm diameter by 15 mm long. The generator needs to be kept separate from the turbine stages due to the high temperatures involved in the thermodynamic processes. The turbine bearings will be hydrodynamic gas bearings – the back of the turbine rotors and the shaft will be plated with a zig-zag pattern, microns thick, designed to created turbulence at high speeds. Dynamically the system will operate below the first critical speed.

The generator bearings will be angle contact bearings (6 mm) using ceramic balls (RHP-INA) and Kluber Isoflex Super LDS 18 grease. The rear bearing is preloaded and free to slide by means of a crinkle spring. The coupling between generator and motor can be a tongue and fork, permitting easy removal, as the torque is below 1 Nm.

The generator losses preheat the air entering the first compressor stage. The generator has laminations of 0.2 mm radiometal with powder coat insulation applied to ensure minimum eddy current losses. The rotor has a one-piece tubular magnet, 22 mm diameter and 15 mm long by 3.5 mm thick, of 'one-five' samarium cobalt. This is glued onto a stainless steel shaft of high

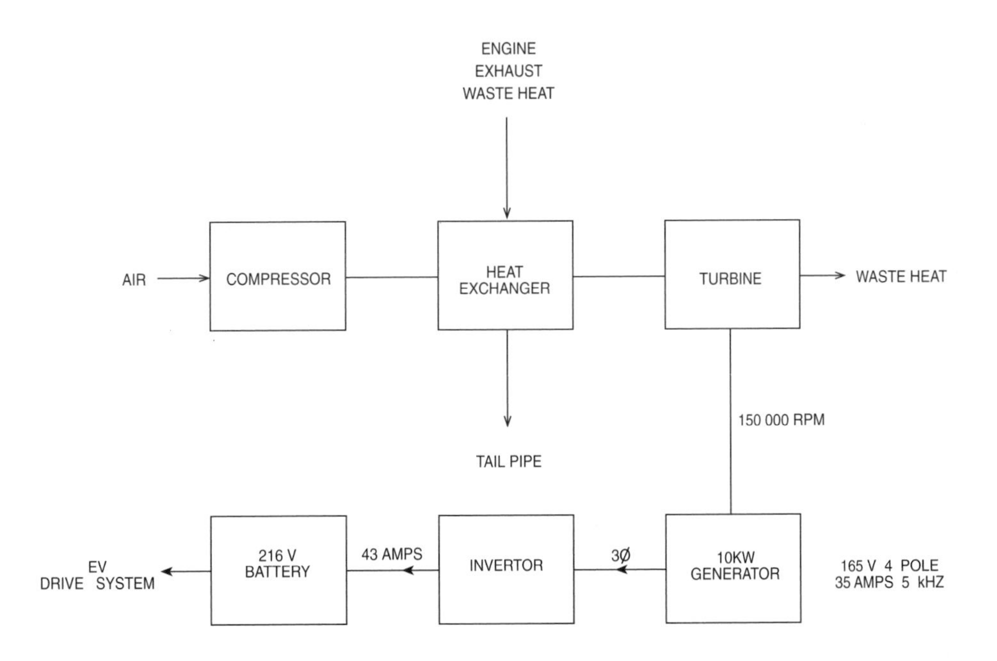

Fig. 2.21 Block diagram of turbine waste heat.

resistivity. The magnets are retained by a prestressed carbon fibre ring of 1.5 mm wall thickness. The generator has a 4 pole configuration and the machine winding is designed to give 165 V (RMS) at 150 000 rpm, resulting in a line current of 42 A at 10 kW. An advantage to this method of construction is that the generator may be built and tested separately from the turbine. The turbine rotors will be only 40 mm outside diameter and machined from aluminium. The rotors are held to the shaft by Loctited nuts and there is a hole down the centre to facilitate temperature measurement. One of the nuts contains the fork for the drive coupling.

The design of these stages with their casing and expanders is confidential. The heat exchanger is an air to air unit rated for 30 kW at a temperature of 600°C. The same unit also functions as an exhaust silencer for the engine and special construction techniques are required to resist the high temperature of the exhaust from a Wankel engine (typically 1000°C). Polaron envisage a battery of 216 V nominal varying between 180 V and 255 V. The speed of the turbine may vary over the entire range but meaningful output will only occur between 120 000 and 150 000 rpm.

The turbine, Fig. 2.22, is started by a transistor bridge connected across the diode bridge and as the compression of inlet air starts, and is expanded, the output turbine takes over supplying rotational power, and the transistor bridge is then used as a switching regulator, to match the generator voltage to the battery voltage. Sensorless timing techniques are possible but it is simpler to use three Hall sensors operating from the rotor field system.

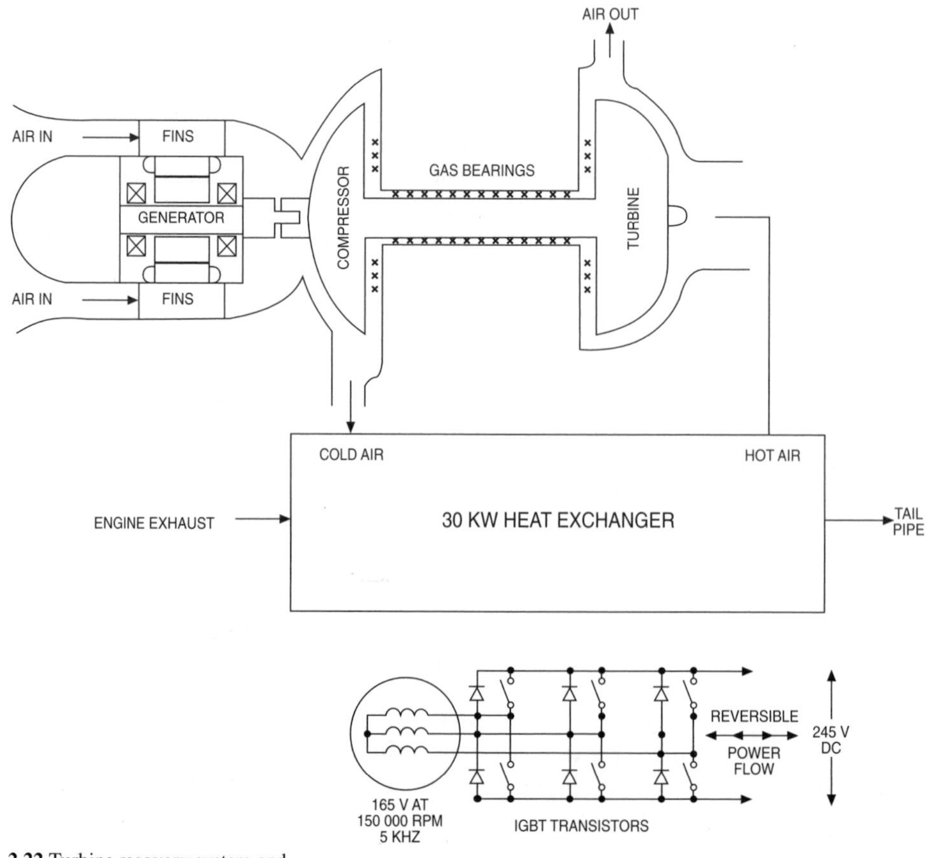

Fig. 2.22 Turbine recovery system and recuperator power control.

You might ask: Why not let the load line of the turbine generator intersect with that of the battery on an open loop basis? The problem is that in most cases one would not obtain the correct operating point. The turbine power is proportional to speed cubed. One obtains the correct operating point at just one speed for a given power, whereas the battery operates from 1.75 to 2.35 V per cell. Consequently it is necessary to have closed loop control of the power flow from generator to battery. But there is a second reason; this mode of control with the transistor bridge permits the turbine to be used as a brake – power flow is reversible between turbine and battery. This is very useful when negotiating long steep gradients, for example.

Overall it is believed that an efficiency of 30% is achievable with such a process and thus the system can make a major contribution to fuel utilization under motorway conditions.

2.5.3 THERMOELECTRIC GENERATOR

The turbine recuperator technique involves some very high technology mechanics to make the system work. It prompts the questions: Is there any other way of achieving the same objective? Is a solid state solution possible?

Thermoelectrical devices were invented in 1821 and are perhaps best known today for the small fridges we have on our cars and boats to cool food and drinks. An array of bismuth telluride chips 40 mm square can produce 60 watts of cooling with a temperature differential of 20°C. If we go back 60 years to the 1930s there were thermopiles which one placed into a fire and the pile provided the current for a vacuum tube radio. It is only very recently that here in the UK a group of engineers started to ask the question 'Why are thermopiles so inefficient?' What happens to the 96% of the energy consumed that does not appear at the output terminals? Why is the output voltage so small – typically microvolts per °C at top temperature?

At Southampton University Dr Harold Aspden soon identified the answer to the efficiency question. The energy was being consumed by circulating currents within the device. It was then realized that if a dielectric was placed between the thermopile layers, and the pile was oscillated mechanically, that an AC voltage could be obtained up to 50 times the amplitude of the original DC voltage, Fig. 2.23. This oscillation has been tested with frequencies from DC to RF and the process holds good across the spectrum. Dr Aspden has concentrated his efforts on producing thermopile arrays for use on the roof of a building, with temperature differentials of 20–40°C.

However, if we return to our waste heat recovery problem we are dealing with top temperatures of 600°C plus and consequently alternative materials will be required and the number of stages in series to produce a given voltage will be reduced. But, with a top temperature of 30°C existing, devices can convert 20 W of power with an efficiency of 25%. It should be emphasized that this work is at an early stage of development at this time.

The thermopile elements suitable are iron and constantin 40% nickel/60% copper (Type J thermocouple material); at 600°C, with mechanical excitation, a voltage of 300–500 mV per stage can be achieved, hence 500 cells in series would produce 216 V DC. The circulating current in each cell is proportional to the temperature difference but the output AC voltage may be controlled by adjusting the amplitude of the mechanical excitation. The most interesting point is that to give 10 kW a suitable unit could be very compact – our calculations suggest about 100 mm cube. We believe the mechanical excitation is best supplied by ultrasonic piezoelectric transducers driven by a HiFi amplifier. The power required is around 200 watts. One interesting point is that the unit offers reversible power flow. How? It can be converted from refrigerator to heater and act as a braking device.

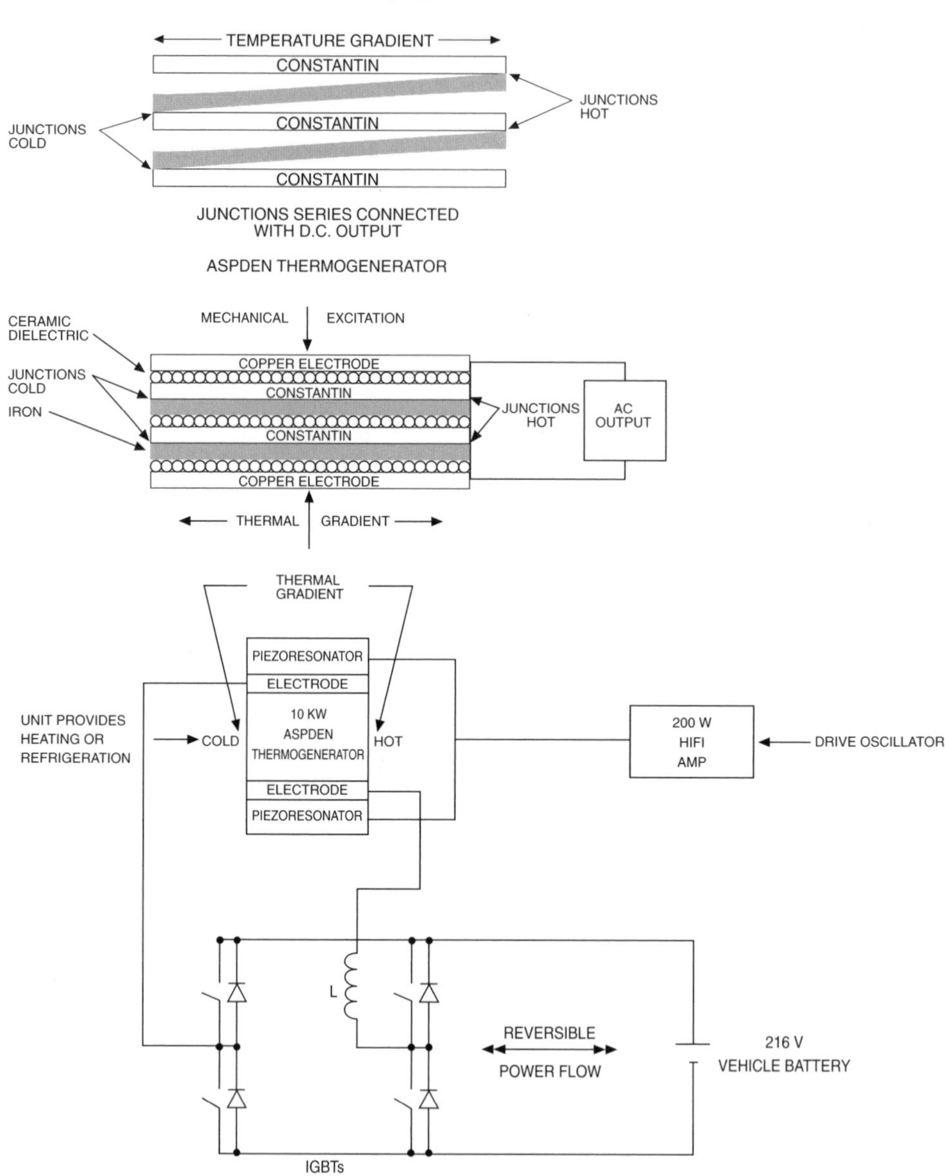

Fig. 2.23 Aspden thermogenerator and its control system (below).

References

1. Hodkinson, R., The electronic battery, paper 98EL004, ISATA31
2. Hodkinson, R., The aluminium battery – a status report, paper 99CPE012, ISATA 32, 1999
3. Zaromb, S. and Faust, R. A., *Journal of the Electrochemical Society,* 109, p. 1191,1962
4. Despic, A. and Parkhutik, V., *Modern Aspects of Electrochemistry,* No. 20, J. O. M. Bockrus, Plenum Press, New York

5. Gifford, P. R. and Palmissano, J. B., *Journal Electrochem. Soc.,* 135, p. 650, 1988
6. Zagiel, A., Natishan, P. and Gileadi, E., *Electrochim Acta,* 35, p. 1019, 1990
7. Rudd, E. J., *Development of Aluminium/Air Batteries for Applications in Electric Vehicles,* Eltech Research Corp. to Sandia Nat Labs, Contract AN091–7066, December 1990
8. ALUPOWER INC, Internal ALUPOWER-Canada Report 1992
9. Gibbons, D. W. and Rudd, E. J., *The Development of Aluminium/Air Batteries for Propulsion Applications*
10. Hodkinson, R., Advanced fuel cell control system, *EVS* 15, Brussels, September 1998
11. Hodkinson, R., Waste heat recovery – a key element in supercar efficiency, paper 94UL004, ISATA 27, 1994

Further reading

Proceedings 28th IECEC 1993

Rand *et al., Batteries for electric vehicles,* Research Studies Press/Wiley, 1998

Berndt, *Maintenance-free batteries,* Research Studies Press/Wiley, 1993

3

Electric motor and drive-controller design

3.1 Introduction

While Chapter 1 introduced the selection and specification of EV motors and control circuits, this chapter shows how system and detail design can in themselves produce very worthwhile improvements in efficiency which can define the viability of an EV project. The section opens with discussion of the recently introduced brushed DC motor, by Nelco Ltd, for electric industrial trucks, then considers three sizes of brushless DC machine for electric and hybrid drive cars, before examining the latest developments in motor controllers.

3.2 Electric truck motor considerations

EV motor makers Nelco say the requirements for traction motors can be summarized as light weight, wide speed range, high efficiency, maximum torque and long life. The company recently developed their diagonal frame Nexus II motor, for general electric truck operation. In this motor, Fig. 3.1, active iron and copper represent 50 and 30% respectively of the motor weight. Holes in the armature lamination, (a), have resulted in some weight reduction and the use of a faceplate commutator, (b), has also helped keep weight down – with only 30% of the copper required for a barrel-type commutator – because the riser forms part of the brush contact face. With use of aluminium alloy for the non-active parts, such as brush holders (c) of the motor, weight of the 132L motor is held to 80 kg, a power to weight ratio of 450 watts/kg. Tolerance of high accelerations comes from perfection of the faceplate commutator to retain brush track surface stability. Usually the constraint on high power at high speeds, particularly when field strengths are reduced, is commutation ability, Nelco maintains.

The patented segmented frame of the Nexus, (d), makes the provision of interpoles quite an easy option – to optimize commutation at all current loadings, so reducing brush heating losses and compensating for interpole coil resistance losses. As output torque is a function of armature current, flux and the number of conductors, all these must be maximized. Short time high current densities, over the constant torque portion of the performance envelope, are possible given adequate cooling. Cost is held down by such measures as use of a segmented yoke/pole assembly, (e); extruded brush holders are also used, (f). Figure. 3.2 shows rating and efficiency curves for the N180L machine.

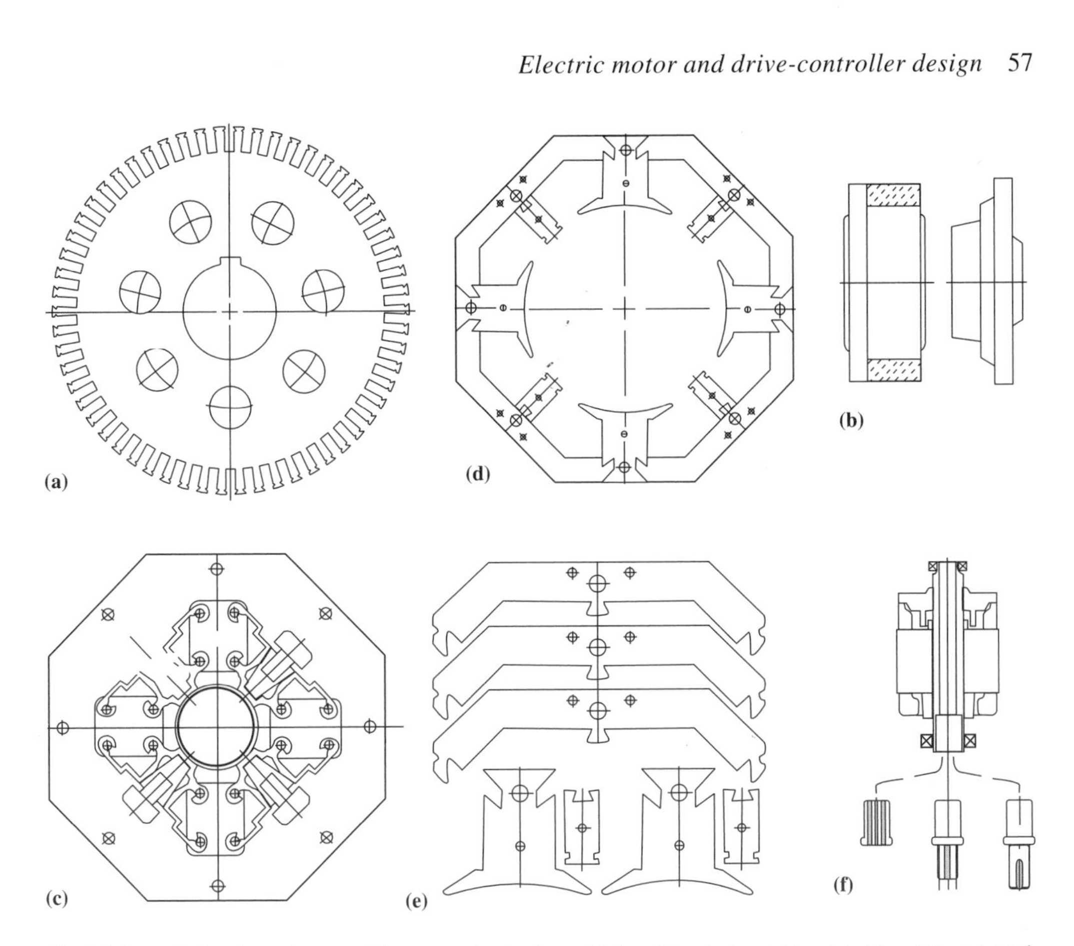

Fig. 3.1 Nexus II electric truck motor: (a) armature laminations; (b) faceplate adaptor; (c) brush holders; (d) segmented frame; (e) segmented yoke/pole assembly; (f) brush holder extrusions.

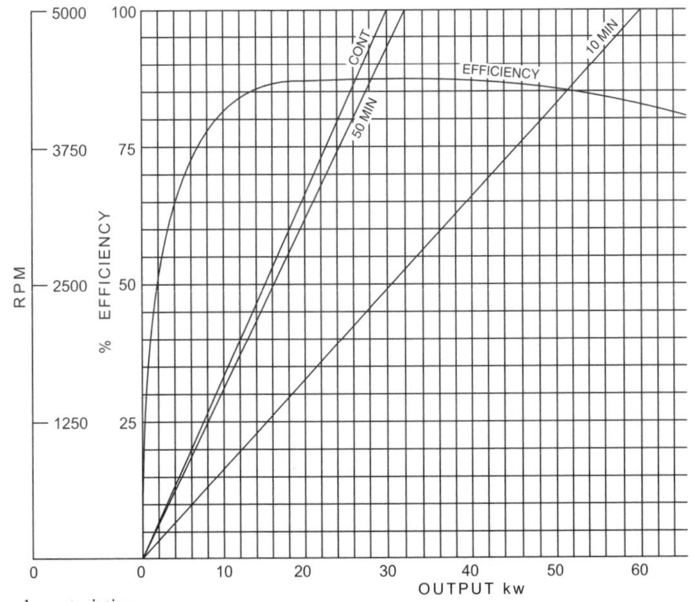

Fig. 3.2 N180L motor characteristics.

3.3 Brushless DC motor design for a small car

In this case study of the design of a 45 kW motor[1] commissioned for a small family hatchback – the Rover Metro Hermes – the unit was to give rated power from 3600–12 000 rpm at a terminal voltage of 150 V AC. The unit has been tested on a dynamometer over the full envelope of performance and methods for improving the accuracy of measurement are discussed below. The results presented show a machine with high load efficiency up to expectations and the factors considered are important in minimizing losses.

3.3.1 BRUSHLESS MOTOR FUNDAMENTALS

A key aspect of motor design for improved performance is vector control, which is the resolution of the stator current of the machine into two components of current at right angles. *Id* is the reactive component which controls the field and *Iq* is the real component which controls the power. *Id* and *Iq* are normally alternating currents. In this example, Fig. 3.3, the machines being considered are of the rare-earth surface-mounted magnet type with a conventional 3 phase stator and a rotor consisting of a magnetic flux return with a number of motor pole magnets mounted on it. The open loop characteristics of the machine are considered as follows: if the shaft of the motor is driven externally to 12 000 rpm a voltage of 260 V will be recorded, (a). In this condition with full field at maximum speed, iron losses will be high and the stator will heat up very quickly. At this operating point the motor could supply about 135 kW of power. However, this is not the purpose of the design, (b).

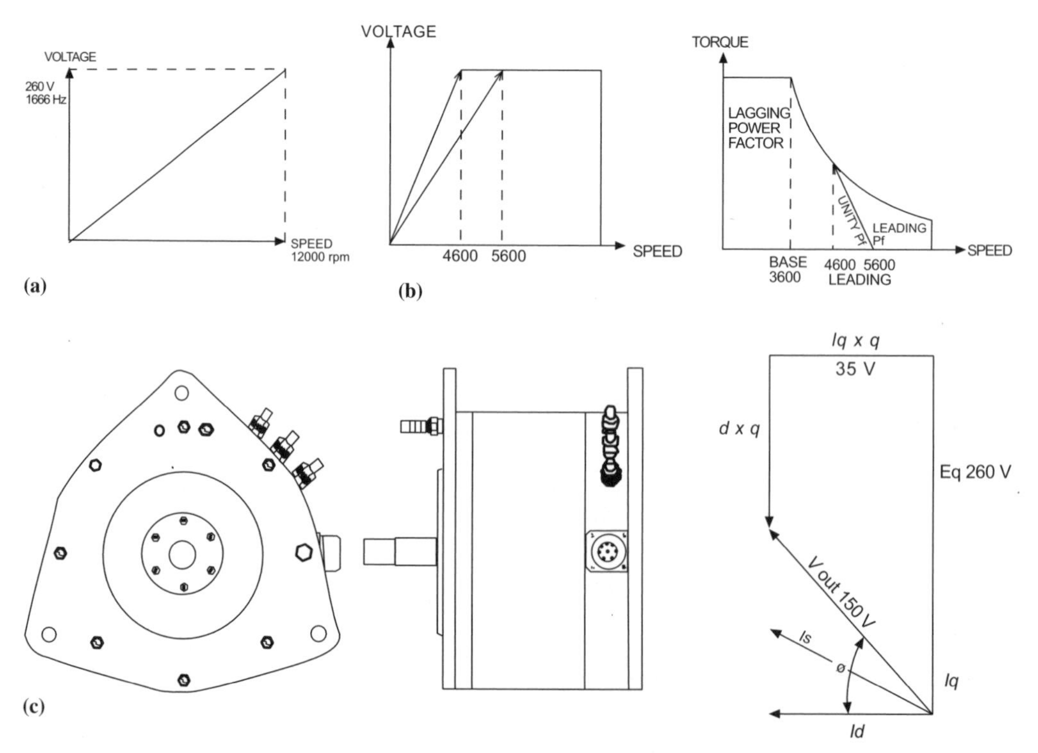

Fig. 3.3 Example brushless motor characteristics: (a) no-load terminal voltage when machine is operated as a generator; (b) variation of machine terminal voltage with torque and speed (left) with variation of power factor with torque and speed (right); (c) vector diagram (right) of PMB DC motor (left), in field weakening condition 12 000 rpm no-load.

The torque–speed requirement for a typical small vehicle is shown to be constant torque to base speed (around 3600 rpm) then constant power to 12 000 rpm. This assumes a fixed ratio design speed reducer. During the first region the voltage rises with speed. In the second region the voltage is held constant at 150 V by deliberately introducing a circulating current $-Id$ which produces 152 V at 12 000 rpm to offset the 260 V produced by the machine, to leave 150 V at the machine terminals. The circulating current produces this voltage across the inductance of the machine winding. It also produces armature reaction which weakens the machine field; total field = armature reaction + permanent magnet field gives a lower air gap flux and lower iron losses. This mode of operation is known as vector control. What happens if we reverse the direction of Id? Theoretically we strengthen the field. However, with a surface mounted magnet motor the machine slows down due to the effect of the circulating current on the machine inductance. However, the torque per amp of Iq current remains constant.

If we supply the motor from a square wave inverter we observe some interesting phenomena when we vary the position of the rotor timing signals. In the correct position the stator current is very small. When the current lags the voltage the motor slows and produces current with sharp spikes and considerable torque ripple. When the current leads the voltage the motor runs faster and produces a near sine wave with smooth torque output. It is the field weakening mode we wish to use in our control strategy, (c).

3.3.2 MOTOR DESIGN: METHOD OF MEASUREMENT

In the following account details are given of the motor design, Fig. 3.4, and of the predicted and measured efficiency maps. The measured efficiency maps were carried out using a variable DC link voltage source inverter. Polaron conducted the trials with two waveforms: a square wave with conduction angle 180° and a square wave with harmonic reduction, conduction angle 150°, the purpose being to assess the effects of the harmonics on motor performance, (a).

(a)

Stack OD	220 mm	Stator mass	14.1 kg
Stack ID	142.5 mm	Rotor mass	4.12 kg
Length	80.5 mm	Total mass	34 kg
Overall length	140.5 mm	Rotor inertia	0.016 kg m^{2}V
Pole number	16	Thermal resistance	0.038°C/Watt
Peak torque	200 Nm	Thermal capacity	6000 joules/°C
Motor constant km RMS	3.03 Nm/sqr (W)	Rotor critical speed	21 000 rpm
Motor constant km DC2	89 Nm/sqr (W)	Nominal speed	12 000 rpm
Electrical time constant	10.4 millisecs	Back EMF at	12 000 rpm = 260 V
Mechanical time constant	1.9 millisecs	Winding resistance	0.096 ohm
Friction	0.171 Nm	Winding inductance	100 microhenries
Motor torque constant	0.3 Nm/A	Vector control voltage	150 V
Winding star connected		RMS line to line	

(b)

a = 150° audible					a = 180° audible			
SPEED	V	I	P	noise	V	I	P	noise
1000	29V	7.3A	75W	52dB	28V	12.4A	72W	54dB
2000	55V	8.1A	216W	54dB	55V	12.8A	216W	54dB
3000	82.6V	8.4A	396W	56dB	84V	13.2A	405W	55dB
4000	113V	9.12A	540W	56dB	110V	13.6A	630W	56dB
5000	138V	9.12A	765W	58dB	137V	13.8A	900W	57dB
6000	150V	25A	990W	59dB	150V	24A	1080W	58dB
8000	150V	87A	1440W	60dB	150V	84A	1800W	63dB
10000	150V	122A	2250W	67dB	150V	123A	2700W	69dB

Fig. 3.4 Motor design data: (a) XP1070 machine data; (b) no-load losses (machine only).

The measurement of electrical input power is accurately achieved using the 'three wattmeter' method. Measurement of mechanical power is more difficult. Polaron found it necessary to mount the motor into a swing frame with a separate load cell to obtain accurate results at low torque. Even so, other problems such as mechanical resonances and beating effects at 50 Hz harmonics require care in assessing results. The operating points were on the basis of maximum efficiency below 150 V AC terminal voltage.

Results are in the form of three efficiency maps which give predicted and measured performance on both waveforms. The losses in this type of motor are dominated by resistance at low speed and iron losses at high speed. What the results show is that low speed performance was accurately predicted but high speed performance was less efficient especially at light load. The reason for this is that the iron loss at 10 000 rpm, no-load, should be about 1000 W, sine wave, (b). With 150 V terminal voltage the measured figure was 2200 W. The following paragraphs discuss the factors affecting this result but it is believed that the main contributors are larger than expected hysteresis losses due to core steel not being annealed, and larger than expected eddy current losses because of lower than specified insulation between laminations.

Annealing causes oxidation of the surface of the steel, leading to improved interlayer insulation. Polaron subsequently coat the laminations with epoxy resin then clamp them in a fixture to form a solid core for winding.

3.3.3 MOTOR DESIGN FACTORS AFFECTING MACHINE EFFICIENCY

For the stator the important factors are: (i) shape of lamination – optimized lamination has a much larger window than 50 Hz induction motor lamination and a bigger rotor diameter relative to the stator diameter; (ii) use of high nickel steels is counteracted by poor thermal conductivity. Thin silicon steel with well-insulated laminations gives best results. Laminations should be annealed and not subjected to large mechanical stresses. The core can be a slide fit in casing at room temperature as expansion due to core heating soon closes the gap. Stator OD should be a ground surface; (iii) winding must be litz wire and vacuum impregnated to ensure good thermal conductivity. Varnish conducts 10 times the heat of air gap.

For the rotor the main ones are: (i) if magnets are thick (10 mm in this case) mild steel flux return is satisfactory; (ii) magnets are unevenly spaced to remove cogging torque; (iii) individual poles must not contain gaps between magnet blocks making up the pole. Such gaps lead to massive high frequency iron losses. This can be checked by rotating the machine at lower speed and observing the back-EMF pattern. If there are sharp spikes in the wave form the user will have problems with losses.

3.3.4 MOTOR CONTROL

Battery operated drives must make optimum use of the energy stored in the battery. To do this, the efficiency of both motor and driveline are critically important. This is especially true in vehicle cruise mode typically two-thirds speed one-third maximum torque, therefore Polaron proposed to build a drive with two control systems: (i) current source control in constant torque region and (ii) voltage source operation in constant power region. At 45 kW 6000 rpm we would expect I_L 175 A, V_{AC} 150 V; inverter switching loss 10 kHz, 1.8 kW; converter saturated loss 0.9 kW, using PWM on the windings and IBGT devices.

If, however, we use a square wave at the machine frequency, Fig. 3.5, and the machine operates with a leading power factor, the switching losses are greatly reduced for additional iron loss, of 225 W, at top speed. The inverter efficiency increases from 94% to 97%. In the low speed constant torque region there is no alternative to using PWM in some form.

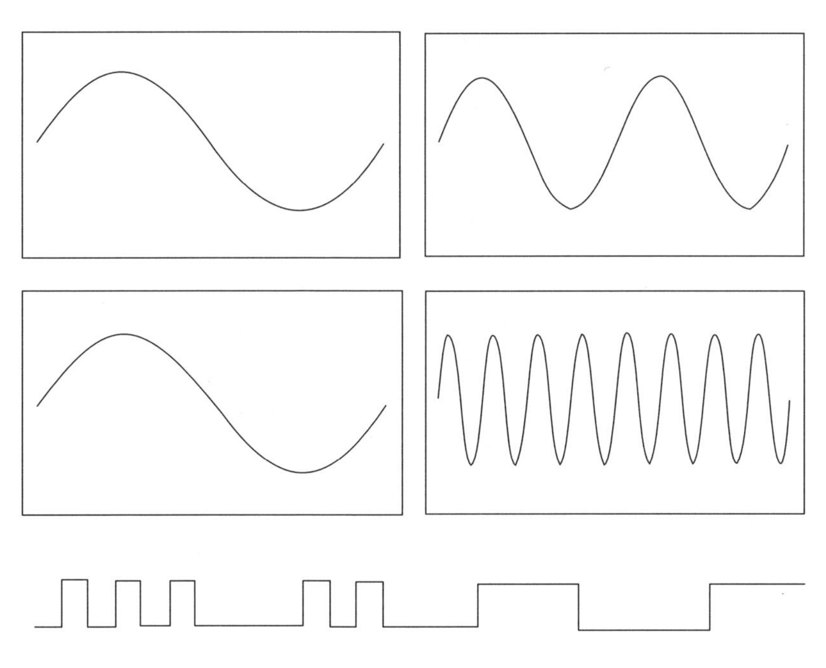

Fig. 3.5 Motor line current waveforms.

3.4 Brushless motor design for a medium car

3.4.1 INTRODUCTION

Here the task is to optimize the 45/70 kW driveline for the family car of the future[2]. This involves improvements in fundamental principles but much more in materials and manufacturing technology. The introduction of hybrid vehicles places ever greater demands on motor performance.

It is the long-term aim of the US PNGV programme to reduce the cost of 'core' electric motor and drive elements to 4 dollars per kW from around 10 dollars charged in 1996 for introductory products supplied in volume. The price may be reduced to 6.5 dollars using new manufacturing methods to be reviewed below. Further savings may come from very high volume production. This will require significant investment which will not occur until there is confidence in the market place and technical maturity in a solution. In terms of design, we may increase speed from 12 000 to 20 000 rpm. For reasons to be explored, a further increase becomes counterproductive unless there is a breakthrough in materials. In the inverter area Polaron believe the best cost strategy is to use a double converter with 300 V battery, 600 V DC link and 260 V motor. This assumes power levels of 70 kW.

The motor can be induction type or brushless DC. Induction is satisfactory in flat landscape/ long highway conditions. For steeper terrain, and shorter highways as exists in Europe brushless DC is more suitable – especially for high performance vehicles and drivelines for acceleration/ braking assistance in hybrid vehicles. Excellent progress has been made in the silicon field. The introduction of high reliability wire bonded packaging in association with thin NPT chip technology for IGBTs is reducing prices and improving performance. Currently a 100 A 3 phase bridge costs around $100 in volume. The arrival of complete 3 phase bridge drivers in a single chip at low cost is a further improvement in this area. Individual driver chips provide better device protection and drive capability at this time.

Great progress has been made in batteries in recent years. However, the time has come for a change in emphasis. Previously the pure battery electric was seen as the desired solution. Even if the remaining technical issues can be addressed, we are still impeded by weight and cost of such a solution. Consequently Polaron believe they should focus on hybrid solutions and this needs batteries optimized for peak power not energy capacity. It requires batteries with geometries optimized for peak power – ultra-low internal resistance and perhaps high capacitance at the same time. It will certainly require new packaging. A capacity of 2 kWh at 2 minute rate would be adequate for the average family car. It will also require a low cost short-circuit device to bypass high resistance cells in long series strings.

There is now little doubt that brushless DC machines offer the best overall performance when used in vector control mode, with high voltage windings, Fig. 3.6. The reason is that the brushless DC motor offers the lowest winding current for the overall envelope of operation. An electric vehicle has to provide a non-linear torque/speed curve with constant power operation from base speed to maximum speed. In a brushless DC motor, the motor voltage may be held constant over this range using vector control. In an induction motor, the motor voltage must rise over the constant power speed range. If V and I are the voltage and current at maximum speed and power the values at base speed are $V \times (\text{Base Speed/Max Speed})^{1/2}$, $I \times (\text{Max Speed/Base Speed})^{1/2}$. If maximum speed / base speed = 3.5 times, the current at base speed is $1.87I$. Consequently the induction motor inverter requires 1.87 times the current capacity of the brushless DC motor inverter.

The most significant improvement recently for brushless DC machines has been the development of the Daido magnet tube in Magnaquench material. This product offers the benefits of high energy magnet and containment tube. This leads to a third benefit which is not immediately obvious but very significant. Surface magnet motors usually employ a containment sleeve which adds several millimetres of air gap to the magnetic circuit. Since magnet tube does not require a sleeve if used within its speed capability, a thinner magnet tube is possible whilst maintaining

Power (3.5:1 CPSR) (kW)	45	70	70	70	150
Speed max	12 000	10 000	13 500	20 000	20 000
Stator OD (mm)	218	200	220	200	225
Rotor OD (mm)	141	113	141	113	145
Active length (mm)	80.5	190	97	110	160
Overall length (mm)	141	260	157	170	230
Stator voltage (V)	150	360	460	360	460
Max Efficiency	96%	96%	98%	96.5%	98.6%
Winding L (mH)	0.1	1.78	1.37	0.85	0.28
Winding R (mW)	9.6	66	116	38	13.4
Poles	16	8	8	8	8
Stator/rotor mass (kg)	19	40	21	24	44

*NOTE: 35 kW continuous, 70 kW short time rated.

Fig. 3.6 Current designs of vector controlled brushless DC machines.

the same air gap flux density. The benefit is reduced magnet weight for a given motor design. For example, 140 mm diameter Daido grade 3F material with a 5 mm wall will operate unsupported to 13 500 rpm.

The rotor of the machine, Fig. 3.7, is assembled with the magnet tube glued to the flux return tube, with the magnets de-energized. The pole pattern is applied with a capacitor discharge magnetizer from inside the flux return tube. The end plates and motor shafts are then fitted using a central bore for precise axial alignment. Use of a solid rotor is not practical unless a rotor material which does not saturate until 3 tesla is used. Since such material costs $50 per kg the hollow tube is the best alternative. The use of magnet tube makes complete automation of rotor construction possible achieving significant savings in labour costs, Fig. 3.7a.

Many designers are attracted by the possibility of running motors faster than the current 12 000 rpm. The objective is to reduce the peak torque requirement in an effort to reduce weight and cost of active materials. One obvious method is to compromise the constant power over the 3.5:1 speed:range requirement. Polaron's own investigations into faster speed suggest any increase above 20 000 rpm will be counterproductive. There are many reasons for this:

(a) The maximum frequency of operation is limited to 1500 Hz using Transil 315 in 0.08 mm thickness (3.15 W/kg at 50 Hz). Most designers are concerned with no load line losses and are endeavouring to optimize this.

(b) Consequent on (a), as the speed rises above 20 000 rpm the pole count has to be reduced from 8 to 6 to 4 poles. This results in thicker magnets and longer flux return paths.

(c) Optimum machine geometry is rotor OD = stator length. The Polaron 70 kW machine has rotor OD = 140 mm and rotor length of 95 mm which is close to optimal. The machine has 8 poles and gives 70 kW from 4000 to 13 500 rpm.

(d) Machines that are below 100 mm rotor diameter are not easy to make as the windings cannot be inserted by automatic machinery. This is especially true of heavy current windings.

(e) Machines with low pole count have poor rotor diameter to stator diameter ratio, which increases the mass of stator iron and results in large winding overhangs increasing copper losses.

(f) Laminations for these machines should have a large number of teeth to reduce the thermal resistance from copper to water or oil jacket. The limitation is when the tooth achieves mechanical resonance in the operating frequency range of the machine. Typically it is the $6f$ component that causes excitation ($6f$ = 6 times motor frequency). Silicon steel (Transil) has good thermal conductivity. High nickel steels such as radiometal exhibit poor thermal conductivity but lower

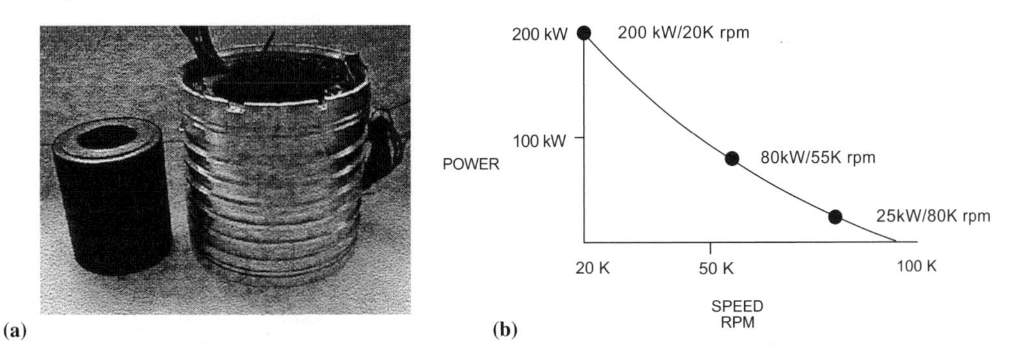

Fig. 3.7 Rotor design and machine performance: (a) a 150 kW, 20 000 rpm brushless DC stator-rotor; (b) power/speed for brushless DC motor with 3.5:1 constant power speed range.

iron losses. Machines with a high peak torque requirement are better in Transil where the copper losses of peak torque can be safely dissipated.

(h) If a better core material at a sensible price were available it would be a real boon. This is one area where there is much room for improvement. Polaron are aware of powder core technology using sintered materials but the tooth tip flux density is only 0.8 tesla. Ferrites are worse at 0.5 tesla.

(i) If makers are prepared to use containment sleeves, a power–speed graph for high speed radial brushless DC machines would look like that in Fig. 3.7a (based on 3.5:1 constant power torque/speed curve). This is the maximum power achievable in consideration of dynamic stability requirements. This graph assumes two point suspension and that the first critical speed must be 20% higher than the top speed of operation (25 kW rotor from 25 000 to 80 000 rpm would be 57 mm OD × 100 mm long).

(j) One problem with high speed machines is the increased kinetic energy stored in the rotor. This can place a severe strain on subsequent speed reducers unless torque limiting devices are provided.

(k) Acoustic noise is often severe at high speed. For a reduction try: (i) impregnation of stator; (ii) removing sharp edges on outside of rotor; (iii) operating rotor at reduced pressure using magnetic seals or (iv) using machine with liquid cooling jacket.

(l) Speed reduction is another difficult area at high speed. Since torques are low, friction speed reducers are quieter than gears by a factor of ten.

(m) Bearings and mechanical stability are challenging problems at turbomachinery speeds.

Polaron believe the best cost/performance ratio can be achieved for 70 kW system by: (1) using a Transil 315 stack 0.08 mm thick made as a continuous helix using the punch and bend technique; (2) using a rotor made from 5 mm magnet tube of surface mount structure mounted on 12.54 mm of 14/4 stainless steel; (3) magnetizing the rotor after assembly to flux density of 3 tesla for 2 millisecs for maximum flux density; (4) choosing a stator frequency of less than 1500 Hz, mean air gap flux density 0.6 tesla; (5) using a liquid cooled stator; (6) insulating the stator from earth with low capacitance coupling; (7) choosing stator of 215 mm OD with 48 teeth stack of 95 mm giving 70 kW from 4000 to 13 500 rpm. Alternatively a stator of 185 mm with 24 teeth and rotor of 110 mm OD × 140 mm long will give 70 kW from 6000 to 20 000 rpm; (8) winding the machine for 460 V in constant power region (460 V at 4000 rpm) with machine driven as a generator open circuit. This gives good efficiency and substantial winding inductance to minimize carrier ripple, Fig. 3.7b.

3.5 Brushless PM motor: design and FE analysis of a 150 kW machine

3.5.1 INTRODUCTION

High speed permanent magnet (PM) machines with rotor speed in the range from 5000 to 80 000 rpm have been developed[3], applications of which include a gas turbine generator with possible application in hybrid electric vehicles. The motor considered below runs at infinitely variable speeds up to 2 kHz at full power and has been designed for different requirements at an output power of 150 kW. Machine parameters have been calculated from software package 141 developed at Nelco Systems Ltd.

The drive system of this design consisted of a brushless DC machine and an electronic inverter (a chopper and DC link) to provide the power. The performance parameters set out are aimed to producing a design specification of the machine shown in Fig. 3.8, Fig. 3.9 showing the machine controller.

Fig. 3.8 150 kW PM brushless machine.

In the initial stage, a detailed specification was set out for the peak torque performance of 150 Nm from 10 000 to 20 000 rpm, the no-load back-EMF at 20 000 rpm of 600 V(RMS); the total number of poles are 8 (1.33 kHz at 20 000 rpm), and the maximum total weight is 45 kg.

3.5.2 ROTOR AND STATOR CONFIGURATION

Main constraints were found to be weight and inductance; in the high speed application it is important to keep the weight to a minimum, therefore a ring design is the most suitable which means a sufficient number of poles is required on the rotor, 8 poles in this case. The main advantage with this configuration is that the return path for the magnetic circuit in the core and yoke is much smaller in cross-section area (the thickness of the ring has been considered within the customer's shaft requirements). While 8 pole design was found to give the best solution, a 16 pole design was also considered which resulted in lower weight, but was rejected because the return path had to be increased, in that area, to give sufficient mechanical strength to the unit. In general a machine of high number of poles, at high frequency, produces high specific core loss and the reduction in the stator mass meant that the total core loss was a few watts more. To achieve the required winding inductance, careful attention had to be made to the shape of the stator lamination so as to reduce the slot leakage. The reduction of current density in the copper conductors has also been considered, but the slot shape and area have had an effect on the winding inductance. The final lamination design has been optimized for minimum slot leakage, to achieve the required performance.

3.5.3 MAGNETIC MATERIAL SELECTION

High energy-density rare earth magnets, of samarium cobalt, have been chosen in this design because of the material's higher resistance to corrosion, and stability over a wide temperature range. Also it has a high resistance to demagnetization, allowing the magnetic length of the block to be relatively small. This shape of block lends itself to being fixed onto the outside diameter of the rotor hub, to produce the field in the *d*-axis, which gives advantages of a greater utilization of the magnet material with lower flux leakage, the low slot leakage resulting in low winding inductance. The magnets in this application have been fitted with a sleeve on the rotor outside diameter, for mechanical protection and to physically hold the magnets in place. A carbon-fibre sleeve was chosen for this application; it offers at least twice the strength of the steel sleeve in tension, so a much greater safety factor can be achieved. The sleeve on the rotor increases the effective air gap but an unloaded air gap flux density of 0.6 tesla was achieved from this high energy density rare earth magnet. The core loss in the stator, due to high frequency, is considered and must be kept to an acceptable level. The grade of material considerable is radiometal 4550. This alloy has a nominal 45% nickel content and combines excellent permeability with high saturation flux density.

3.5.4 MAGNETIC CIRCUITS

The magnetic circuit for this design was calculated using the Nelco software. The most important parameters in the design of the magnetic circuit were weight and to keep the core losses down to a minimum whilst reducing the slot leakage to minimize the winding inductance. This is achieved when a compromise has been reached in which the flux density in the teeth is 1.15 tesla, the density in the core is 0.8363 tesla, and the yoke flux density is 0.78 tesla.

3.5.5 BRUSHLESS MACHINE DRIVE

The machine drive consists of a polyphase, rotating field stator, a permanent magnet rotor, a rotor position sensor, and the electronic drive. During operation the electronic drive, according to the signals received from the rotor position sensor, routes the current in the stator windings to keep the stator field perpendicular to the rotor permanent field, and consequently generates a steady torque. Conceptually, the drive operates as the commutator of a DC machine where the brushes are eliminated. The main advantage here is that no current flow is needed in the rotor. As a result, rotor losses and overheating are minimal, the input power factor approaches unity and maximum efficiency is obtained. This is especially relevant in continuous duty applications, where the limiting factor of traditional induction drives is invariably the difficulty of removing rotor losses.

Fig. 3.9 Machine controller.

3.5.6 MOTOR DESIGN: FINITE-ELEMENT MODELLING

3D finite-element modelling (FEM) was not required, as the topology of the machine in x–y plane is the same along the axial length, except at each end where the end turns winding exists. However, a 2D finite-element model has been employed for the machine to calculate and analyse the flux distribution in it, Fig. 3.10. This is done to facilitate the rotor movement relative to the stator, so that the characteristics of interest such as the flux modulation due to slot ripple effect on the magnet and the rotor hub can be examined. To carry out this kind of analysis, several meshes have to be created, one for each rotor position, and then each solved in turn. The software program has a facility for coupling meshes, using Lagrange multipliers. This technique has been used to join the independent rotor and stator meshes at a suitable interface plane, a sliding Lagrange interface being placed in the middle of the air gap. The view at (a) shows a close-up view of the joined meshes for the machine, and in (b) is the rotor of the machine at 45° from base (half of the rotor mesh is missing for clarity).

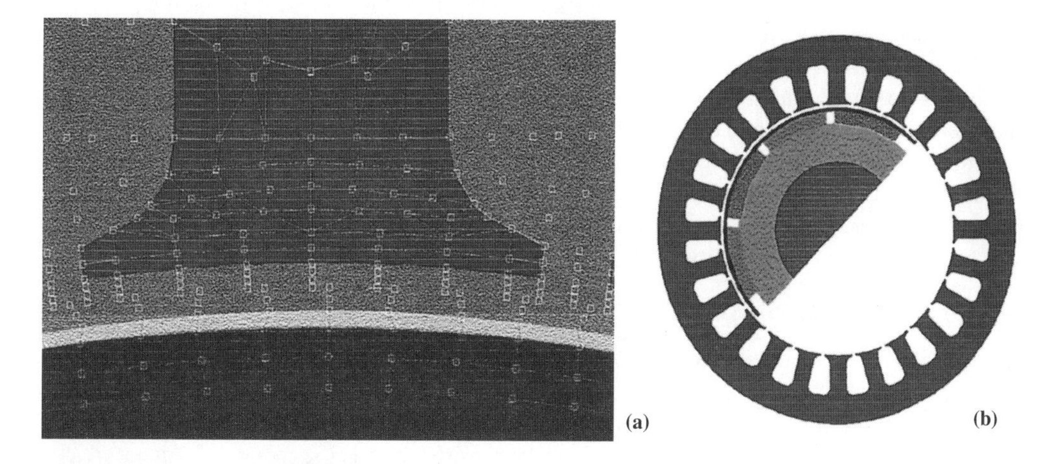

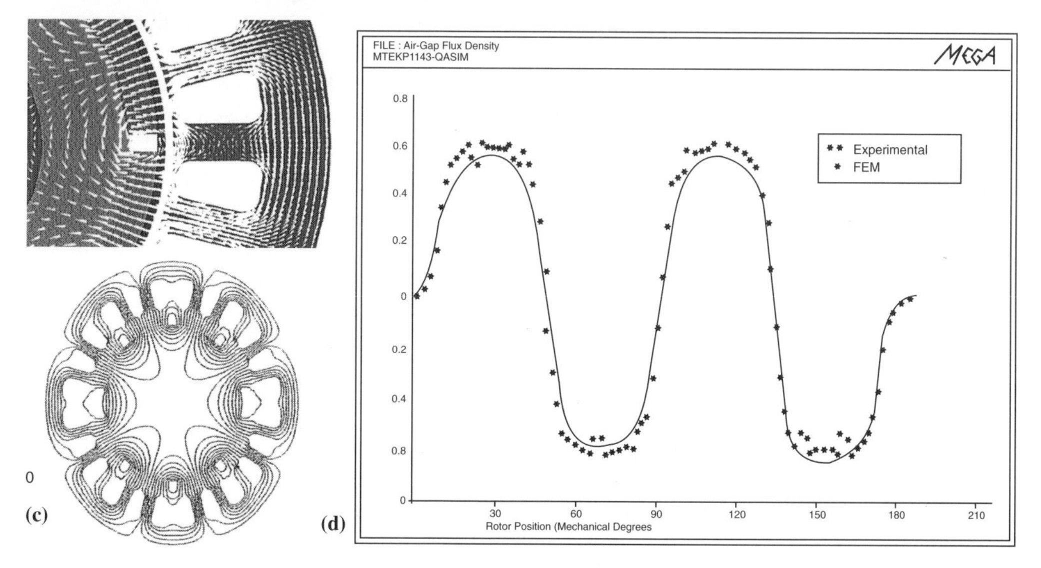

Fig. 3.10 Finite-element modelling: (a) the coupling meshes; (b) rotor at 45° from base; (c) air gap flux waveforms; (d) contour and vector flux.

The stator winding flux linkage waveforms of the machine have been calculated from the time transient solution, as the rotor speed is dynamically linked to the program, at 20 000 rpm. The experimental phase flux linkage has been deduced by integration of the phase EMF generated from the machine at no-load. These EMFs are shown to be within 8% difference, the value calculated by FEM being the higher. The flux in the air gap was measured using a search coil that is inserted on the stator side. From this search coil, a flux waveform was recorded and it is shown together within the flux calculated from FEM in (c). The flux plot, as contours and vectors at 0° rotor position for the machine, is shown at (d).

3.6 High frequency motor characteristics

In the 1970s motor designers were introduced to Bipolar Darlington transistors which permitted switching up to 2 kHz at mains voltage. In the 1980s insulated packaging was mastered and motor costs have been reduced. In the 1990s we have the IBGT which permits operation to 16 kHz for the first time at high power. This gives the designer a new freedom[4]. Hitherto the market sector has been dominated by 50 Hz machine designs. Now we can choose our operating point so the question must be asked: what is the optimum point and which is the best type of motor?

There is no simple answer to this question. We have several types of machine each with characteristics which are good in particular tasks. What is certain is that whatever type of machine is used, it can be made smaller than its 50 Hz counterpart by using a high frequency design. During the next ten years lies the challenge of the hydrogen economy with an increased demand for electric drives. IBGTs make new inverter topologies possible. The inverter on a chip in the back of the motor is now a reality.

3.6.1 HF MACHINE PROPERTIES

Motors designed for high frequency operation are of many types; however, they all share common design attributes. The 50 Hz motor designer will be used to the idea that at the full-load operating point copper loss = iron loss. This is not true for HF machines – iron loss dominates, accounting for up to 80% of the losses. Another factor is the power density which is in general 5–20 times greater. The use of HF windings means that the number of turns on a winding is reduced. So a high frequency motor can be expected to have much lower winding resistance and inductance than a 50 Hz machine.

For good loss management it is necessary to minimize the weight of core material. Generally, the flux density at the tooth is greater than in the main body of the core. It is common for all 50 Hz machines to use 2 or 4 pole windings; on HF machines, 8–32 poles are much more common.

Machines with a high pole number have a much smaller diameter build-up on the rotor; for a given stator OD the designer achieves a bigger rotor diameter which gives more torque and reduces stator mass. Machines with large numbers of poles are much easier to wind with only short winding overhangs. This is important because the overhangs contain the winding hot spots. See the example below.

DI60 frame IM 380 V 50 Hz motor

	1500 RPM	**12 000 RPM**
Power	11 kW (air cooled)	60 kW (water cooled)
Frequency	50 Hz	400 Hz
Resistance	0.5 ΩL/L	0.2 ΩL/L
Inductance	2.5 mH	400 mH
Stator flux density	1.5 tesla	0.75 tesla

Currently 500–1000 Hz represents the optimum operating point for stator iron. HF machines are very suitable for use in non-linear torque speed regimes because it is possible to operate at much higher flux densities at low speed. We therefore need to investigate vector control characteristics.

3.6.2 VECTOR CONTROL

Understanding of this subject has been delayed by years with torturous mathematical explanations of how it is achieved. In practice, vector control is a powerful technique because: (1) the full power of the stator controller can be brought to bear on the field system; (2) only a single winding set is involved. The stator current has two components: (a) field component *Id* and (b) real power component *Iq*. As these axes are at right angles, they may be independently controlled so long as the field is capable of supporting the demanded torque.

Vector control is nothing more than power factor control. The reactive element controls the field and the real power element controls the generated torque. In induction motors there is an added complication; there has to be slip between the rotor and stator to create rotor current for producing the field. This involves an axis transformation which makes for all the difficult mathematics. Synchronous motors are much easier; vector control only involves manipulation of phase shift.

Permanent magnet machines offer great flexibility because it is possible to manipulate the field with vector control currents. This has no damaging effect on the magnets so long as the material has a recoil permeability of unity (or a linear 2nd quadrant demagnetization curve) such as ferrite and samarium cobalt.

However, the level of ampere turns needed to control the field varies dramatically between different types of machines in accordance with the magnetic reluctance in the *d*-axis. It may be seen that this becomes more critical in HF machines which have smaller numbers of turns on the stator, for example a machine with four sets of windings per phase. If windings are arranged in star, Fig. 3.11a, generated back-EMF is 380 V at 2800 rpm, or by letting the circulating current at 100% field be 1 and rearranging the windings in parallel delta, as at (b), an alternative situation arises. Now 380 V is produced at 30 000 rpm and the current for 100% field increases to $4(3)^{1/2}I$ or $6.82I$. To give some idea, I is approximately 30 A for a 500 nm surface mounted PM magnet machine.

It may seem attractive to do away with the permanent magnets altogether. In practice this is not a good idea because the machine has a poor power factor and requires an oversize inverter. However, there is a variation on the concept which is possible, called the switched reluctance motor. This machine ignores Fleming's LH rule and instead relies on the attraction forces between an electromagnet and soft iron. The problem is that the production of torque is not smooth; however, they are suitable for use in difficult environments.

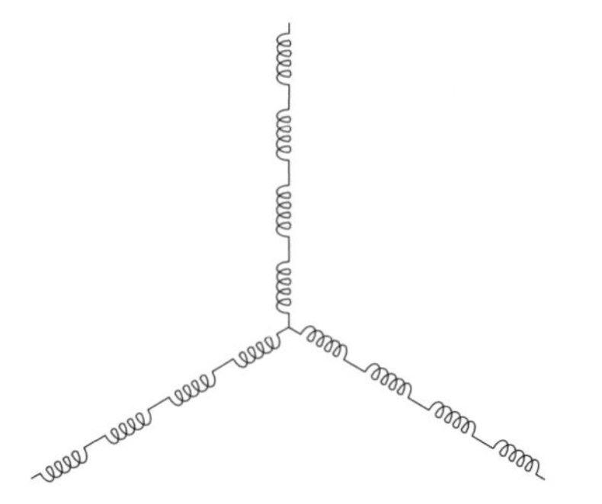

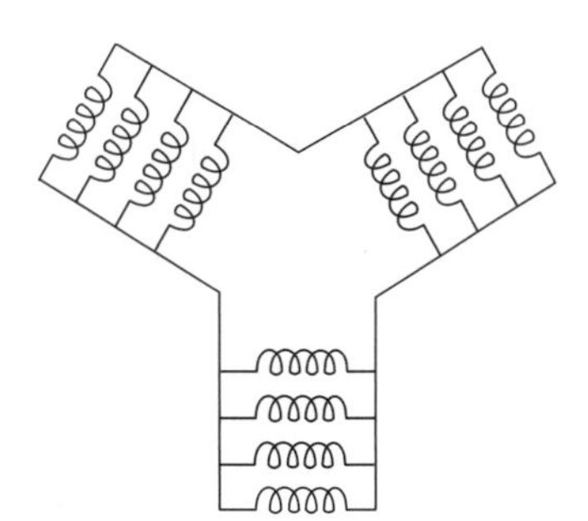

Fig. 3.11 Star, left, and parallel-delta, right, winding.

3.6.3 OPEN LOOPS OR CLOSED LOOPS FOR INDUCTION MOTORS

In the early days of inverter drives, open loop operation of induction motors was the main objective. Generally this is satisfactory over a 10:1 speed range but is problematical at slow speed due to: harmonic torques; stability problems – especially with low load inertias; and lack of rotor cooling.

Vector control may be used to improve stability and can be applied on an open loop basis. To do this, estimates are used for the load inertia/rotor current and lead to errors where fast dynamics are involved. However, Jardin and Hajdu wrote one of the leading papers on this subject whilst developing the Budapest Tramcar drive system[6].

As the motor frequency is increased, the low winding resistance makes eddy current losses, induced by DC circulating currents in the windings, a problem. Voltage source inverters without active balancing are unlikely to be satisfactory.

One machine which gives excellent performance on an open loop control is the buried PM motor developed by Brown Boveri/CEM/Isosyn and now also produced by GEC/Alsthom Parvex. Ken Binns at Liverpool University is a well-known authority in this area.

For fast dynamics and tight control there is no substitute for proper closed loop operation of a permanent magnet machine using vector control. Such an arrangement can give a constant power torque/speed range of 4:1 and this can be increased by using winding switching up to 70:1. Such systems are ideal for traction drives in vehicles with torque bandwidths of up to 1 kHz.

3.6.4 INDUCTION MOTORS

Of the various types of motor, Fig. 3.12, induction motors (IMs) are practical up to about 30 kW and 15 000 rpm. Beyond these limits exhausting the rotor losses is generally a problem (at 1500 rpm megawatt level machines are commonly constructed). At 15 kW IMs run satisfactorily up to 100 000 rpm but special motor construction techniques are needed to give strength to the cage. Water cooling is used at high power. A typical specification, for example, might be 36 kW, 12 000 rpm, PF 0.9 at 400 Hz, 36 kW – 4 pole: efficiency 0.9 at 400 Hz, 36 kW, 380 V, 68 A line current; slip (pure aluminium cage) 50 rpm cold, 70 rpm hot – torque 29 Nm (0.7 tesla), hot rotor diameter 6 in, active length 4 in, stator 9 in OD, peak torque at low speed 100 Nm (1.5 tesla), rotor cooling 8 CFM compressed air at 17 Psi, stator cooling 4 litres water/minute.

3.6.5 SURFACE MOUNTED MAGNET SYNCHRONOUS MACHINES

This design is first choice for high power drives. The rotor consists of a steel sleeve to which the magnets are glued and a containment band fitted on the outside. This fits inside a standard stator with water jacket. This design is practical up to 0.5 megawatts at up to 100 000 rpm and is used for traction drives.

Another benefit is that the output frequency is no longer related to shaft rpm and multipole designs/speeds over 3000 rpm may be considered. Using vector control, voltage and frequency may be separately controlled and much faster speed of response can be achieved. Many people are wary of PM designs because of concern about high temperature performance. The latest Nitromag alloys operate up to 250° C. These use nitrogen as the alloying element and are being investigated as part of the Joint European Action on Magnets Programme.

Most commercial motors use samarium cobalt of the 1/5 variety which has superior mechanical properties to the 2/17. Generally speaking, alloys of 20 MGO are in common use and the trick is to design rotors around standard size blocks, $1 \times 1/2 \times 6$ inches thick. Modern high coercivity magnets need very large currents to demagnetize the magnets and typically 3 tesla are needed to achieve full initial magnetization for about 1 millisecond.

(a) INDUCTION MOTOR

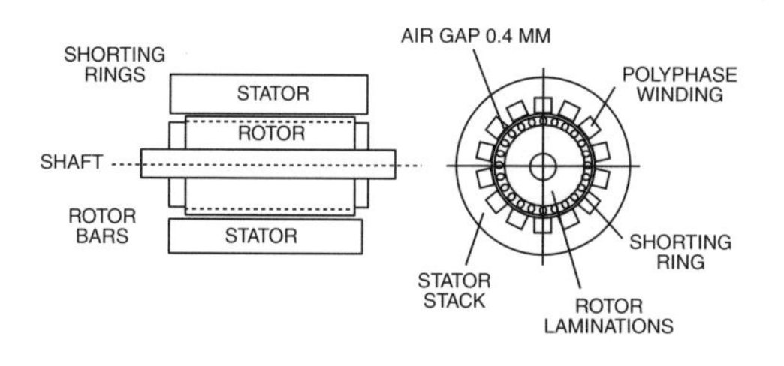

(c) CLAW TYPE PM MAGNET MOTOR

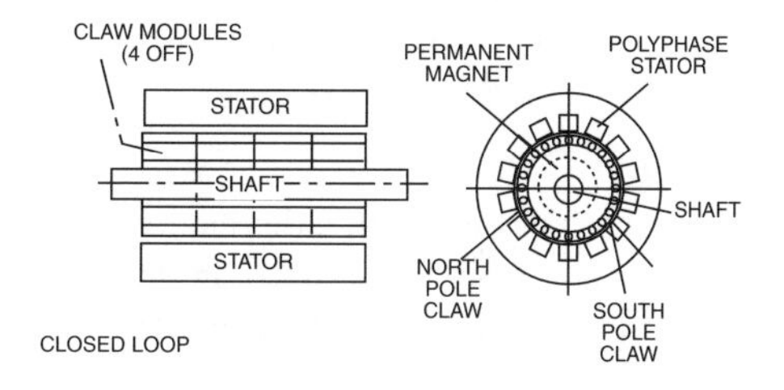

(b) SURFACE MOUNTED PM MAGNET MOTOR

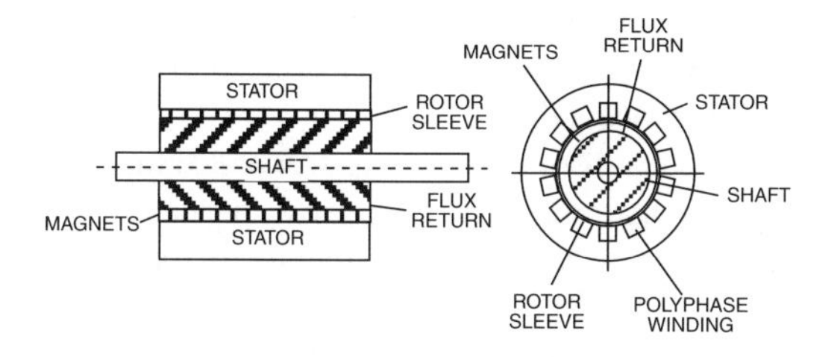

(d) UNIO TYPE MOTOR (Contentional construction)

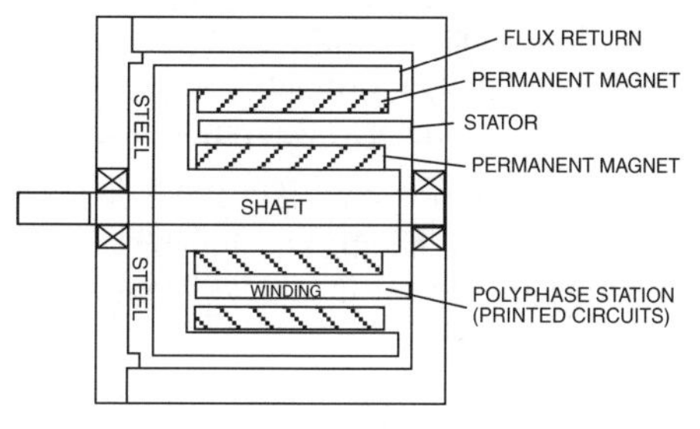

(e) BURIED MAGNET MOTOR (4 POLE)

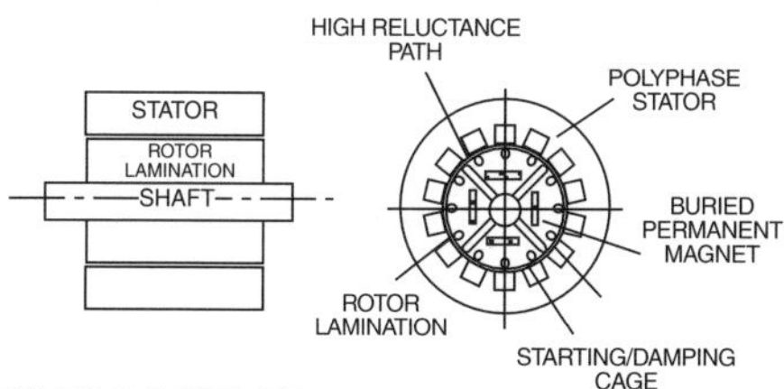

(g) HOMOPOLAR MACHINE

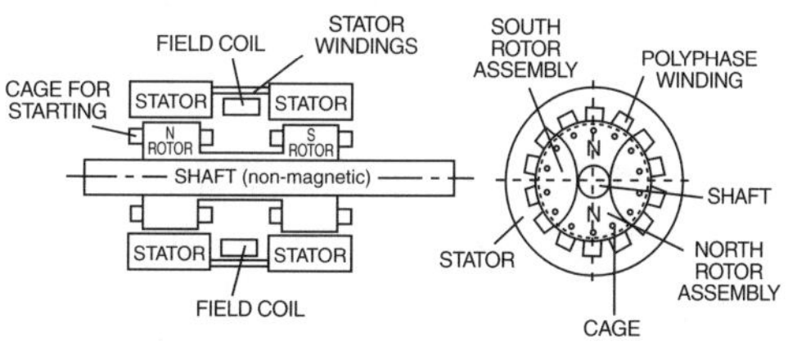

(f) SWITCHED RELUCTANCE MOTOR
(8 STATOR POLES, 6 ROTOR POLES)

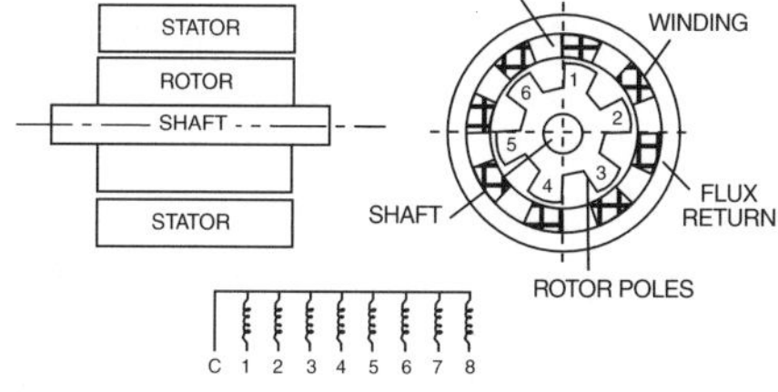

(h) RELUCTANCE MOTOR

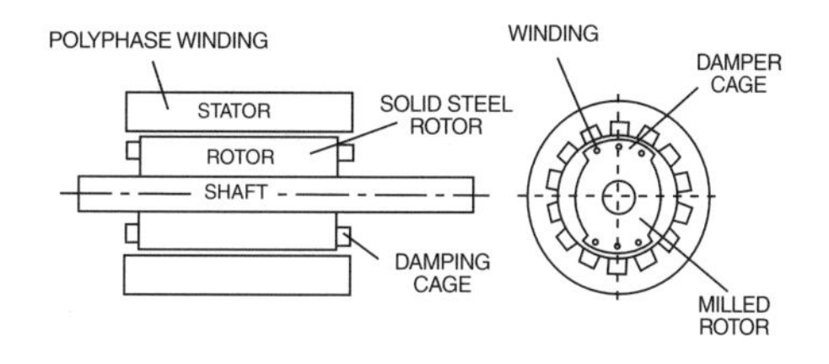

Fig. 3.12 Motor types.

Typical machine specification for 60 kW, 10 000 rpm (surface mounted) would be: stator OD 10 in, rotor diameter 7 in, active length 3 in; operating point 0.7 tesla at 666 Hz, 8 poles, 380 V, 103 A, efficiency 0.97, power factor 1; winding resistance 0.015 Ω L/L, winding inductance 300 mH L/L; iron loss 1.5 kW at 666 Hz, core Transil 270 0.35 mm non-orientated; load torque 57 Nm, peak torque 150 Nm, vector control current 100 amps for 0.7 tesla.

3.6.6 IRONLESS PM SYNCHRONOUS MOTOR

This machine has been developed by UNIQ (USA) for hub mounted motors for use in electric vehicles. It consists of a machine with both an internal and external rotor which are mechanically linked and a thin stator winding which is usually fabricated using printed circuit techniques. The result is a lightweight machine with a very high power density and low winding inductance since there is no stator iron. Performance is largely determined by the quality of permanent magnet used. The *d*-axis reluctance is high due to the double air gap so that the currents needed for vector control can be large compared with a conventional PM machine. Such machines have been built up to 40 kW rating at 7500 rpm with epicyclic speed reducers that are wheel-mounted.

At present such machines are costly to manufacture because of the large amount of PM material involved, which has to be of the cobalt/neodynium variety to achieve good performance. Losses are all due to stator copper which is generally operated at extremely high current density to give a very thin stator.

3.6.7 WOUND ROTOR SYNCHRONOUS MACHINE WITH BRUSHLESS EXCITATION

This machine is sometimes used for inverter drives in addition to the well-known use as an electricity generator. The presence of the exciter/rectifier means that this solution is applied at higher powers. The rotor can be salient pole or of surface slot construction at high speed. Whichever solution is chosen, the full field thermal loss in the motor is significant and a particular problem if the machine is to be run slowly at high load torques. This type of machine is used in traction drives using thyristor-based converters.

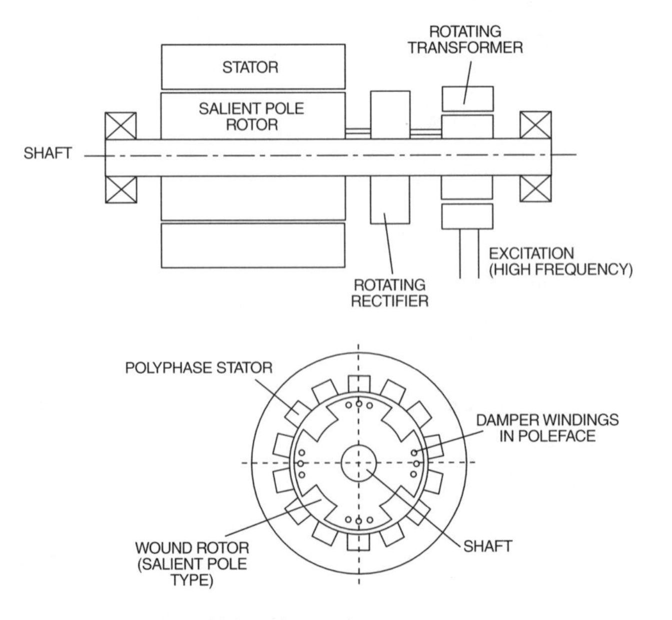

Fig. 3.13 Wound rotor synchronous machine with brushless exciter.

3.7 Innovative drive scheme for DC series motors

Many DC brushed motor drive schemes for EVs use a DC shunt motor and it has been suggested that such a solution is the most appropriate[5]. This section investigates an alternative solution. There are many railway locomotives which successfully use series wound motors and we hope to establish that indeed this is the best solution for electric vehicles.

3.7.1 MOTOR DRIVES: WHY CHANGE THE SYSTEM?

Because the system is already subject to change brought about by new requirements and developments. First, we have the introduction of sealed battery systems. These will permit much higher peak powers than hitherto possible and consequently will run at high voltages. 216 V DC is a common standard working with 600 V power semiconductors. Second, we have the introduction of hybrid vehicles. This will result in the need for drives and motors to operate for long sustained periods – previously batteries did not store enough energy. Third, the DC series motor has the right shape of torque–speed curve for traction, constant power over a wide speed range. Fourth, DC series field windings make much better use of the field window than high voltage shunt windings where much of the window is occupied by insulation. The series field winding is a splendid inductor for use in battery charging mode. Losses in series mode are significantly reduced.

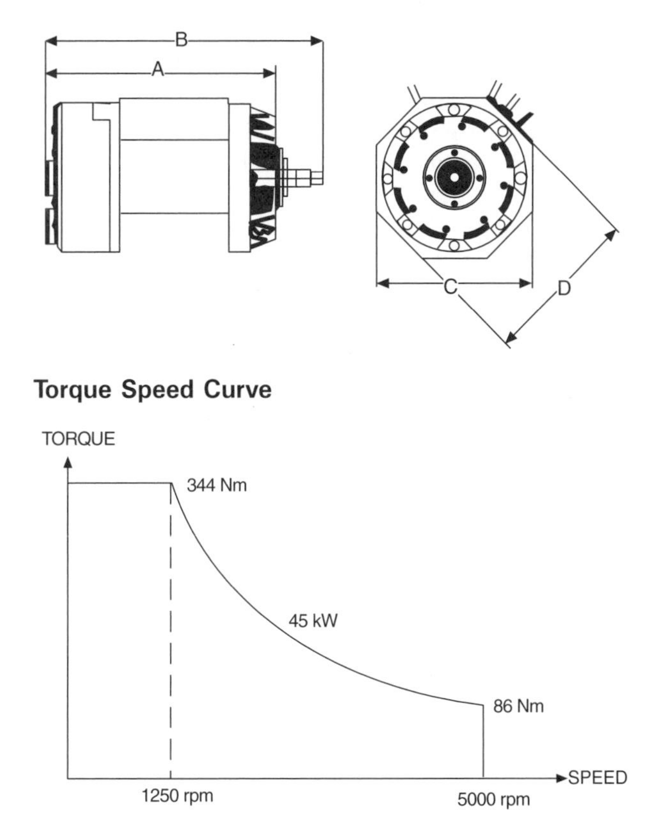

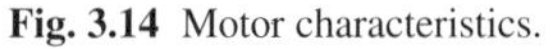

Fig. 3.14 Motor characteristics.

An example specification is typified by the Nelco N200, Fig. 3.15(a), which compares with a 240 mm stack, Fig. 3.15(b):

Shunt field	Series field
$N = 227$	$N = 12$
Hot resistance 7 Ω	Hot resistance 0.014 Ω
Watts 700 at 10 A	Watts 500 at 189 A

So why hasn't somebody attempted to use series motors in EVs before? They have for single quadrant low voltage systems but not on multi-quadrant, high voltage schemes. This account proposes a new control concept akin to vector control for AC machines. We will show how it is possible to achieve independent control of field current I_f and armature I_a, with very fast response, using a transistor bridge.

3.7.2 VEHICLE DYNAMICS AND MOTOR DESIGN

A vehicle represents a large inertia load with certain elements of resistance some of which increase with speed; see Chapter 8. For a small family car, mass = 1250 kg at 60 mph (26.8 m/sec) typical cruising speed. Windage accounts for 6 kW, rolling resistance 2 kW and brake drag 2 kW, a total of 10 kW in steady state conditions. Windage varies as the 3rd power of vehicle relative velocity with respect to the wind.

Kinetic Energy = $1/2\ MV^2$, where M = mass = 1250 kg and V = velocity in metres/sec. So we have:

SPEED	(MPH)	10	20	30	40	50	60	70	80
	(m/sec)	4.5	8.9	13.4	17.8	22.3	26.7	31.2	35.6
KE	(kilojoules)	12.5	49.5	111	198	309	446	607	792

What this illustrates is that recovered energy below 20 mph is small, consequently regeneration only matters at high speed. It also illustrates that the inertia load, not the static resistance, is the main absorber of power during acceleration.

3.7.3 MOTOR CHARACTERISTICS

These are shown in the following table:

Voltage	216 V
Rated power	45 kW, 1250–5000 rpm
Frame D 200 M-	4 pole with interpoles
Weight	170 kg

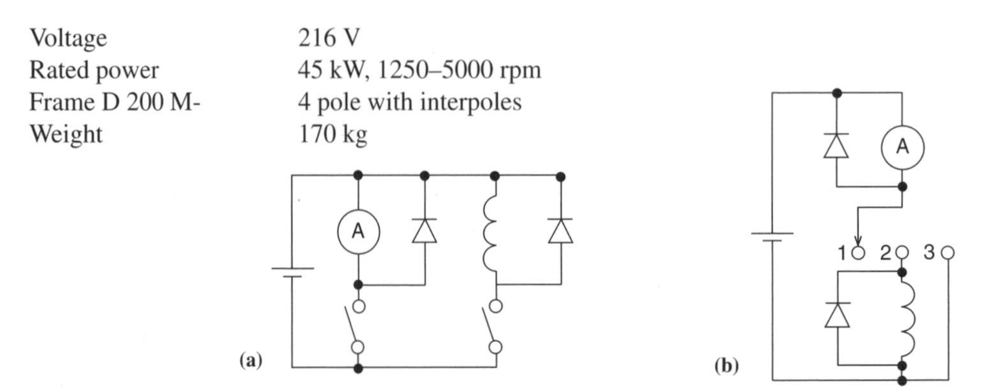

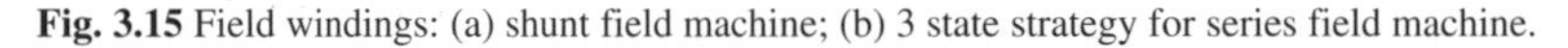
Fig. 3.15 Field windings: (a) shunt field machine; (b) 3 state strategy for series field machine.

Cooling	air forced, separate fan
Winding, series field	245 A/216 V full load
Efficiency at full load	85%
Field	Resistance 10 milliohm, inductance 1.2 mH
Armature inc. brushgear interpoles	Resistance 30 milliohm, inductance 260 mH
Dimensions	A = 490 mm, B = A + shaft, C = 335 mm, D = 350 mm; see Fig. 7.14

This illustrates that when the field current is strengthened in the constant power region, the armature voltage can be made to exceed the battery voltage and regenerative braking will take place. Below 1250 rpm plug braking must be used; however, the energy stored at this speed is small.

3.7.4 SWITCHING STRATEGY (SINGLE QUADRANT), FIG. 3.15

Figure 3.15(a) shows the arrangement for a 216 V, 45 kW shunt field machine with separate choppers for field and armature. There are some disadvantages with this scheme: (a) field is energized when not needed; (b) forcing factor of field is small – for a 45 kW shunt field, $R = 7$ ohm, $I = 10$ A nominal, $L = 1.2$ henries, $t = 0.17$ seconds; (c) when extended to multi-quadrant design two bridge chopper systems are needed if contactor switching is to be avoided; (d) extensive modifications are needed to provide for high power sine wave battery charging; (e) field power losses are significant (3 kW at max field).

Figure 3.15(b) illustrates the proposed new circuit which has a single 3 state switch: state (1) open-circuit; state (2) armature + series field; state (3) armature. So as an example, consider the following situation:

Full load torque at standstill	
Field voltage for 245 A	= 2 V
Armature voltage for 245 A	= 16 V

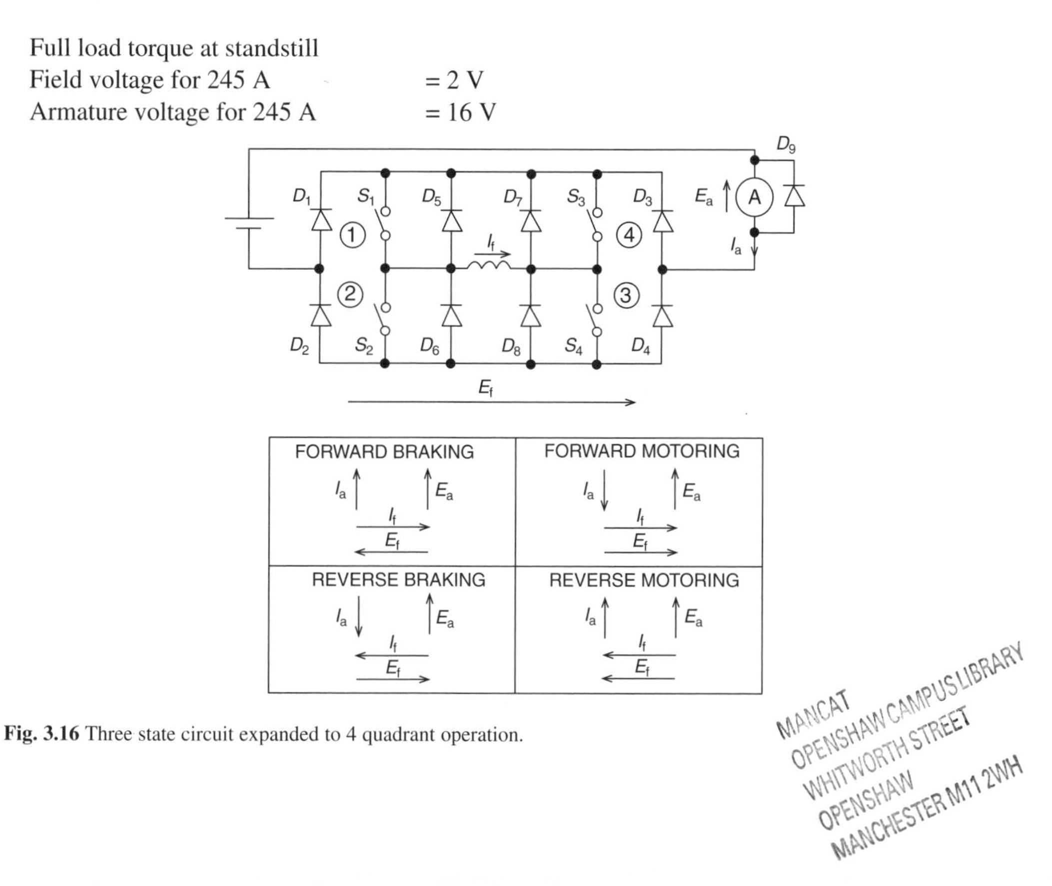

Fig. 3.16 Three state circuit expanded to 4 quadrant operation.

so with 216 V battery:
$D = 2/216$ in state 2
$D = 16/216$ in state 3

The balance of the time will be off (D = duty cycle ratio for chopper).

It can be seen that by manipulating the relative times spent in each of the states, separate control of field and armature currents may be exercised.

When the speed of the motor exceeds the base speed (1250 rpm) the back-EMF is equal to the battery voltage and the switch henceforth operates only in states (2) and (3).

Let D = duty cycle for single quadrant chopper, then $V_{out}/V_{in} = D$, hence

$$D_2\,(V_B - 5 - K_A\omega I_f - I_a R_a - L_a dI_a/dt) = I_f R_f + L_f\, dI_f/dt \quad \text{and}$$

$$V_B - 5 = (K_A\omega I_f + I_a R_a + L_a\, dI_a/dt) \times (D_2 + D_3)$$

where
ω = motor speed, rads/sec
V_B = battery voltage
K_A = armature back-EMF constant V/amp/rad/sec $(D_2 + D_3)$
D_2 = duty cycle state 2
D_3 = duty cycle state 3

Other symbols are self-explanatory.

3.7.5 MULTI-QUADRANT STRATEGY

Figure 3.16 illustrates the 3 state circuit when expanded to 4 quadrant operation: state 1 is all switches off; state 2 either S_1/S_4 or S_2/S_3 on and state 3 is either S_1/S_2 or S_3/S_4 on. As is clear, the third state is produced by having a controlled shoot-through of the transistor bridge. It may be considered that with two transistors and two diodes in series, voltage drops in the power switching path make the circuit inefficient. In fact with the latest devices: $V_{ce\ sat}$ for switches = 1.5 V at 300 A; V_f for diodes = 0.85 V at 300 A, giving a total drop = 4.7 V. So (4.7/216) × 100 = 2.3% power loss.

When the motor loses 15% this is a small deficiency. It represents 1.2 kW at full power. As the table illustrates in Fig. 3.16, all states of motoring and braking can be accommodated. The outstanding feature of this scheme is that the full power of the armature controller can be used to force the field, giving very fast response. From Fig. 3.16, it will be seen that the 4 quadrant circuit

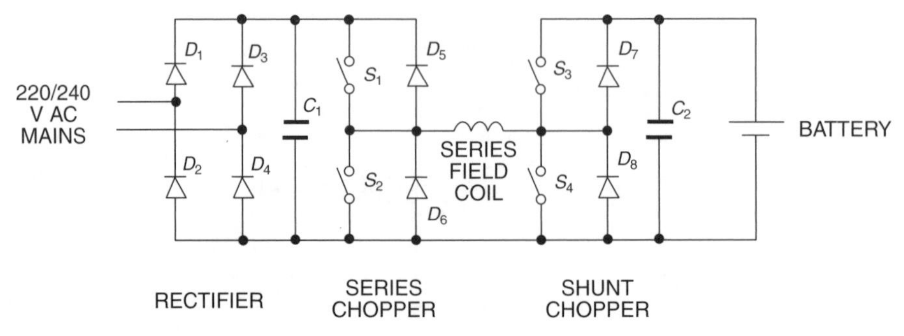

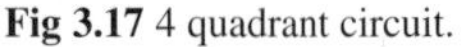
Fig 3.17 4 quadrant circuit.

consists of a diode bridge D_1–D_4 and a transistor bridge S_1–S_4 (D_5–D_8). D_9 acts as a freewheel diode when the transistor bridge is operated in shoot-through mode. Bridge D_1/D_4 is required because the direction of armature current changes between motoring and braking. Control in braking mode is a two-stage process. At high speed the armature voltage exceeds the battery voltage and the battery absorbs the kinetic energy of the vehicle. At low speed the field current is reversed and plug braking of the armature to standstill is achieved via D_9.

3.7.6 DEVICE PROTECTION IN A MOTOR CONTROLLER

Switches S_1–S_4 form a bridge converter and the devices require protection against overvoltage spikes from circuit inductances. The main factors are: (1) minimize circuit inductances by careful layout. The key element is the position of D_9 and associated decoupling capacitor relative to D_1–D_4; (2) fit 1 mF of ceramic capacitors across the DC bridge S_1 /S_4 plus varistor overvoltage protection.

D_1–D_4 can be normal rectification grade components but D_9 must be a fast diode with soft recovery. D_5–D_8 are built into the transistor blocks.

3.7.7 SINE WAVE BATTERY CHARGER OPERATION

With little modification the new circuit, Fig. 3.17, can be used as a high power (fast charge) battery charger with sine wave supply currents. The circuit exploits the series field as an energy storage inductor. S_1 and D_6 are used as a series chopper with a modulation index fixed to give 90% of battery volts. This creates a circulating current in the storage inductor. Switch S_4 and diode D_7 function as a boost chopper operating in constant current mode and transfer the energy of the storage inductor into the battery. Charging in this manner is theoretically possible up to 250 amps but will be limited by: (a) main supply available and (b) thermal management of the battery.

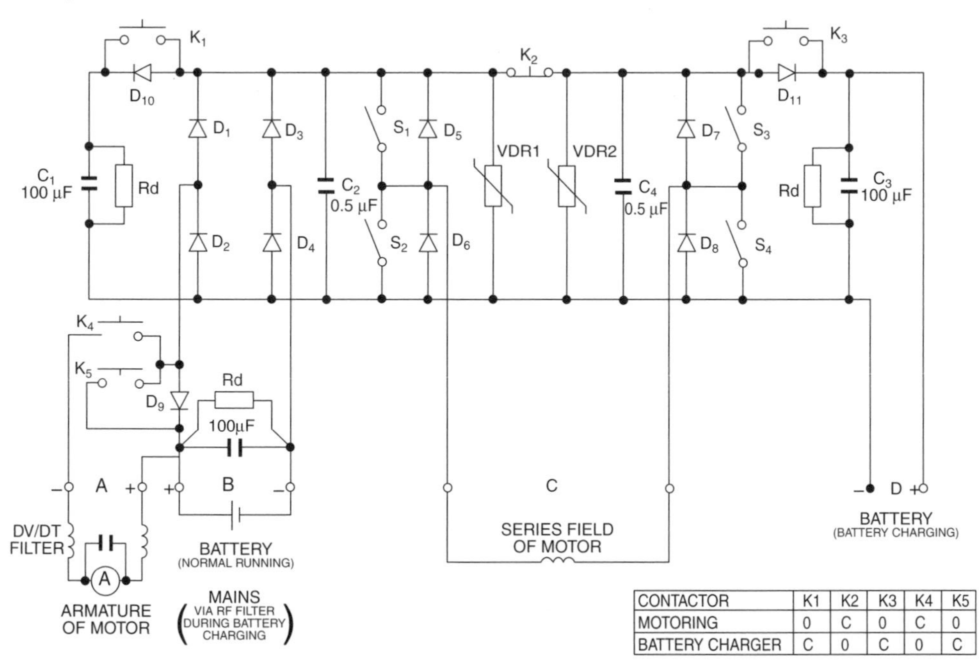

CONTACTOR	K1	K2	K3	K4	K5
MOTORING	0	C	0	C	0
BATTERY CHARGER	C	0	C	0	C

Fig. 3.18 Full circuit diagram of combined chopper/battery charger.

Experience shows that charging at 30 amps is possible on a 220 V, 30 A, USA-style house air conditioning supply. Charging at greater currents will require special arrangements for power supply and cooling. One advantage of the scheme presented is that it may be used on any supply from 90 V to 270 V.

It is also possible to adopt the circuit for 3 phase supplies in one of two ways: (1) add an additional diode arm – this would produce a square wave current shape on the supply; (2) fit a 3 phase transistor bridge on the supply – this would permit a sine wave current in each line at a much increased cost.

3.7.8 POWER DIAGRAM FOR MOTORING AND CHARGING

Figure 3.18 presents the combined circuit diagram for motoring and battery charging. Reservoir capacitors and mode contactors have been added. The capacitors function as snubbers when running in motoring mode. As drawn, to adapt to battery charging, the battery plug is moved to outlet D and the mains inserted into plug B, alternatively contactors could be used to do the job. Battery safety precautions comprise: (1) the battery is connected via a circuit breaker capable of interrupting the full short-circuit current of a charged battery; (2) this circuit breaker is to contain a trip to disconnect battery by mechanical means only; (3) battery/motor/controller are each to contain 'firewire' to disconnect the circuit breaker; (4) circuit breaker is to be tripped by 'G' switch when 6G is exceeded in any axis.

3.7.9 CONTROL CIRCUIT IN MOTORING MODE

Figure 3.19 shows the block diagram of the controller for motoring mode. The heart of the system is a memory map which stores the field and armature currents for the machine under all conditions

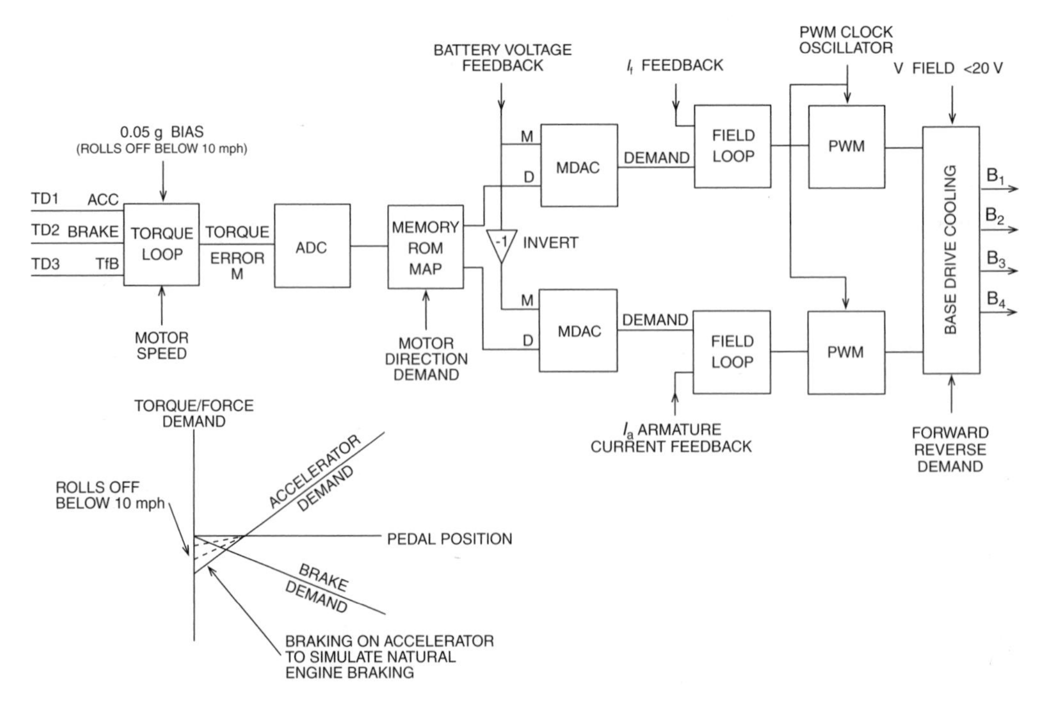

Fig. 3.19 Control system in motoring mode.

of operation. These demands for I_f and I_a are then compensated for in accordance with the battery voltage before conversion into analogue form, to be passed to operational amplifier loops which drive the modulators. Current feedback is provided by Hall effect CTs. The torque loop has input from two pedals and a feedback from a torque arm attached to the motor. Above the base speed there is no open circuit condition and the armature loop error is used to control the field.

3.7.10 CONTROL CIRCUIT IN BATTERY CHARGING MODE

The control circuit for battery charging is shown in Fig. 3.20. When the battery is below 2.1 V per cell and 40°C it is charged at the maximum current obtainable from the supply. Above 2.1 V/cell the battery is operated at reduced charging up to 2.35 V per cell, compensated at –4 mV/°C for battery temperature. This data assumes lead–acid cells.

As can be seen from the block diagram there are two separate loops for the buck and shunt choppers. The fast current loops stabilize the transfer function for changes in battery impedance. The current limit function must be user-set in accordance will supply capabilities.

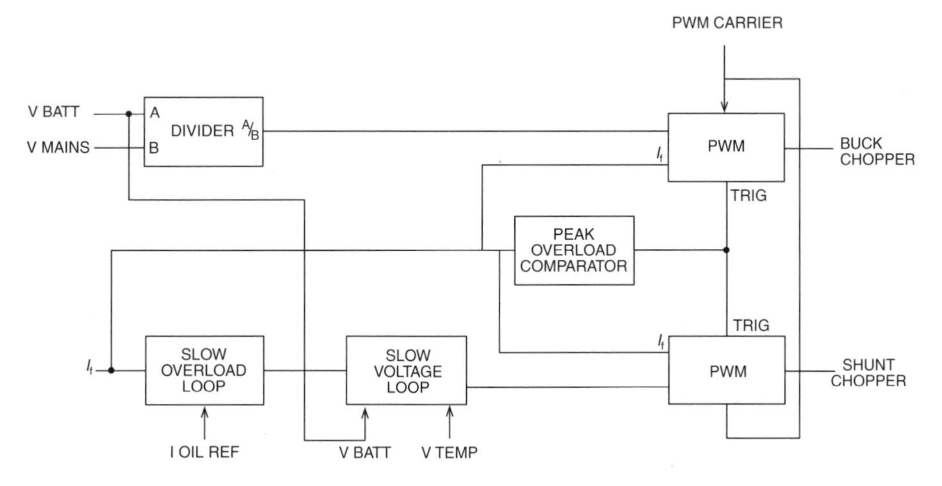

Fig. 3.20 Block diagram of battery charging controller.

References

1. Hodkinson, R., Operating characteristics of a 45 kW brushless DC machine, *EVS* 12, Aneheim, 1995
2. Hodkinson, R., Towards 4 dollars per kilowatt, *EVS* 13, Osaka, 1996
3. Al'Akayshee *et al.,* Design and finite element analysis of a 150 kW brushless PM machine, *Electric Power Transactions,* IEEE, 1998
4. Hodkinson, R., The characteristics of high frequency machines, Drives and Controls Conference, 1993
5. Hodkinson, R., A new drive scheme for DC series machines, *ISATA* 24, Aachen, 1994
6. Jardin and Hajdu, Voltage Source Inverter with Direct Torque Control, IEE PEPSA, 1987

Further reading

Alternative transportation problems, *SAE,* 1996
The future of the electric vehicle, *Financial Times Management Report,* 1995
Battery electric and hybrid vehicles, *IMechE,* 1992
Electric vehicle technology seminar report, *MIRA,* 1992
Electric vehicles for Europe conference report, *EVA,* 1991

4

Process engineering and control of fuel cells, prospects for EV packages

4.1 Introduction

The first three sections of this chapter will give a history of fuel cells; describe the main types of fuel cells, their characteristics and development status; discuss the thermodynamics; and look at the process engineering aspects of fuel-cell systems. It is based on a series of lectures given by Roger Booth to undergraduates at the Department of Engineering Science at the University of Oxford, under the Royal Academy of Engineering Visiting Professor Scheme in 1999. The assistance of Dr Gary Acres of Johnson Matthey in preparing this chapter is greatly appreciated.

The remaining sections deal with the control systems for fuel cells that turn them into 'fuel-cell engines' and considers the problems of package layout for all EVs as an introduction to the package design case studies reviewed in the following two chapters.

4.1.1 WHAT IS A FUEL CELL?

The easiest way to describe a fuel cell is that it is the opposite of electrolysis. In its simplest form it is the electrochemical conversion of hydrogen and oxygen to water, as shown in Fig. 4.1. Hydrogen dissociates at the anode to form hydrogen ions and electrons. The electrons flow through the external circuit to the cathode and the hydrogen ions pass through the electrolyte to the cathode and react with the oxygen and electrons to form water. The theoretical electromotive force or potential of a hydrogen–oxygen cell operating at standard conditions of 1 atm and 25°C is 1.23 V, but at practical current densities and operating conditions the typical voltage of a single cell is between 0.7 and 0.8 V. Commercial fuel cells therefore consist of a number of cells in series.

4.1.2 TYPES OF FUEL CELL

Fuel cells are described by their electrolyte:

Alkaline – AFC
Phosphoric acid – PAFC
Solid Polymer – SPFC (also referred to as proton exchange membrane – PEMFC)
Molten carbonate – MCFC
Solid oxide – SOFC.

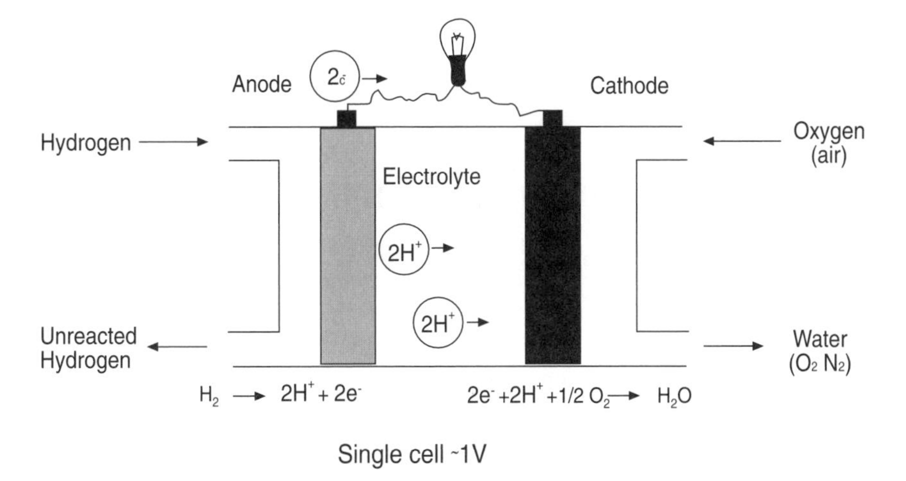

Fig. 4.1 Basic chemical reactions in a fuel cell.

The reaction shown in Fig. 4.1, with hydrogen ion transfer through the electrolyte, is only applicable to fuel cells with acid electrolytes and solid polymer fuel cells. The reactions in each of the fuel cell types currently under development[1] are:

Cell	Anode	Cathode
AFC	$H_2 + 2OH^- \longrightarrow 2H_2O + 2e^-$	$O_2 + 2H_2O + 4e^- \longrightarrow 4OH^-$
PAFC	$H_2 \longrightarrow 2H^+ + 2e^-$	$4e^- + 4H^+ + O_2 \longrightarrow 2H_2O$
SPFC	$H_2 \longrightarrow 2H^+ + 2e^-$	$4e^- + 4H^+ + O_2 \longrightarrow 2H_2O$
MCFC	$H_2 + CO_3^= \longrightarrow H_2O + CO_2 + 2e^-$	$O_2 + 2CO_2 + 4e^- \longrightarrow 2CO_3^=$
	$CO + CO_3^= \longrightarrow 2CO_2 + 2e^-$	
SOFC	$H_2 + O^= \longrightarrow H_2O + 2e^-$	$O_2 + 4e^- \longrightarrow 2O^=$
	$CO + O^= \longrightarrow CO_2 + 2e^-$	
	$CH_4 + 4O^= \longrightarrow CO_2 + 2H_2O + 8e^-$	

4.1.3 HISTORY

The concept of the fuel cell was first published in 1839 by Sir William Grove when he was working on electrolysis in a sulphuric acid cell. He noted a passage of current when one platinum electrode was in contact with hydrogen and the other in contact with oxygen. In 1842 he described experiments with a stack of 50 cells, each with one quarter of an inch wide platinized platinum electrodes and he noted the need for a 'notable surface of action' between the gases, electrolyte and electrodes. Over the next 90 years a number of workers published papers on both acid and alkali fuel cells, including the development of three dimensional electrodes by Mond and Langer in 1889. But it was not until 1933, when F. T. (Tom) Bacon (an engineer with the turbine manufacturers C. A. Parsons & Co. Ltd.) started work with potassium hydroxide as the electrolyte and operating at 200°C and 45 atm, that significant progress was made. The main thrust for development of fuel cells was the space programme of the early 1960s, when NASA placed over 200 contracts to study and develop fuel cells. The first major application was the use of solid polymer fuel cells developed by General Electric for on-board power in the Gemini programme. By 1960 Bacon had transferred to the Pratt and Whitney Division of United Aircraft Corporation (now United Technologies Corporation) in the USA, and led the development of the on-board power system for the Apollo lunar missions. Ninety-two systems were delivered and 54 had been used to power nine moon shots by 1965. This was followed by UTC's development of a 7 kW stack which is used in the Space Shuttle. During the 1990s fuel-cell development accelerated, with particular interest in automotive and small distributed power systems as the main target applications[2,,3,7].

4.2 Reforming and other hydrogen feedstocks

The prime source of hydrogen for fuel cells is by reforming a hydrocarbon, particularly natural gas. The hydrogen production can be carried out either at a central large-scale facility, or immediately upstream of the fuel cell, including on board for transport applications.

Steam methane reforming is the main commercially applied process for hydrogen production, with an annual global production capacity in excess of 40 million t/a and single units with capacities greater than 400 t/d. Overall thermal efficiencies (LHV) are over 65% and there is the potential to exceed 70%. The process is carried out over a nickel catalyst at temperatures of about 800°C on a sulphur-free feedstock. The main reaction is:

$$CH_4 + H_2O = CO + 3H_2 + 205.1 \text{ kJ}$$

with the endothermic heat of reaction provided by combustion of natural gas (note this chapter follows the convention of enthalpy changes as +ve when energy is absorbed by the system).

Partial oxidation is also applied on a significant commercial scale with both natural gas and liquid hydrocarbons as feedstock:

$$CH_4 + (1/2)O_2 = CO + 2H_2 - 37.7 \text{ kJ}$$

Both catalytic (nickel based) and non-catalytic processes are applied.

Autothermal reforming processes are being developed which combine steam methane reforming with partial oxidation, the latter providing the energy for the former. This application is of particular interest to developers of on board reforming systems for road vehicles.

The reactions are significantly more complex than given above and one of the main problems is caused by the Bouduoard reaction:

$$2CO = CO_2 + C$$

which results in soot production and lower hydrogen yield.

Water gas shift reaction is used to increase the yield of hydrogen from all the above processes[9]

$$CO + H_2O = CO_2 + H_2 - 41.9 \text{ kJ}$$

Hydrogen from electrolysis is also a potential fuel, but unless the electricity is produced from solar energy, the overall process is not attractive (second law of thermodynamics).

Feedstocks other than natural gas can also be used for hydrogen production.

Coal was of particular interest at the turn of the century via partial oxidation (gasification) and water gas shift:

$$(CH_{0.8}) + (1/2)O_2 = CO + 0.4H_2 \qquad + H_2O = CO_2 + 1.4H_2$$

Biomass, particularly from sustainable plantations, can also be used by gasification plus shift:

$$(CH_2O) \longrightarrow CO + H_2 \qquad + H_2O = CO_2 + 2H_2$$

Methanol has been of interest for a number of years, both via reforming and in the direct Methanol fuel cell (DMFC), where most current developments focus on a variant of the SPFC. Basic reactions are:

$$\text{anode: } CH_3OH + H_2O \longrightarrow CO_2 + 6H^+ + 6e$$
$$\text{cathode: } (3/2)O_2 + 6H^+ + 6e^- \longrightarrow 3H_2O$$

Gasoline (and diesel) on-board autothermal reforming is being developed for transport applications, as it would mean fuel-cell powered vehicles could make use of the existing fuel infrastructure. Fuel cells operating on a host of exotic fuels, including ammonia and hydrazine, have been studied, but the above are the most likely fuels to be used on a large commercial scale.

4.3 Characteristics, advantages and status of fuel cells

The main characteristics of fuel cells, arranged in ascending order of operating temperature, are given in the following table:

Type	Electrolyte	Typical °C	Electrode	Status	Efficiency
AFC	KOH	80	Pt on C	100 kW	> 50
SPFC	Polymer e.g. 'Nafion'	85	Pt on C	250 kW	45 – 55
PAFC	Phosphoric Acid	200	Pt on PTFE/C	0.2–11 MW	~40
MCFC	Li or K Carbonate	650	Ni or Ni alloy	2 MW	50 – 60
SOFC	Yttria stab'd zirconia	1000	Ni/zirconia cermet	100 kW	50 – 55

The quoted status and efficiencies are typical of systems in 1999 and are based on electricity as a percentage of the lower heating value of hydrogen consumed. Efficiency is discussed in more detail in the next section.

The alkaline fuel cell has the advantages of using cheap materials; quick to start and high power density. The main disadvantages are the corrosive electrolyte and the intolerance to carbon dioxide – current limits are about 50 ppm, which means CO_2 removal would be required for transport applications. The technology has been proven for space applications by IFCC (International Fuel Cells Corporation), which is a joint venture between UTC and Toshiba, and one developer (ZEVCO, who took over the work of the Belgian company Elenco) has developed a London taxi.

The phosphoric acid fuel cell has the main advantages of CO_2 tolerance (up to about 20%) and moderate CO tolerance. The main disadvantages are the cost of the noble metal catalyst, lower efficiency and a current system cost of about $3000/kW. PAFC is the most commercially advanced fuel-cell technology, with over 140 ONSI (a division of IFC) 200 kW systems in operation for distributed power systems. In addition, Tokyo Electric Power have an 11 MW system, supplied by IFC and constructed by Toshiba, which started operation in 1991. Other developers include Fuji, Hitachi and Mitsubishi Electric.

The solid polymer fuel cell has many advantages, including solid non-corrosive electrolyte; quick start; long life; produces potable water; ease of volume manufacture. The main disadvantages are the use of noble metal catalyst and intolerance to sulphur and carbon monoxide. In recent years development has focused successfully on reducing the platinum loading (now less than 10% of the loading in the late 1980s) and improving tolerance to CO, which is now up to about 50 ppmv. A further disadvantage is the limited scope to use waste heat which results from the low operating temperature. The SPFC is the leading contender in the automotive market and has potential in the cogeneration and battery replacement markets. Consequently the list of active companies is large and includes Ballard, Alstom, IFC, Toyota, Plug Power, Dais Corporation, Warsitz Enterprises, Advanced Power Sources, Siemens, DeNora, Sanyo.

The Molten Carbonate fuel cell is being targeted at the large scale decentralized power generation and industrial Combined Heat and Power (CHP) markets and has the advantages of being able to support internal reforming of natural gas and produces high grade heat for CHP applications. The main disadvantages are the stability of the electrolyte; poisoning by sulphur and halogens; and there are questions about the achievable operating life. The largest unit is the 2 MW Energy Research Corporation facility commissioned in 1997. M-C Power have demonstrated at the capacity of 250 kW and have announced plans for commercial production at the rate of 2–3 MWpa, targeting customers needing between 500 kW and 3 MW CHP systems. Other developers include IFC, IHI, MTU, Ansaldo (100 kW demonstration) and ECN.

The solid oxide fuel cell has similar advantages to the MCFC, but has easier electrolyte management and is less corrosive. The main disadvantage is the high cost, the 100 kW Westinghouse unit in the Netherlands which was delivered in 1998 is reported to have cost $10 million. Development is not as advanced as with the PAFC and MCFC, but active programmes are being carried out by Westinghouse, Sulzer, Siemens, Sofco, Mitsubishi, Eltron, Zetek, Global Thermoelectric amd Ceramic Fuel Cells.

Automotive company activities were triggered by the Californian legislation for, so-called zero emission vehicles (ZEVs) which were seen as being required to improve urban air quality. With the exception of Zevco who are using AFCs, all other manufacturers' activities are based on SPFCs and are summarized in the following table.

Company	Comments
Daimler-Chrysler	Probable leader (Necar I-IV); Joint venture with Ballard (Daimler Benz/Ballard), Ford, Mazda, Shell
Mazda	In-house cell development, metal hydride storage, but now with Daimler-Benz/Ballard
Toyota	In-house, hybrid vehicles both H_2 (hydride storage) and MeOH
Honda	In-house programme, also used Ballard
Nissan	In-house programme, also used Ballard
Ford	PNGV; cooperating with Plug Power and Daimler-Benz/Ballard
GM	PNGV; cooperating with Ballard and Arco/Exxon for reformer
Chrysler	AD Little gasoline reformer; Plug Power cell, Merged with Daimler-Benz
Renault/PSA/Volvo/VW	Joule programmes with DeNora cell

There is also development work on marine applications using SPFC, SOFC and MCFC technology[3,6,9,10].

4.4 Thermodynamics of fuel cells

The efficiency of the heat engine is limited by the Carnot cycle and equals:

$$(T_h - T_l)/T_h$$

A Carnot efficiency of about 70% could be achieved theoretically with upper and lower temperatures of 1000 K and 300 K, which would require a compression ratio of about 20:1. However, fuel and material restrictions limit the practical efficiency to about 50%, which is achieved by modern, large, low speed diesel engines, but automotive gasoline and diesel engines achieve much lower efficiencies, particularly when averaged over a standard driving cycle.

A true direct energy conversion device is one which can convert the Gibbs free energy of a chemical reaction directly into work. A fuel cell converts the Gibbs free energy of a chemical reaction into a stream of electrons under isothermal conditions. The change in Gibbs free energy of a reaction is given by:

$$\Delta G_r = H_r - TS_r$$

Fuel-cell reactions which have negative entropy change (e.g. $H_2 + (1/2)O_2 = H_2O$) generate heat and those with positive entropy change (e.g. $C_2H_6 + 3.5\ O_2 = 2CO_2 + 3H_2O$ and $CH_3OH + 1.5O_2 = CO_2 + 2H_2O$) extract heat from the surroundings. For a fuel cell operating at constant temperature and pressure, the maximum electrical energy is given by the change in Gibbs free energy:

$$W_{el} = -\Delta G = nFE \qquad (1)$$

Where n = the number of electrons in the reaction, F = Faraday's constant (96 500° C/equivalent) and E = the reversible potential. If all reactants are at standard conditions of 1 atm and 25°C:

$$\Delta G^o = -nFE^o \qquad (2)$$

For the reaction

$$H_2(g) + (1/2)O_2(g) = H_2O(l)$$

the Gibbs free energy change[12] is –237 kJ, $n = 2$, and therefore the maximum reversible potential, $E^o = 1.23$ V. The maximum reversible potential under actual fuel-cell operating conditions can be calculated from the Nernst equation. For the general reaction:

$$aA + bB = cC + dD$$

The free energy change can be expressed:

$$\Delta G = \Delta G^o + RT\ \ln([C]^c[D]^d/[A]^a[B]^b)$$

Substituting equations (1) and (2) gives:

$$E = E^o + (RT/nF)\ \ln([A]^a[B]^b/[C]^c[D]^d$$

For the hydrogen/oxygen fuel cell this can be simplified[1] to:

$$E = E^o + (RT/2F)\ \ln[P_{H_2}/P_{H_2O}] + (RT/2F)\ \ln[P_{O_2}{}^{1/2}]$$

Normal practice for conventional power generation is to use the thermal efficiency, expressed as the electrical output as a percentage of the heat of combustion of the fuel. It is common practice in Europe to use the lower heating value (LHV) or lower calorific value (LCV) (water as gas), whereas in the United States it is common practice to use the higher heating value (HHV) or Gross Calorific Value (GCV) (water as liquid). The heat of combustion is equal to $-\Delta H$, the change in enthalpy. The thermal efficiency of a fuel cell is given by:

$$\text{Thermal efficiency} = \frac{\text{Gibbs free energy converted to electricity}}{\text{Enthalpy change (–heat of combustion)}}$$

For a hydrogen/oxygen fuel cell with liquid water as product, the Gibbs free energy change is –237 kJ per mole, equivalent to –118.5 MJ/kg hydrogen and the higher and lower heats of combustion are[2] 142.5 and 121.0 MJ/kg, resulting in maximum theoretical efficiency of 83% basis HHV which most accurately represents the reaction, but for comparison to conventional power systems, would be equivalent to 98% basis LHV.

A number of definitions of fuel-cell efficiency are used in the literature, without always stating which the author means, and care is therefore required in applying the information. The voltage efficiency being one of the most frequently used:

$$\eta E = E_I / E_O$$

Where E_I is the cell potential at current I and E_O is the open circuit at the cell operating conditions. If a fuel cell operated reversibly, then the efficiency would be:

$$\eta_{rev} = \Delta G / \Delta H$$

Where ΔG and ΔH are the changes in Gibbs free energy and enthalpy (both –ve). As a fuel cell does not operate reversibly, the efficiency is given by:

$$\eta = -nFE_I / \Delta H$$

Rearranging and multiplying numerator and denominator by $\Delta G/\Delta H$ gives:

$$\eta = -(\Delta G/\Delta H)E_I / \{(\Delta G/\Delta H)(\Delta H/nF)\}$$

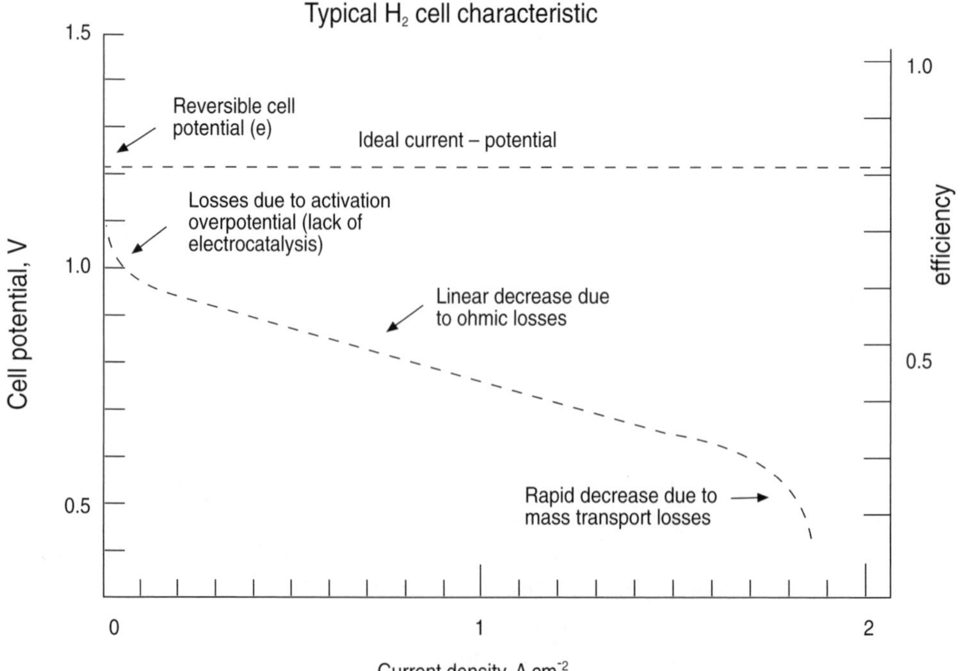

Fig. 4.2 Voltage/current relationship for hydrogen/oxygen cell.

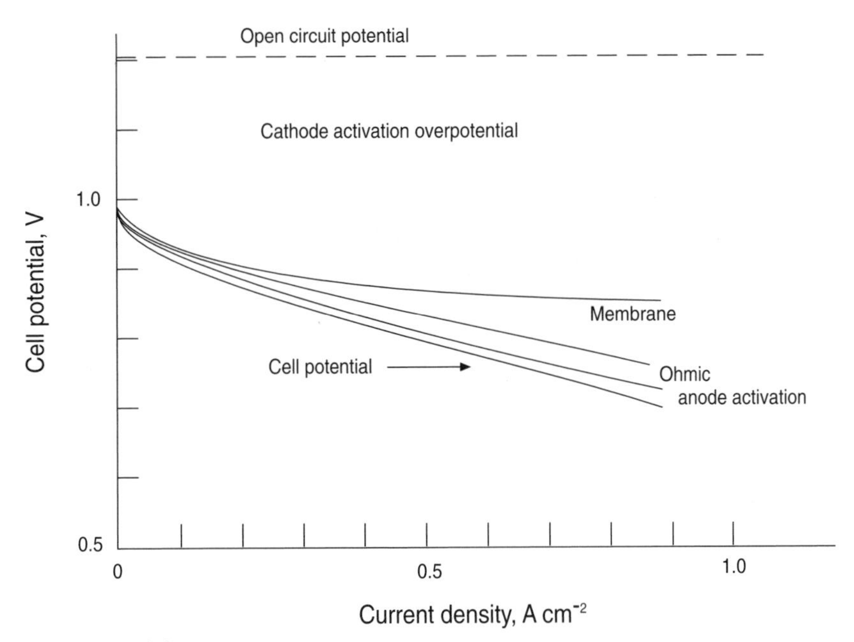

Fig. 4.3 SPFC characteristics.

As $\eta_{rev} = \Delta G/\Delta H$ and $\Delta G = -nFE_O$, then:

$$\eta = \eta_{rev}(E_I/E_O) = \eta_{rev}\eta_E$$

As the current efficiency (η_I) is not unity and there are further system inefficiencies due to feed pretreatment, cooling, gas recycle, inversion from DC to AC, etc. (η_s), the overall thermal efficiency is:

$$\eta = \eta_{rev}\eta_E\eta_I\eta_s$$

Figure 4.2 shows[11] the potential versus current relationship for a typical hydrogen–oxygen cell. A further reduction in achievable efficiency results from less than total hydrogen utilization and use of air rather than oxygen.The contribution of cathode activation overpotential, membrane, ohmic electrode and anode activation losses for a specific SPFC are shown[11] in Fig. 4.3.

The above figures show why considerable R&D is being focused on improving the cell voltage efficiency η_E via improved electrocatalysis. Further extensive development is being made to improve the system efficiency η_s by improved design of the stack and the ancillary systems. Overall thermal efficiencies of about 50% can be achieved at current densities of 0.7 A cm^2 with state of the art SPFC stacks.

One major advantage of fuel cells is the lower sensitivity of efficiency to scale than is achievable with thermal power systems, which means that fuel cells in the kW range have very similar overall system efficiencies to those in the MW range. A further benefit is the relatively flat efficiency versus load curve compared to internal combustion engines.

4.5 Process engineering of fuel cells

The discussion in the last section refers to the system efficiency η_s and an indication of the engineering role in the development of a practical fuel cell system can be seen in Fig. 4.4, which

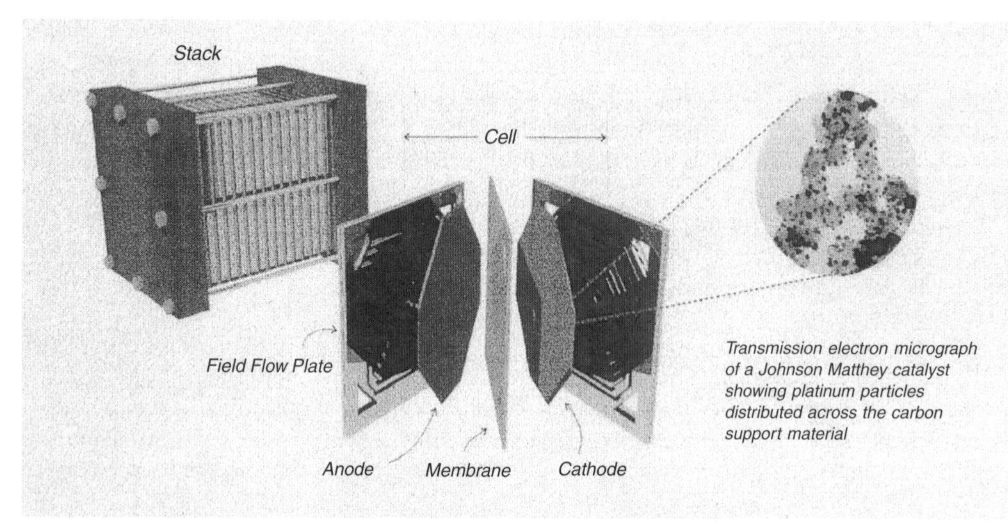

Fig. 4.4 PEM fuel cell.

shows how a single SPFC cell is designed to ensure even distribution of the reactants over the electrodes and the importance of catalyst distribution. The figure also shows a group of cells built into a small stack, but before addition of the manifolding for the supply of reactants and collection of products from the stack. An indication of the pipework complexity is shown in Fig. 4.5, which is the Ballard 240 kW system used in a bus.

During the 1990s Ballard and Daimler-Benz improved the system design to the extent that in the Necar I, the whole of the normal load carrying space of the van was filled with the fuel cell system, but in the small car, called Necar III, the system takes little more space than in the internal combustion powered version. Necar IV is based on the A-class Mercedes, with space for five passengers and luggage.

Fig. 4.5 Ballard 240 kW system.

Standard test conditions used by Ballard for SPFC stack systems are a pressure of 3 atm, stoichiometry of 2.0 for air and 1.5 for hydrogen. Their MK 513 stack is reported[9] to give 13 kW at 0.58 V and a power density of 300 W/kg and 300 W/l. The same reference quotes the performance of the stack developed for Daimler-Benz as achieving 32.3 kW at 0.68 V, and power densities of 1000 W/l and 700 W/kg, which meet the target set in 1990 by the US DOE for automotive fuel cells of 560–1100 W/kg.

The major challenges for fuel cells are (a) to develop systems which require similar mass and volumes to the engine plus fuel tank of current gasoline or diesel cars, at competitive prices when mass produced and (b) development of the required fuel infrastructure.

4.6 Steps towards the fuel-cell engine

Earlier sections of this chapter, contributed by Roger Booth, have dealt with process engineering of the fuel-cell stack. Hereafter the steps leading to the development of viable fuel-cell engines are considered. While hybrid drive vehicles, using conventional battery-electric and thermal-engine power sources, provide improved fuel economy and a viable solution for urban operation, the fuel-cell powered vehicle is now seen as the long-term option. Already it is realized that thermal-engine driven vehicles can never provide the necessary fuel economy and emissions control required by world governments, primarily because the thermal engine is running at 10% of its total power potential for most of its time and there is no known way of eliminating CO_2 emissions from it. Over the next 20 years, people using HC vehicles are going to face increasing fuel scarcity, increasing fuel cost and ever increasing restrictions on use and size of vehicles, because of the emissions they produce. The high operating efficiency and zero emission characteristic of the fuel-cell vehicle are strong arguments for its adoption. But the passage into the period of the hydrogen-fuel economy has to be a gradual and peaceful one, requiring considerable changes in attitude by the motoring publics worldwide.

4.6.1 FUEL CELLS: TOWARDS A PAINLESS TRANSITION

The only way to break out of the cycle of increasing fuel costs and heavier restrictions is for motorists to accept the necessity to move over to a hydrogen-fuel economy in the shorter term and to lessen its impact by replacing their current vehicles, when the time comes, with much more fuel-efficient types. There are two advantages to the hydrogen economy: if early hydrogen-fuelled vehicles are not very efficient it does not matter because pollution-wise they have zero emissions, and more importantly, the existing cost of expensive exhaust after-treatment is removed.

The publication by the OECD International Energy Agency in 1998 of 'World Energy Outlook', a year after the Kyoto Earth Summit, was a pivotal point in understanding world energy and pollution problems. The basic message was that, as people now live longer, energy usage and pollution rise exponentially and a 'brick-wall situation threatening in ten years' time means that we cannot stay as we are'. It was also from this point that the major G7 economies took global warming seriously. China, too, takes it seriously, realizing that nine-tenths of its population live in its southeastern corner delta region, which could be subject to flooding if global warming is not seriously addressed. World oil supply is expected to peak in 2010 and then tail off; this is unless the cost can be met of tapping into the vast oil deposits beneath the polar icecaps. But North American motorists remain blissfully unaware of these threatening situations. While the UK and Europe pay what gasoline and diesel actually costs, in total, US users have an enormous effective subsidy which hides the governmental costs, and keeps them oblivious to the problems involved. Fuel/vehicle taxes in the UK pay for road building, health care related to accidents, also defence costs relating to naval protection of oil rigs, whereas in the USA the petrol price paid at the pumps is the direct cost of the fuel at world market prices. Even the cost of the US fleet in the Middle East is reckoned by some to be equivalent to a subsidy of $1/gallon and of course there is no contribution to the health-care costs associated with exhaust pollution. In the US such costs are covered by road tolls and other forms of local/central taxation. It is considered that Americans pay pump prices which cover only one-third of the real costs, the other two-thirds being borne by the state and, in the USA, local oil extraction has been declining since 1975 to the point where nowadays some 70% is imported.

4.6.2 FUEL-EFFICIENT VEHICLES

The US PNGV programme described later in this chapter is an effort to provide the technology for very low fuel consumption vehicles which industry could adopt if motorists accept the need. Researchers such as Lovins at the Rocky Mountain Institute have shown that 500 mpg hypercars are possible, paving the way for some intermediate value to the 30 mpg average consumption vehicle typical in America. The 10 kW absorbed at 60 mph alone could be halved by the adoption of underfloor aerodynamics. In combination with an aerodynamic superstructure, double-acting brake cylinders and low rolling-resistance tyres, it could be cut to one-third.

In the deal that was struck after the Kyoto summit in 1997 hydrogen is to become available one day at filling stations; both Shell and Esso are already committed in Europe and Japan, and will start operating only when they consider the fuel markets are operated on an orderly basis. On the other hand, in America they are operated on a subsidized basis. In Europe, too, at least one manufacturer, VW, has shown with the Lupo that a 1.3 litre 80 mpg car can be built to USA PNGV requirements. This uses a high technology diesel engine with common rail fuel injection and is an instant starter without the use of glow-plugs or comparable devices. It has top speed of over 100 mph and nippy accelerative performance. But this performance could not be achieved on larger cars typical in the American market, that is not without hybrid drive technology as now made available on the Toyota Prius, which has interior accommodation comparable with some American cars. The US version has a 21 kW battery to improve acceleration and features full air-conditioning. In 2000, this car sells at $18 000 compared with $10 000 for an American 'base model' car. So far some 40 000 are in use worldwide but the UK version is two-thirds dearer at £18 000, for a much higher performance vehicle required in this market.

Such fuel-efficient vehicles should permit an increase in gasoline prices that would allow oil extraction from the polar regions, permit other fuels to compete with gasoline and permit other transport systems to compete with cars and aeroplanes. The GM Precept (Fig. 4.6) is the corporation's PNGV car and similar ones have been developed by Ford and Chrysler. GM also intends to make the vehicle in hybrid and fuel-cell versions. It is important to note that it is an ultra-low weight vehicle made primarily in aluminium alloy with underbody streamlining, a TV camera in lieu of wing mirrors, low rolling-drag tyres and double-acting brake pistons. The latter

Fig. 4.6 GM Precept PNGV car.

overcome the problem of drag due to brake hang-on with conventional systems when the single-acting piston fails to withdraw, and some 2 kW can typically be lost at motorway speeds. Of course, hydro-mechanical brakes only operate rarely on EVs, which can rely on regenerative braking for light-duty work. The Precept is an order-of-magnitude change in technology compared with ordinary American-produced vehicles, but of course would cost considerably more than the $10 000 dollars now charged.

4.6.3 FUEL-CELL VEHICLE PROSPECTS

The real future is with fuel-cell cars because the Precept version so fitted will have a fuel-cell stack volume of just 1.3 ft^3 and produce 70 kW continuously, 95 V at 750 A, Fig. 4.7. Plate current density is 2 A/cm^2; the cell is currently world leader in PEM-stack design and if used intelligently has higher power density than a thermal engine. This is, then, the turning point, and it is underlined by Mercedes whose work on the Necar IV is showing that 38% efficiency is obtained from hydrogen in the fuel tank to power at the road wheels. This compares 13–15% for a standard thermal-engine car. With this three times improvement in fuel economy GM believes that they are going to develop fuel-cell vehicles. The major challenge over the next five years is making it at a low enough cost to be attractive.

4.6.4 BATTERY ELECTRICS/HYBRIDS IN THE INTERIM

EV technology will be needed for the future fuel-cell car but currently it cannot be proven with current traction batteries which are too heavy, too expensive and made from materials that are not in plentiful enough supply. Most good quality batteries are based on nickel technology and the reason for being in this difficulty is not because the battery research

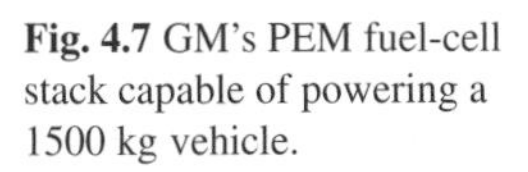

Fig. 4.7 GM's PEM fuel-cell stack capable of powering a 1500 kg vehicle.

programme was a failure, but rather the reverse. The outcome is that people are using batteries for communications, camcorders and computers; they pay far more for their batteries than car manufacturers could sensibly afford. The '3Cs' are prepared to pay three times what the EV builder can pay.

A better alternative to nickel technology is aluminium. This is because the nickel–metal hydride battery, to give 80 kWh needed for a 3–400 mile range (on a PNVG car), would weigh 850 kg and cost $25 000, at year 2000 prices from Ovonic. The same 80 kWh can be obtained at a weight of 250 kg with aluminium, and at a cost of just $5000; and of course the material from which it is made is the most abundant on earth, next to hydrogen. The electric car is not a failure from a performance point of view but merely waiting for its time to come when high performance batteries can be produced economically. Energy density is not the problem with aluminium, rather it is the corrosion problem which causes aluminium batteries to degrade rapidly because of the formation of aluminium hydroxide jelly. However, two years ago scientists in Finland put forward a raft of patents which overcame some of the problems; these were revealed at the 2000 ISATA conference.

For the PNVG programme, Dr Alan Rudd built an aluminium battery for the US government; this was a pump-storage one, that was totally successful apart from the above-mentioned corrosion problem – the factor which persuaded the Americans to discontinue development. The technical committees took the decision to ignore the lower voltage couples on the grounds that the higher voltage couples were sold on the basis of having fewer cells in series. This gave the lightest batteries for portable power applications but the materials involved could never be cheap enough for an electric car.

A key advantage of the aluminium battery is its ability to operate at temperatures down to –80°C, overcoming the disadvantage of many existing types. The corrosion problem is now thought to be soluble and effective EV batteries are foreseen in 5 years' time. It is considered best therefore to hold back on battery electrics, while hybrids hold the fort, until such time as the most cost-effective high performance solution is found. If the current European development programme is successful EV makers will have D-cells (like torch batteries), each handling 150 Ah at 1.5 V DC and able to be discharged at about 500 A maximum. Batteries will be formed from matrices of such cells, as discussed in Chapter 2. This is the most effective solution to the battery-electric vehicle problem. Existing technology batteries are going to be used for energy stores in hybrid-drive vehicles where the capacity needed will be less than 5 kWh.

4.6.5 POLLUTION CONTROL MEASURES

Hybrid drive cars present an early stepping point to fuel-efficient cars but very few car makers have produced the full gamut of drag-reducing measures that could transform fuel economy, and hence CO_2 emissions, with existing thermal-engine technology cars. To do so could give huge improvements at modest cost, with immediate fuel saving and emission control benefits. The other big potential benefit would be catalytic converters for diesel engines. Such engines produce as much pollution now as petrol engines did in the 1960s and are the worse by far for PM10 particulates. Converters would dramatically cut diesel engine emissions and it is only their poisoning by sulphur in conventional diesel fuel that has prevented their widespread use. Now that clean 'City Diesel' is beginning to be seen at filling stations there is real hope for a positive step forward. This new fuel has around 10 ppm sulphur instead of 250 ppm with conventional fuel. The final stage in pollution control is, of course, the zero emission vehicle, the main contender being fuel-celled vehicles which GM intends to introduce in 2004.

4.6.6 OVERALL ENERGY POLICY

The global perspective is that over the next 20 years, road vehicles and aircraft will switch to hydrogen fuel. The exact time will depend on how consumers react to the problems outlined above. Should consumers adopt a helpful attitude and accept the certain introduction of fuel-efficient cars soon? This would allow fuel prices to rise sufficiently for the extraction of polar oil deposits, then gasoline could still be used in 100 years' time. Conversely by doing nothing, and continuing to drive 30 mpg cars, we shall be subject to a crash introduction of the hydrogen economy in ten years' time. From an industry and cost perspective it would obviously be better to have a gradual transition; a sudden transition could have an economic effect similar to that of a world war. It is really vital that the G7 economies, at least, introduce fuel-efficient cars within the next five years. By staying with conventional HC fuels, and avoiding the hydrogen transition, will only lead to more regulation, slow strangulation and severe restrictions – which would be very hard to impose on the US market, for example.

When the transition does occur hydrogen will be produced first from the reformation of natural gas and then by the electrolysis of water using electricity from fusion reactors. In 2010 it is estimated there will be dual-fuel aircraft with paraffin in the wing tanks and liquefied hydrogen in tanks over the passenger compartment. Over the next 20 years, many airlines will prefer to transfer directly to hydrogen solely, as they gain major operational benefits. They will, at a stroke, double range or payload and thus be prepared to pay a higher price for hydrogen. The fuel will be supplied as a liquid at –180°C and 20 bar pressure for the aerospace market and thus high quality hydrogen will also be available in quantity for road vehicles. The reformation of natural gas will be carried out in central facilities (much more efficiently than in on-board installations) with the important proviso that energy must be extracted from the carbon in the methane, as well as from the hydrogen molecules, since there is three times the energy in carbon over hydrogen. Something like an Engelhard ion–thermal catalytic process is thus required. It is also important that after the carbon has been burnt out it should be in the form of a carbonate or a carbide (and not in the form of CO_2 which would revert to the atmosphere). Although energy is released in this way during the conversion, overall, energy is consumed during the process. Some 90% of the initial energy is retained in the form of pure clean hydrogen fuel. From the point of production hydrogen can be distributed as a gas through existing natural gas pipelines. This is the likely scenario until 2050 when the exhaustion of natural gas will dictate the need for fusion reactors.

4.7 Prospects for EV package design

Electric traction was viable even before 1910 when Harrods introduced their still familiar delivery truck, with nickel–iron battery, which is still in daily use and speaks volumes for the longevity and reliability of the electric vehicle. But very important structural changes have taken place since. At the turn of the Twentieth century EVs did not have solid-state controls and sophisticated control was achieved using primitive contactor technology. Amazingly successful results were obtained with contactor-changing and field-weakening resistor solutions, as is seen in the Mercedes vehicle described at the end of the Introduction.

But the 'writing on the wall' for the first generation of electric cars appeared in the World War I period with the development of electric starters for thermal engines. This was followed by unprecedented improvement, by development, of the piston engine and the success of the Ford Model T generation vehicles in the 1920s which substantially outperformed early electric cars. From then on until 1960 when high-power solid-state switching devices were developed and EVs were basically used for delivery and other secondary applications. Between 1960–1980, a new generation of EVs was developed, of which mechanical-handling trucks and golf-carts were the most notable. These were based on Brushed DC motors with lead–acid batteries and some millions of golf-carts are in use today. However, with growing pollution, and fuel-availability problems in the 1970s, there was an impetus to try to build a successful passenger car, with the realization that it would take an order-of-magnitude improvement in technology to make this happen. There was, however, a significant advance with the coming of power transistors in place of thyristors, and big improvements in drive controllers resulted, epitomized by the successful Curtis controller. This was a field-effect transistor chopper that was to become almost universally used in low power DC vehicles. High performance AC drives also came into being, and four machines thus came to do battle for the EV market.

4.7.1 MOTOR CONSTRAINTS

There is now general acceptance that the brushless DC motor will be the one used now and in the future. At a conference in Toronto in June 2000, GM gave details of its latest version, with inverter, in the Precept car. Compared with their earlier induction motor drives, they have halved the size by going to the permanent-magnet motor as well as reducing current consumption, for equivalent performance, by a factor of 1.8. They thus also have an inverter of half the size of that required for an induction motor, and this has resulted, too, in substantial manufacturing cost reduction.

The disadvantage of brushed DC motors is the high unit weight for the performance obtained (and the commutator is moisture sensitive, perhaps beyond the capability of tolerating a high-pressure car wash). This is quite acceptable in industrial trucks where extra weight is often required to counterbalance handling of the payload at high moment arms, but not of course in cars. Because the frequency of the commutator in a brushed DC machine is 50 Hz, compared with 1 kHz for a brushless PM machine, the latter is smaller and lighter; also the electronics can switch at 20 times the equivalent rate of mechanical brushes, which is the basis for the weight advantage. A 45 kW brushed machine weighs 140 kg, and typically runs at 1200–5000 rpm, while an equivalent powered brushless machine operates at 12 000 rpm and weighs less than 20 kg. There can of course be a 5 kg weight penalty for a reduction gearbox but even so there is a 75% weight reduction overall. The inverter is also the cheapest of those used with any ‘AC-type’ motor.

The other contenders were the switched reluctance motor (SRM) and AC induction motor, plus the permanent magnet Pancake motor (by Lynch) which serves the specialized light vehicle market but is non-scaleable technology. Key result of the investigation into comparative methods was that one had to look at the development of the whole drive package and not just the traction motor. While once people balked at the $200 required for the permanent magnets required in brushless motors, in 2000 they appreciate that some $1000 of power electronics is saved in the inverter. The induction motor was put ‘out of court’ because traction operation requires constant power over a 4:1 speed range. Since voltage, current, speed (*V, I, N*) characteristics show that to do this an induction motor goes from 0.5 V at double current at the bottom end of the speed range, to *V* and *I* at the top end, so an inverter that can supply double current is required. With a brushless DC machine it can be designed to require *V* and *I* at both minimum and maximum speeds, so only half the size of converter is required. It also has a further advantage that scratches SRM from the equation. Because the SRM is force commutated (current interruption is very dissipative , a ‘hard-switching’ turn-off), power losses in the inverter are very significant compared with that associated with a brushless machine which at high speed operates with a leading power factor and so has hardly any switching losses in the inverter (which may be of very compact construction). The SRM also exhibits significant acoustic noise due to the magnetostriction of its operating dynamics. With the 4 or 6 coils in the machine which have to be moved relative to one another forces of attraction between them are up to ten times greater than the developed torque of the motor; the framework of the motor can physically distort and considerable noise thus generated.

4.7.2 WHEEL MOTORS AND PACKAGE DESIGN

While wheel motors are ideal for low speed vehicles the problem of high suspended mass rules them out for cars. Road damage can be caused at wheel hop frequencies and the perceived threat of losing traction on one wheel, by a single motor failure, would prevent any safety authority from issuing a certificate of roadworthiness. Use of such devices as active suspension makes them possible on medium speed urban buses where road wheel tyres can be as much as one metre in diameter and large brake assemblies reduce the relative weight of wheel motors. Motors driving individual back wheels are a possibility in commercial vehicles, where traction and steer forces are not shared by individual tyres, and 4×4 drives with a single motor power source are ideal for more expensive cars, which could tolerate the cost of multiple control systems. Wheel motors could see wider application if steels with adequate magnetic properties could be developed for lighter-weight PM motors at reasonable cost. Expensive military vehicles use such a steel, called Rotalloy, but it costs some £15 per kg in 2000. Such vehicles sometimes have individually steered and driven wheels which enable them to move sideways so perhaps cheaper future alloys of this type will improve parking manoeuvres.

At the present time safety authorities are unlikely to certificate cars with electrical rather than mechanical differential gears but a number of drive-by-wire solutions may become more feasible on EVs. Introduction of 5 kW, 42 V electrical systems is a strong possibility, that could see the replacement of many hydraulic, pneumatic and mechanical controls by electrical ones monitored electronically. These drive-by-wire systems will be a prelude to convoy control of vehicles on motorways. Many development pains have yet to be cured, however, though EV technology will be helpful in the implementation. Other advances such as starter-alternators are likely to be found on thermal-engined hybrid vehicles; these use a new kind of power electronics with silicon-carbide switching devices cooled by hot water from the engine cooling system, allowing semiconductors to operate safely at 250°C. First is due on the Mercedes 500 to be introduced in 2001 and fitment to all European and American cars is expected in two years' time.

4.7.3 ALTERNATIVE AUXILIARY POWER

Consideration of photovoltaic power is often a pastime of EV promoters but 10–15% light to electricity conversion efficiency has precluded serious traction usage so far, though use as an auxiliary power source is important. Even at high noon in the tropics solar radiation can only generate 1 kW/m^2 which means that the solar cell will produce only 150 W for each square metre. In the Honda Solar Challenger, 8 m^2 of solar cells generates 1–1.5 kW, which would be nowhere near enough to provide propulsion and hotel loads ('parasitic' loads such as lighting and air conditioning) for a conventional car. The most hopeful traction application is for electric scooters operating in the tropics where a reasonable size photocell array, carried in the panniers then unfolded and left out in the sun, could charge the battery of a Honda 50 electric scooter in 6 hours and provide traction for 50 miles without energy being drawn from the grid important in isolated areas.

Photocells are also useful on battery-electric vehicles to ensure that the battery never gets fully discharged (particularly important with lead–acid types). They are valuable sources of auxiliary power for cooling purposes, either lowering interior temperatures on cars parked in the sun or providing refrigeration power to keep gaseous fuels in liquid form. The transformation in usage that could follow an increase in conversion efficiency may be realizable before long if the reported intensity of research bears fruit. BP and Sanyo are both world leaders in this and already enjoy market success with static arrays of low efficiency cells in tropical countries.

4.8 Fuel-cell vehicles and infrastructure

Fuel cells are preferred as a primary traction power source because theoretical stack efficiency on the Carnot cycle is 83%, which is more than double that of the thermal engine, and unlike the thermal engine they become more efficient (90%) at light load operation. Stack EMF drops from 1 V at no load to 0.6 V at full load; stack efficicency thus increases at light load since auxiliary losses do not go down in the same proportion. Whether hydrogen is reformed from fossil fuels on board the vehicle (Fig. 4.8), as an interim approach to carrying compressed liquid hydrogen, is still under debate. This approach is being championed by Chrysler and is attractive in America where gasoline is sold at a subsidized price. Even with this technique overall efficiency is much higher than with a thermal engine. However, the heavy on-cost to the vehicle makes it no more than a transitory solution. The military have used hydrogen propulsion for nuclear submarines, space-craft and specialized assault vehicles for many years now and they have a complete infrastructure in place from which the civil transport market can learn, so that 'critical mass' has been reached in terms of knowledge base and experience. Now the development is directed at moving from cost plus to cost effective. The challenge is to make parts out of plastic that were previously made from stainless steel and achieve one-tenth of existing costs.

It was once said that on-board liquid hydrogen storage was a big problem but the latest C16 carbon fibre has resulted in a 60 litre storage tank with a weight of only 7 kg that will store 16 in^3 of hydrogen. What is not widely understood is how the gas is compressed in the liquefaction process. The method of approach can be the difference between success and failure since a two-stage process is involved, a Stirling cycle stage down to –200 °C and a Linde cycle from –200 to –269 °C, the latter using nearly all the energy. So it is necessary not to liquefy the gas at ambient pressure as the Linde cycle will be involved, burning up 30% of energy within the fuel getting it down to –273°C. The approach is to have pressure tanks operating at –160 to –180 °C, with a metal inner wall then glass fibre in a vacuum for an inch radial thickness followed by a carbon fibre outer wall. By putting hydrogen into a tank with no additional cooling it takes about two weeks before the liquid becomes a gas, and it blows off. Under normal conditions motorists would use a

(a)

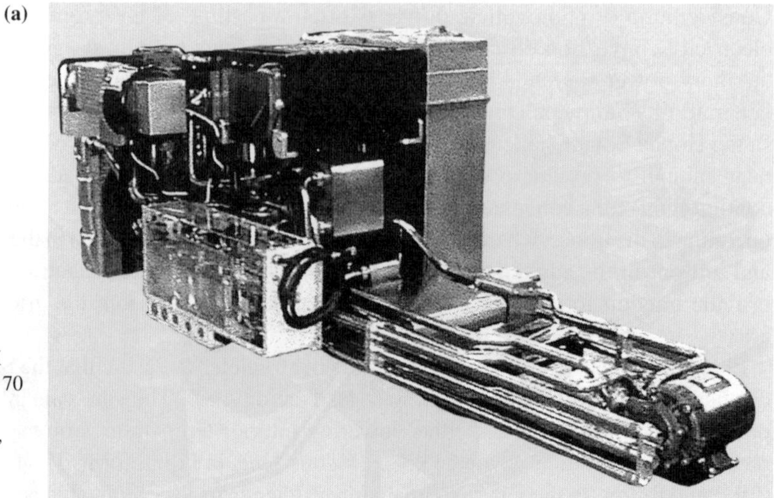

Fig. 4.8(a) Complete GM fuel-cell chassis with POX converter capable of up to 70 kW output at 300 V DC (including AC drive train), **(b)** Gasoline to hydrogen (POX) converter close up.

tank-full every two weeks so that no additional refrigeration would be required. However, if it is necessary to use refrigerated gas the basic Stirling cycle refrigerator of 10 W would keep it in liquid form indefinitely, the 10 W coming readily from a photocell array charging a small auxiliary battery. Such units have been made in Israel for 30 years and are extensively used by the military. For cars a 15 litre fuel storage tank will probably be used, having capacity for 50 cubic metres of hydrogen to provide a 400–500 mile range. Larger vehicles, trucks and buses will probably store the gas as a liquid, in view of the larger gaseous volumes otherwise involved. One should also not forget the bonus that natural gas comes out of the ground at 300–400 bar therefore not so much energy is needed to compress the gas, with the recovery of energy in the reformation process.

Generation zero fuel cells cost $8000/kW for the entire system including pumps, power conversion, controls and fuel storage. Autumn 2000 saw the construction of second generation fuel cells and are the first serious attempt at cost reduction. Many separate components will be integrated by such techniques as manifolding, for example. Plastic pumps will be employed instead of metal ones and for the first time custom-engineered chips will be used instead of standard PLCs. This will yield $2000/kW which it is hoped will be reduced to $1000/kW for the third generation by 2002. For example, an air pump used by the author's company once weighed 18 kg and by careful production development this was reduced to 10 kg; in the latest stage of conversion from metal to plastic, it is hoped to achieve 3.5 kg, as well as the reduced cost benefits. The company build all electrical and control systems for Zetec Power's fuel-cell engines. Zetec are opening a new factory in Cologne which will make 2000 stacks per year and thus cost effectiveness is paramount.

4.8.1 HYDROGEN DISTRIBUTION

Currently natural gas is distributed through 600 mm diameter pipes at a pressure of 500 psi. Many of these pipes can be used for hydrogen distribution and energy transport factor will increase significantly as a result with the higher energy density of hydrogen. Many other end-products and processes can be fuelled, as well as road vehicles, by this means. It has to be remembered, however, that the gas is explosive at extremely low levels of hydrogen/air mixture and it must be stored near the roof of vehicles, since the gas is lighter than air. Vehicles themselves must also be stored in well-ventilated areas. Explosive energy is considerably lower than natural gas, however, and the main requirement is to install low level concentration hydrogen sensors in the storage vicinities.

(b)

Although considerable change and expense is necessary to move to a hydrogen-fuel economy, it will be a much easier experience if it can be implemented over a substantial time period, as suggested earlier, but with minimum delay in starting the process. A hydrogen economy has the advantage that one grade and type of fuel replaces the five or six currently on offer at filling stations and domestic heating too is likely to turn to fuel cells so hydrogen will also be their source of supply.

4.9 The PNGV programme: impetus for change

On 29 September 1993, the Clinton Administration and the US Council for Automotive Research (USCAR), a consortium of the three largest US automobile manufacturers, formed a cooperative research and development partnership aimed at technological breakthroughs to produce a prototype 'super-efficient' car. The 'Big Three' (Chrysler, Ford, and General Motors), eight federal agencies, and several government national defence, energy, and weapons laboratories have joined in this Partnership for a New Generation of Vehicles (PNGV). It is intended to strengthen US auto industry competitiveness and develop technologies that provide cleaner and more efficient cars. The 1994 PNGV Program Plan called for a 'concept vehicle' to be ready in about six years, and a 'production prototype' to be ready in about 10 years. Research and development goals included production prototypes of vehicles capable of up to 80 miles per gallon – three times greater fuel efficiency than the average car of 1994.

Background drivers of the initiative include a combination of high gasoline prices, and government fuel economy regulation caused new car fuel efficiency to double since 1972. However, fuel economy standards for new cars peaked at 27.5 miles per gallon (mpg) in 1989 and the average fuel efficiency of all on-road (new and old) cars peaked at 21.69 mpg in 1991, then dropped slightly in 1992 and again in 1993. Further, the large drop in real gasoline prices since 1981 and the increasing number of cars on the road are eroding the energy and environmental benefits of past gains in auto fuel efficiency. The public benefits that could derive from further improvements in auto fuel efficiency include health benefits from reduced urban ozone, 'insurance' against sudden oil price shocks, reduced military costs of maintaining energy security, and potential savings from reduced oil prices.

The Declaration of Intent for PNGV emphasizes that the programme represents a fundamental change in the way government and industry interact. The agreement is seen as marking a shift to a new era of progress through partnership and cooperation to address the nation's goals, rather than through the confrontational and adversarial relationship of the past. Its intent is to combine public and private resources in programmes designed to achieve major technological breakthroughs that can make regulatory interventions unnecessary. The partnership agreement is a declaration by USCAR and the government of their separate, but coordinated, plans to achieve goals for clean and efficient cars. A further objective is to curb gasoline use by 7 billion gallons per year in 2010 and 96 billion gallons per year in 2020, while creating 200 000 to 600 000 new jobs by 2010.

At the time the agreement was struck, the president and executives from the Big Three said they hoped that PNGV research breakthroughs would ultimately make auto emissions and mileage regulations unnecessary. Chrysler's former PNGV director, Tim Adams, noted that the partnership represents the opportunity to address more efficiently fundamental national objectives than the regulatory mandate approach. Further, car-makers say the Supercar's advanced technologies are outside their short-term research focus, and unjustified by fuel costs or market demand for fuel efficiency. They argue that the North American market forces alone would not drive them to create an 80 mile per gallon mid-sized sedan.

Examples of applied technology would be the development of lightweight, recyclable materials, and catalysts for reducing exhaust pollution; research that could lead to production prototypes of vehicles capable of up to three times greater fuel efficiency. Examples would be lightweight materials for body parts and the use of fuel cells and advanced energy storage systems such as ultracapacitors. Using these new power sources would produce more fuel-efficient cars. Further initiatives included lightweight, high-strength structural composite plastics that are recyclable, that can be produced economically in high volume, and that can be repaired. Hybrid drive control electronics and hardware were also cited alongside regenerative braking systems to store braking energy instead of losing it through heat dissipation; also fuel cells to convert liquid fuel energy directly into electricity with little pollution.

Such advances are aimed at more efficient energy conversion power sources, viable hybrid concepts as well as lighter weight and more efficient vehicle designs. The contributions of US government agencies include the following: at its ten National Laboratories, the Department of Energy has technical expertise, facilities, and resources that can help achieve the goals of the partnership. Examples include research programmes in advanced engine technologies such as gas turbines, hybrid vehicles, alternative fuels, fuel cells, advanced energy storage, and lightweight materials. The DOE's efforts are implemented through cost-shared contracts and cooperative agreements with the auto industry, suppliers, and others. Technologies covered include fuel cells, hybrid vehicles, gas turbines, energy storage materials and others. The Department of Defense's Advanced Research Projects Agency (ARPA) is focused on medium-duty and heavy-duty drivetrains for military vehicles which could, in the future, be scaled down to light-duty vehicles. ARPA funds research on electric and hybrid vehicles through the Electric/Hybrid Vehicle and Infrastructure (EHV) Program and the Technology Reinvestment Project (TRP). EHV is a major source of funding for small companies interested in conducting advanced vehicle research that is not channelled through the Big Three auto-makers. NASA will apply its expertise to PNGV in three ways: by applying existing space technologies such as advanced lightweight, high strength materials; by developing dual-use technologies such as advanced batteries and fuel cells to support both the automotive industry and aerospace programmes; and by developing technologies specifically for the PNGV such as advanced power management and distribution technology. The Department of Interior involvement in PNGV-related research includes research to improve manufacturing processes for lightweight composite materials and recycling strategies for nickel–metal hydride batteries. The DOI's Bureau of Mines has developed a system for tracking materials and energy flows through product life cycles. Life-cycle assessment of advanced vehicles and components can help to anticipate problems with raw materials availability, environmental impacts, and recyclability. This includes the worldwide availability of raw materials, environmental impacts of industrial processes, and strategies for recycling of materials.

The US OTA considers that the most likely configuration of a PNGV prototype would be a hybrid vehicle, powered in the near term by a piston engine, and in the longer term perhaps by a fuel cell. It notes that there is no battery technology that can presently achieve the equivalent of 80 mpg. Thus, the proton exchange membrane (PEM) fuel cell is seen as the more likely candidate. The DOE further stresses that meeting the fuel economy goal will require new technologies for energy conversion, energy storage, hybrid propulsion, and lightweight materials.

4.9.1 PARALLEL EUROPEAN UNION AND JAPANESE INITIATIVES CITED BY THE US GOVERNMENT

According to TASC, the European Union (EU) has formed the European Council for Automotive Research and Development (EUCAR) in response to both the US PNGV programme and accelerated vehicle development in Japan. EUCAR's objectives are technology leadership, increased competitiveness of the European automotive industry and environmental improvements. With a leader appointed from industry, EUCAR has requested a budget of over $2.3 billion from the EU over 5 years, representing a 50% EU government cost share. This includes $866 million for vehicle technology, $400 million for materials R&D, $400 million for advanced internal combustion engine (ICE), $333 million for electric/hybrid propulsion, and $333 million for manufacturing technology and processes. An additional $638 million is targeted for control and traffic management, and $267 million is targeted for management and organization structures. The annual EU budget is expected to include $173 million for vehicle technology, $80 million for advanced ICE, $80 million for materials, $67 million for manufacturing, and $67 million for electric and hybrid vehicles. Member companies of the EUCAR cooperative R&D partnership include BMW, Daimler-Benz AG and Mercedes-Benz AG, Fiat SpA, Ford Europe, Adam Opel AG, PSA Peugeot-Citroen, Renault SA, Rover, Volkswagen AG, and Volvo AB. National initiatives include fleet purchases and demonstrations, subsidies and cooperative R&D.

OTA notes that about $700 million of the EUCAR programme is focused specifically on automotive projects. The EUCAR programme is similar in some ways to PNGV, but the research proposed in its Master Plan is broader in scope, encompassing sustainability concerns in the longer term, though with no mention of a timetable for a prototype vehicle. The Master Plan proposes work focused on product-related research on advanced powertrains and materials, manufacturing technologies to match new vehicle concepts, and the total transport system, including vehicle integration into a multimodal transport system. The primary source of funding will be the EU's 5-year Framework IV programme. Also, in 1995, to stimulate R&D on advanced vehicles using traction batteries, the EU initiated a task force named 'Car of Tomorrow' that will collaborate with industry, ensure R&D coordination with other EU and national initiatives, and encourage the use of other funding such as venture capital. OTA also notes that some European nations, such as France, may be a more promising market for advanced vehicles, especially EVs, since it has more compact urban areas with shorter commute distances. France, Germany and Sweden have significant EV and other advanced vehicle programmes under way.

TASC reports that Japan has utilized the Ministry of International Trade and Industry (MITI) as the focus of industry–government cooperation to execute a similar activity with funding expected to reach $250 million per year. Its strategy is focused on market share and electric/hybrid vehicles for the California market. Reduction of nitrous oxide emissions is also an environmental goal of the programme. The annual government share of budget is expected to include $29 million or more for vehicle technology, $40 million for advanced ICE, $20 million for materials, $5 million or more for manufacturing, and $57 million for electric and hybrid vehicles. An infrastructure project is under way at nine major sites located close to industry and covering a wide range of climates. Industry manufacturers gearing up for the 1998 California zero emission vehicle (ZEV) programme include Honda, Mazda, Nissan, and Toyota. Other Japanese manufacturers participating in the cooperative activity include Daihatsu, Mitsubishi, Isuzu, and Suzuki.

OTA notes that the Japanese programme to develop PEM fuel cells began slowly under the MITI's New Energy and Industrial Technology Development Organization, but it is rapidly catching up with US programmes. PEM fuel cells are being actively developed and tested by some of the most powerful companies in Japan. Japanese auto manufacturers have performed research on EVs for more than 20 years, but the effort was given low priority due to problems with traction battery performance and doubts about EV consumer appeal. However, California's adoption of the ZEV regulations raised this priority.

References

1. Appleby and Foulkes, *Fuel cell handbook,* Van Nostrand Reinhold, 1989
2. Blomen and Mugwera, *Fuel cell systems,* Plenum Press, 1993
3. Hart and Bauen, *Fuel cells: clean power, clean transport, clean future,* Financial Times Energy, 1998.
4. Prentice, *Electrochemical engineering principles;* Prentice-Hall Inc., 1991
5. *Fuel cells, a handbook,* US Dept of Energy 1988, DOE/METC-88/6096 (DE88010252)
6. Platinum 1991, Johnson Matthey
7. Appleby, *Journal of Power Sources,* 29, pp. 3–11, 1990
8. Dicks, J. L., *Journal of Power Sources,* 61, pp. 113–124, 1996
9. Prater, *Journal of Power Sources,* 61, pp. 105–109, 1996
10. Ledjeff and Heinzel, *Journal of Power Sources,* 61, pp. 125–127, 1996
11. Acres and Hards, *Phil Trans R. Soc. Lond. A,* pp. 1671–1680, 1996
12. *Blomen or Perry's Chemical Engineers' Handbook,* Sixth Edition, pp. 3–150
13. Shibata, *Journal of Power Sources,* 37, pp. 81–99, 1992

Further reading

Maggetto *et al. (eds), Advanced electric drive systems for buses, vans and passenger cars to reduce pollution,* EVS Publication, 1990

PART TWO

EV DESIGN PACKAGES/DESIGN FOR LIGHT WEIGHT

5
Battery/fuel-cell EV design packages

5.1 Introduction

The rapidly developing technology of EV design precludes the description of a definitive universal package because the substantial forces which shape the EV market tend to cause quite sudden major changes in direction by the key players, and there are a number of different EV categories with different packages. For passenger cars, it seems that the converted standard IC-engine driven car may be giving way to a more specifically designed package either for fuel-cell electric or hybrid drive. While the volume builders may lean towards the retention of standard platform and body shell, it seems likely that the more specialist builder will try and fill the niches for particular market segments such as the compact city car. It is thus very important to view the EV in the wider perspective of its market and the wider transportation system of which it might become a part.

Because electric drive has a long history, quite a large number of different configurations have already been tried, albeit mostly only for particular concept designs. As many established automotive engineers, brought up in the IC-engine era, now face the real possibility of fuel-cell driven production vehicles, the fundamentals of electric traction and the experience gained by past EV builders are now of real interest to those contemplating a move to that sector. A review of the current 'state of play' in sole electric drive and associated energy storage systems is thus provided, while hybrid drive and fuel-cell applications will be considered in the following chapter.

5.2 Electric batteries

According to battery maker, Exide, the state of development of different battery systems by different suppliers puts the foreseeable time availability for the principal battery contenders, relative to the company's particular sphere of interest, lead–acid – as in Fig. 5.1a.

5.2.1 ADVANCED LEAD–ACID

The lead–acid battery is attractive for its comparatively low cost and an existing infrastructure for charging, servicing and recyclable disposal. A number of special high energy versions have been devised such as that shown at (b), due to researchers at the University of Idaho. This battery module has three cells, each having a stack of double-lugged plates separated by microporous glass mats. High specific power is obtained by using narrow plates with dual current collecting lugs and a 1:4 height to width aspect ratio. Grid resistance is thus reduced by shortening conductor lengths and specific energy is improved by plates that are thinner than conventional ones. They have higher active mass utilization at discharge rates appropriate to EV use. At an operating temperature of 110°F specific energy was 35.4 Wh/kg and specific power 200 W/kg. Over 600 discharge cycles were performed in tests without any serious deterioration in performance. The table at (c) lists the main parameters of the battery. The US company Unique Mobility Inc. have compared advanced lead–acid batteries with other proposed systems. In carrying out trials on an advanced EV-conversion of a Chrysler Minivan the company obtained the comparisons shown at (d). The graphs also show the extent to which the specific energy content of batteries is reduced as specific power output is increased. Trojan and Chloride 3ET205 are commercial wet acid batteries

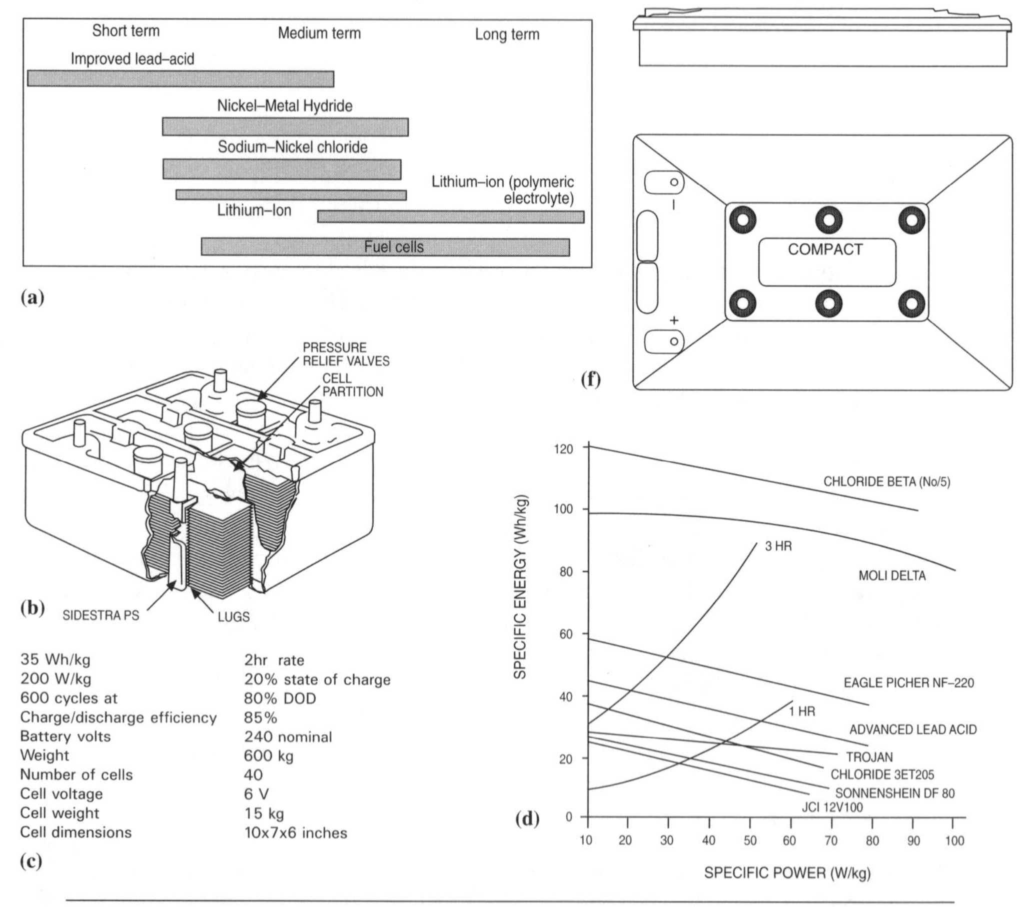

Battery Type	Module weight (lb)	Average module voltage (volts)	C/3 AMP hour	Nominal Vehicle capacity kWh	Energy density Wh/kg	Calculated EV energy density C-cycle Wh/kg	D-cycle Wh/kg	35 mph Wh/kg	55 mph Wh/kg
GC 12 V 100	68.0	11.1	72	16.0	25.9	22.1	18.0	24.9	17.0
DF8D	141.1	11.4	150	16.5	26.7	23.7	19.7	26.5	18.6
MET205	70.5	6.0	187	21.7	35.2	31.2	26.9	34.1	25.8
NI–Fe	75.0	6.3	225	33.8	54.8	54.5	50.8	57.0	49.8
NaS	1102.0	220.0	250	67.9	110.0	115.6	111.6	118.4	110.6
Adv Ph acid	141.1	11.4	225	24.7	40.1	42.9	38.7	45.9	37.5

(e)

Fig. 5.1 The lead–acid battery: (a) development time spans compared; (b) high energy lead–acid battery; (c) parameters of H-E battery; (d) battery characteristics; (e) energy-storage comparisons.

while the Sonnenshein DF80 and JCI 12V100 are gelled electrolyte maintenance-free units which involve an energy density penalty. The Eagle pitcher battery is a nickel–iron one taking energy density up to 50 Wh/kg at the 3 hour rate. The Beta and Delta units are sodium–sulphur batteries offering nominal energy density of 110 Wh/kg. Unique Mobility listed the characteristics of the batteries as at (e).

Exide's semi-bipolar technology has both high electrical performance and shape flexibility. The very low internal resistance allows high specific peak power rates and the electrode design permits ready changes in current capacity. The flat shape of the battery aids vehicle installation. The battery is assembled in a way which allows reduced need for internal connections between cells and a lightweight grid. Coated plates are stacked horizontally into the battery box. Performance is 3.9 Ah/kg and 7.4 Ah/dm^3 and shape profile is at (f).

5.2.2 SODIUM–SULPHUR

For the sodium–sulphur battery, Fig. 5.2, as used in the Ford Ecostar, the cathode of the cell is liquid sodium immersed in which is a current collector of beta-alumina. This is surrounded by a sulphur anode in contact with the outer case. The cells are inside a battery box containing a heater to maintain them at their operating temperature of 300–350°C. This is electrically powered and contained within the charge circuit. When discharging, internal resistance produces sufficient heat for the electrode but some 24 hours are required to reach running temperature from cold. In a

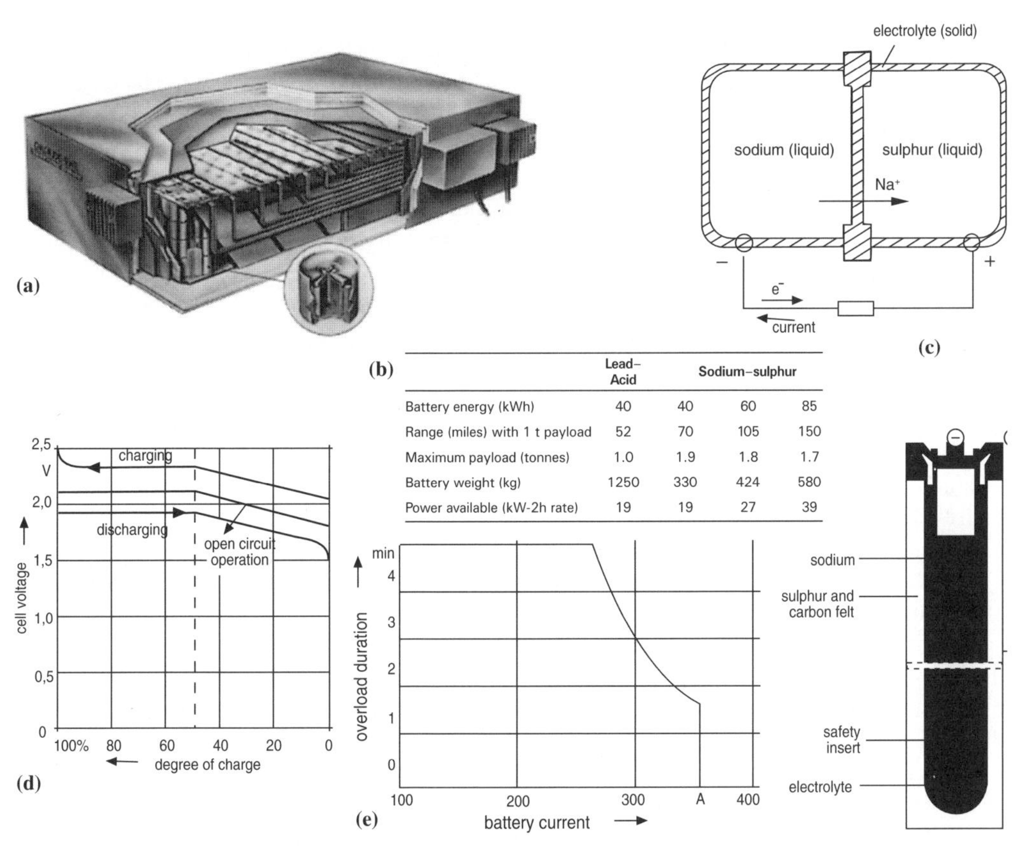

	Lead–Acid	Sodium–sulphur		
Battery energy (kWh)	40	40	60	85
Range (miles) with 1 t payload	52	70	105	150
Maximum payload (tonnes)	1.0	1.9	1.8	1.7
Battery weight (kg)	1250	330	424	580
Power available (kW-2h rate)	19	19	27	39

Fig. 5.2 Sodium–sulphur battery: (a) battery assembly; (b) projected performances; (c) ABB s-s cell; (d) energy capability; (e) overload capability.

typical EV application 100 cells would be connected in series to obtain 100 V and give a battery of 300 Ah, 60 kWh. In use, batteries would typically be charged nightly to bring them up to voltage after daily discharge and to keep the electrode molten. A typical battery installation of chloride cells is seen at (a) with expected vehicle performance, compared with lead–acid, shown at (b). The chloride cells are based on an electropheritic process while those from the Asea Brown Boveri company, used in Ecostar, are made by isostatic pressing.

The ABB cell is seen at (c); the electronic current flowing through the external load resistor during discharge corresponds to a flow of sodium ions through the electrolyte from the sodium side to the sulphur side. Voltage is from 1.78 to 2.08 V according to the degree of discharge involved. A cell with a capacity of 45 Ah has a diameter of 35 mm and length of 230 mm. Its internal resistance is 7 milli ohms and 384 cells of this type can be installed in a battery of 0.25 litre volume. An example produced by ABB has external dimensions $1.42 \times 0.485 \times 0.36$ metres. The cells account for 55% of the total weight of 265 kg. By connecting the cells in four parallel strings of 96, the battery has an open circuit voltage of 170–200 and a capacity of 180 Ah.

The electrical energy which can be drawn from the battery is shown at (d) as a function of the (constant) discharge power. With a complete discharge in 2 hours, energy content is 32 kWh, corresponding to a density of 120 Wh/kg. Associated discharge efficiency is 92%. Complete discharge at constant power is possible in a minimum of 1 hour, and an 80% discharge in less than three-quarters of an hour. The graph at (e) shows that the battery can cope with a load of up to two-thirds of the no-load voltage for a few minutes. This corresponds to a rating of about 50 kW or 188 W/kg. The portion of the heat loss not removed by the cooling system which is incorporated into the battery is stored in the heated-up cells – and covers losses up to 30 hours. Additional heat must be supplied for longer standstill periods either from the electric mains or from the battery itself. Effective, vacuum-type, thermal insulation maintains the power loss at just 80 W so that when fully charged it can maintain its temperature for 16 days. In order to maintain the battery in a state of readiness, the battery must be held above a minimum temperature and it takes about 4–10 hours to heat up the battery from cold – but a limit of 30 freeze–thaw cycles is prescribed. Life expectancy of the battery otherwise is 10 years and 1000 full discharge cycles, corresponding to an EV road distance of 200 000 km.

5.2.3 NICKEL–METAL HYDRIDE

As recently specified as an option on GM's EV1, the nickel–metal hydride alkaline battery, Fig. 5.3, was seen as a mid-term solution by the US Advanced Battery Consortium of companies set up to progress battery development. According to the German Varta company, they share with nickel–cadmium cells the robustness necessary for EV operation; they can charge up quickly and have high cycle stability. The nickel–metal hydride however, is superior, in its specifications relative to vehicle use, with specific energy and power some 20% higher and in volumetric terms 40% higher. Unpressurized hydrogen is taken up by a metallic alloy and its energy then discharged by electrochemical oxidation. The raw material costs are still signalling a relatively high cost but its superiority to lead–acid is likely to ensure its place as its associated control system costs are lower than those of sodium sulphur. Specific energy is 50–60 Wh/kg, energy density 150–210 Wh/litre, maximum power more than 300 W/kg; 80% charge time is 15 minutes and more than 2000 charge/discharge cycles can be sustained.

The negative electrode is a hydrogen energy-storage alloy while nickel hydroxide is the positive electrode. An optimum design would have weight around 300 kg, and capacity of 15 kWh, with life of 2000 discharge cycles. For buses Varta have devised a mobile charging station, in cooperation with Neoplan, which will allow round-the-clock operation of fleets. This removes the need for

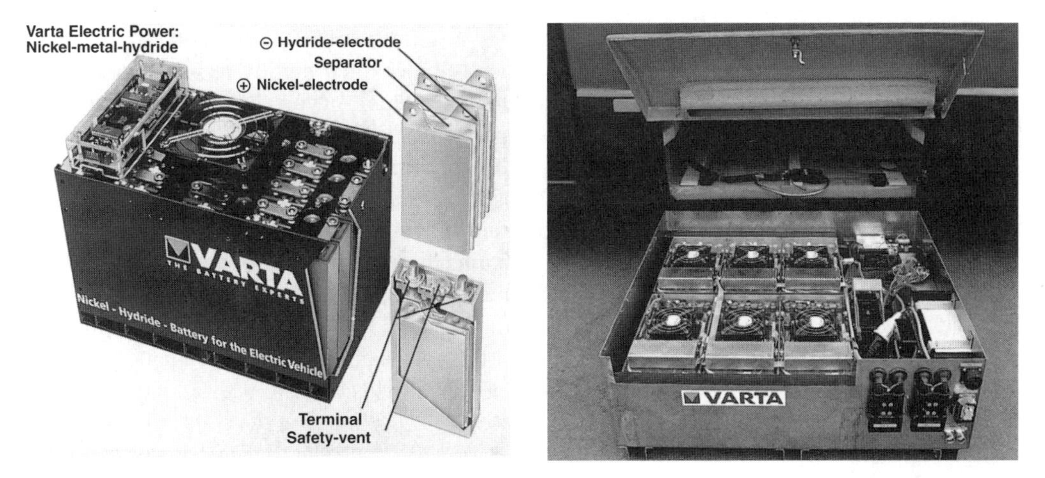

Fig. 5.3 Nickel–metal hydride battery.

fixed sites and allows battery charging and changing to be carried out by the bus driver in a few minutes. The mobile station is based on a demountable container which can be unloaded by a conventional truck. Trials have shown that a bus covering a daily total distance of 75 miles on a three-mile-long route needs to stop at the station after eight journeys. Discharged batteries are changed semiautomatically on roller-belt arms, by a hand-held console.

5.2.4 SODIUM CHLORIDE/NICKEL

Sodium chloride (common salt) and nickel in combination with a ceramic electrolyte are used in the ZEBRA battery, Fig. 5.4, under development by Beta Research (AEG and AAC) and Siemens. During charging the salt is decomposed to sodium and nickel chloride while during discharge salt is reformed. Its energy density of 90 Wh/kg exceeded the target set by the USA Advanced Battery Consortium (80 Wh/kg energy density, to achieve 100 miles range under any conditions and 150 W/kg peak power density to achieve adequate acceleration) and can achieve 1200 cycles in EV operation, equivalent to an 8 year life, and has a recharge time of less than 6 hours. The USABC power to energy ratio target of 1.5 was chosen to avoid disappointing short-range high power discharge of a ZEV battery and for a hybrid vehicle a different ratio would be chosen.

Each cell is enclosed in a robust steel case with electrodes separated by a β-ceramic partition which conducts sodium ions but acts as a barrier to electrons, (a). The melt of sodium/aluminium chloride conducts sodium ions between the inner ceramic wall and into the porous solid $Ni/NiCl_2$ electrode. As a result, the total material content is involved in the cell reaction. Apart from the main reversible cell reaction there are no side reactions so that the coulometric efficiency of the cell is 100%. The completely maintenance-free cells are hermetically sealed using a thermal compression bond (TCB) ceramic/metal seal.

The cell type SL09B presently produced in the pilot production line has an open-circuit voltage of 2.58 V at 300°C with a very low temperature coefficient of 3×10^{-4} V/K, a capacity of 30 Ah and an internal resistance that varies between 12 and 25 mW, dependent on temperature, current and rate of discharge. This variation is because, during the charging and discharging process, the electrochemical reaction zone moves from the inner surface of the β-ceramic electrolyte into the solid electrode. During this process the length of the sodium ion path and the current-density in the reaction zone increases and so the internal resistance increases. In principle this effect is used

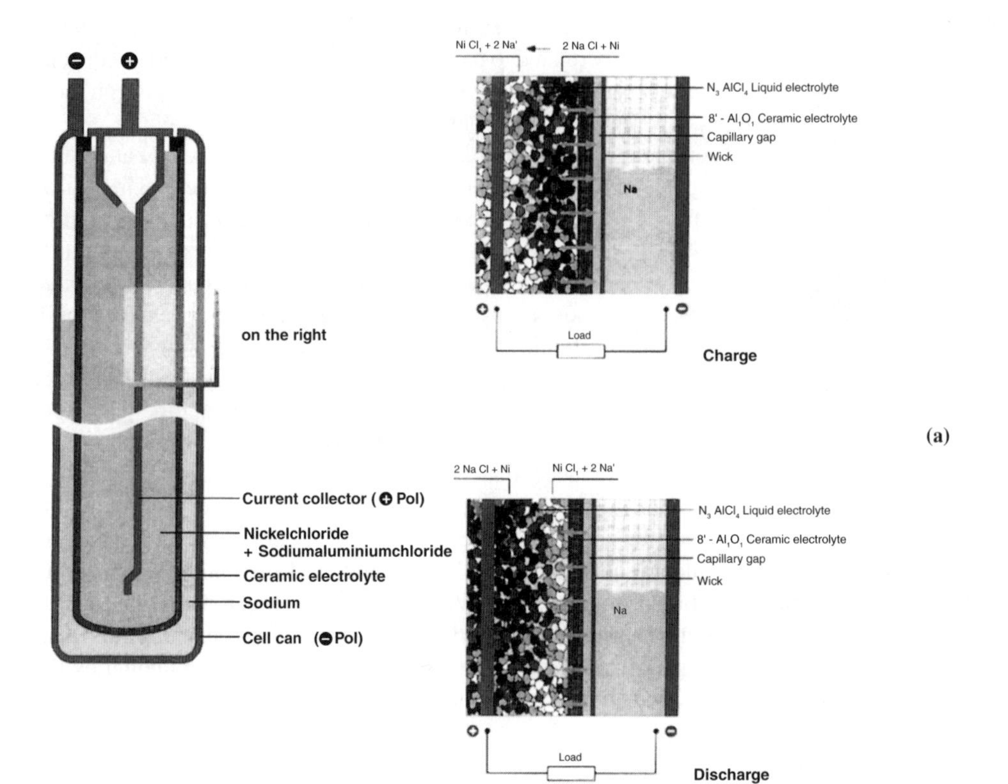

	Z5	Z11 (preliminary)
Dimensions 1 x w x h[1]	730 x 541 x 315 mm	933 x 665 x 315 mm
Weight battery [1]	194 kg	310
Weight accessories[2]	6 kg	about 10
Cell type	SL09B	ML1
Thermal losses at 270°C	max. 125 W	170 W
Rapid charging	75% in 45 min	75% in 45 min
Rated energy	17 kWh	29 kWh
Energy density	88 Wh/kg	94 Wh/kg
Peak Power (80% DOD, $^2/_3$ OCV, 30s)	15 kW	42 kW
Peak Power Density (80% DOD, $^2/_3$ OCV)	75 W/kg	135 W/kg
OCV	284/188/142 V	302 V
Capacity	60/90/120 Ah	96 Ah

(c)

Cooling
Vaccum insulation
R_1
+
Current terminals
–
Electric heater
(b)

Fig. 5.4 ZEBRA battery: (a) cell; (b) cell-box; (c) performance comparison.

to enable a stable operation of parallel connected strings of cells. But from the vehicle point of view the available power which is directly related to the internal cell resistance should not depend on the battery charge status. The redesigned cell type ML1 is a good compromise between these two requirements. The battery is operated at an internal temperature range of 270–350° C.

The cells are contained in a completely sealed, double walled and vacuum-insulated battery box as shown at (b). The gap between the inner and outer box is filled with a special thermal insulation material which supports atmospheric pressure and thus enables a rectangular box design to be utilized. In a vacuum better than 1.10^{-1} mbar this material has a heat conductivity as low as 0.006 W/mK. By this means the battery box outside temperature is only 5–10° C above the ambient temperature, dependent on air convection conditions. Cooling systems have been designed, built and tested using air cooling as well as a liquid cooling. The latter is a system in which high temperature oil is circulated through heat exchangers in the battery with an oil/water heat exchanger outside the battery. By this means heat from the battery can be used for heating the passenger room of the vehicle.

In the ML1 cell, internal resistance is reduced to increase power. The resistance contribution of the cathode is due to a combination of the ion conduction between the inner surface of the β-aluminium ceramic with the reaction zone (80%) and electric conduction between the reaction zone and the cathode current collector (20%). The ML1 has a cloverleaf section shape ceramic to enlarge its surface area over the normal circular section, with resultant twofold reduction in cathode thickness and 20% reduction in resistance. Based on this form of cell construction a new, Z11, battery has been produced with properties compared with the standard design as shown by the table at (c) and the battery is under development for series production.

5.2.5 SOLAR CELLS

According to Siemens, solar technology is a probable solution for Third World tropical countries. Solar modules are available from the company to supply 12 V, 100 Ah batteries from a 50 W solar module. The company recently installed a system on the Cape Verde Islands with a collective power output of 550 kW at each of five island sites. Even in Bavaria, the village of Flanitzhutte, which has an average 1700 hours annual sunshine period, has severed its links with the national grid with the installation of 840 solar modules, with a total area of 360 square metres, to provide peak power of 40 kW. Maintenance-free batteries provide a cushion.

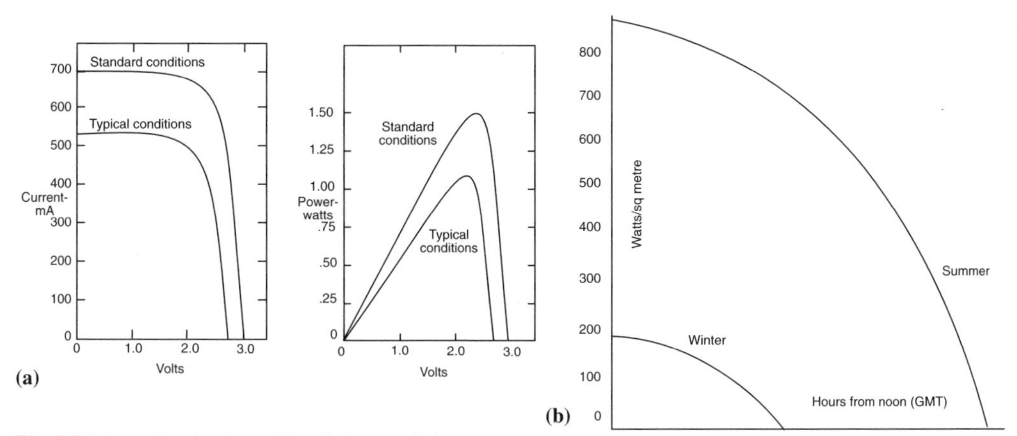

Fig. 5.5 Solar cell technology: (a) cell characteristics; (b) solar energy variation.

The technology of solar cells, Fig. 5.5, has been given a recent boost by the Swiss Federal Institute of Technology who claim to have outperformed nature in the efficiency of conversion of sunlight to electricity even under diffuse light conditions. The cell has a rough surface of titanium dioxide semiconductor material and is 8% efficient in full sunlight rising to 12% in diffuse daylight. For more conventional cells, such as those making a Lucas solar panel, these are available in modules of five connected in series to give maximum output of 1.3 watts (0.6 A at 2.2 V). Some ten modules might be used in a solar panel giving 13 watts output in summer conditions. Power vs voltage and current vs voltage are shown at (a) for so-called 'standard' and 'typical' operating conditions. 100 mW/cm^2 solar intensity, 0° C cell temperature at sea level defines the standard conditions against 80 mW/cm^2 and 25° C which represent 'typical' conditions at which power output per cell drops to 1 W. Temperature coefficients for modules are 0.45% change in power output per 1° C rise in temperature, relative to 0° C; cell temperatures will be 20° C above ambient at 100 mW/cm^2 incident light intensity. Variation of solar energy at 52° north latitude, assuming a clear atmosphere, is shown at (b). On this basis the smallest one person car with a speed of 15 mph and a weight of 300 lb with driver would require 250 W or 50 ft^2 (4.65 m^2) of 5% efficient solar panel – falling to 12.5 ft^2 (1.18 m^2) with the latest technology cells. A 100 Wh sealed nickel–cadmium battery would be fitted to the vehicle for charging by the solar panel while parked.

The future, of course, lies with the further development of advanced cell systems such as those by United Solar Systems in the USA. Their approach is to deposit six layers of amorphous silicon (two identical n-i-p cells) onto rolls of stainless steel sheet. The 4 ft^2 (0.37 m^2) panels are currently 6.2% efficient and made up of layers over an aluminium/zinc oxide back reflector. The push to yet higher efficiencies comes from the layer cake construction of different band-gap energy cells, each cell absorbing a different part of the solar spectrum. Researchers recently obtained 10% efficiency in a 12 in^2 (0.09 m^2) module.

Rapid thermal processing (RTP) techniques are said to be halving the time normally taken to produce silicon solar cells, while retaining an 18% energy conversion efficiency from sunlight. Researchers at Georgia Institute of Technology have demonstrated RTP processing involving a 3 minute thermal diffusion, as against the current commercial process taking 3 hours. An EC study has also shown that mass production of solar cells could bring substantial benefits and that a £350 million plant investment could produce enough panels to produce 500 MW annually and cut the generating cost from 64 p/kWh to 13p.

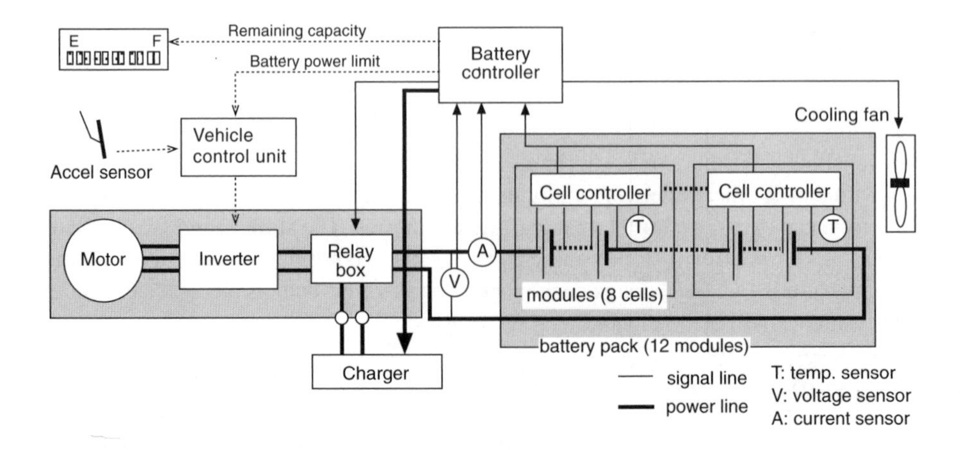

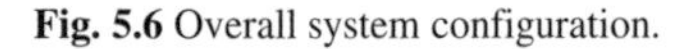
Fig. 5.6 Overall system configuration.

5.2.6 LITHIUM–ION

A high energy battery receiving considerable attention is the lithium–ion cell unit, the development of which has been described by Nissan and Sony engineers[1] who point out that because of the high cell voltage, relatively few cells are required and better battery management is thus obtained. Accurate detection of battery state-of-charge is possible based on voltage measurement. In the battery system developed, Fig. 5.6, cell controllers and a battery controller work together to calculate battery power, and remaining capacity, and convey the results to the vehicle control unit. Charging current bypass circuits are also controlled on a cell-to-cell basis. Maximizing lifetime performance of an EV battery is seen by the authors to be as important as energy density level. Each module of the battery system has a thermistor to detect temperature and signal the controllers to activate cooling fans as necessary.

Nissan are reported to be launching the Ultra EV in 1999 with lithium–ion batteries; the car is said to return a 120 mile range per charge. Even further into the future lithium–polymer batteries are reported to be capable of giving 300 mile ranges.

5.2.7 SUPERCAPACITORS

According to researchers at NEC Corp.[2], the supercapacitor, Fig. 5.7, will be an important contributor to the energy efficient hybrid vehicle, the absence of chemical reaction allowing a durable means of obtaining high energy charge/discharge cycles. Tests have shown for multi-stop vehicle operations a 25–30% fuel saving was obtained in a compact hybrid vehicle fitted with regenerative braking. While energy density of existing, non-automotive, supercapacitors is only about 10% of that of lead–acid batteries, the authors explain, it is still possible to compensate for some of the weak points of conventional batteries. For effective power assist in hybrids, supercapacitors need a working voltage of over 100 V, alongside low equivalent series resistance and high energy density. The authors have produced 120 V units operating at 24 kW fabricated from newly developed activated carbon/carbon composites. Electric double layer capacitors (EDLCs) depend on the layering between electrode surface and electrolyte, (a) showing an EDLC model. Because energy is stored in physical adsorption/desorption of ions, without chemical reaction, good life is obtained. The active carbon electrodes usually have a specific surface area over 1000 m^2/g and double-layer capacitance is some 20–30 $\mu F/cm^2$ (activated carbon has capacitance over 200–300 F/g). The EDLC has two double layers in series, so it is possible to obtain 50–70° F using a gram of activated carbon. Working voltage is about 1.2 V and storable energy is thus 50 J/g or 14 Wh/kg.

The view at (b) shows a cell cross-section, the conductive rubber having 0.2 S/cm conductivity and thickness of 20 microns. The sulphuric acid electrolyte has conductivity of 0.7 S/cm. The view at (c) shows the high power EDLC suitable for a hybrid vehicle, and the table at (d) its specification. Plate size is $68 \times 48 \times 1$ mm^3 and the weight 2.5 g, a pair having 300 F capacity. The view at (e) shows constant power discharge characteristics and (f) compares the EDLC's energy density with that of other batteries. Fuji Industries' ELCAPA hybrid vehicle, (g), uses two EDLCs (of 40 F total capacity) in parallel with lead–acid batteries. The stored energy can accelerate the vehicle to 50 kph in a few seconds and energy is recharged during regenerative braking. When high energy batteries are used alongside the supercapacitors, the authors predict that full competitive road performance will be obtainable.

5.2.8 FLYWHEEL ENERGY STORAGE

Flywheel energy storage systems for use in vehicle propulsion has reached application in the light tram vehicle discussed in the Introduction (pages xiii, xiv). They have also featured in pilot-production vehicles such as the Chrysler Patriot hybrid-drive racing car concept. Here, flywheel energy storage is used in conjunction with a gas turbine prime-mover engine, Fig. 5.8. The drive was developed by Satcon Technologies in the USA to deliver 370 kW via an electric motor drive to the road wheels. A turbine alternator unit is also incorporated which provides high frequency current generation from an electrical machine on a common shaft with the gas turbine. The flywheel is integral with a motor/generator and contained in a protective housing affording an internal

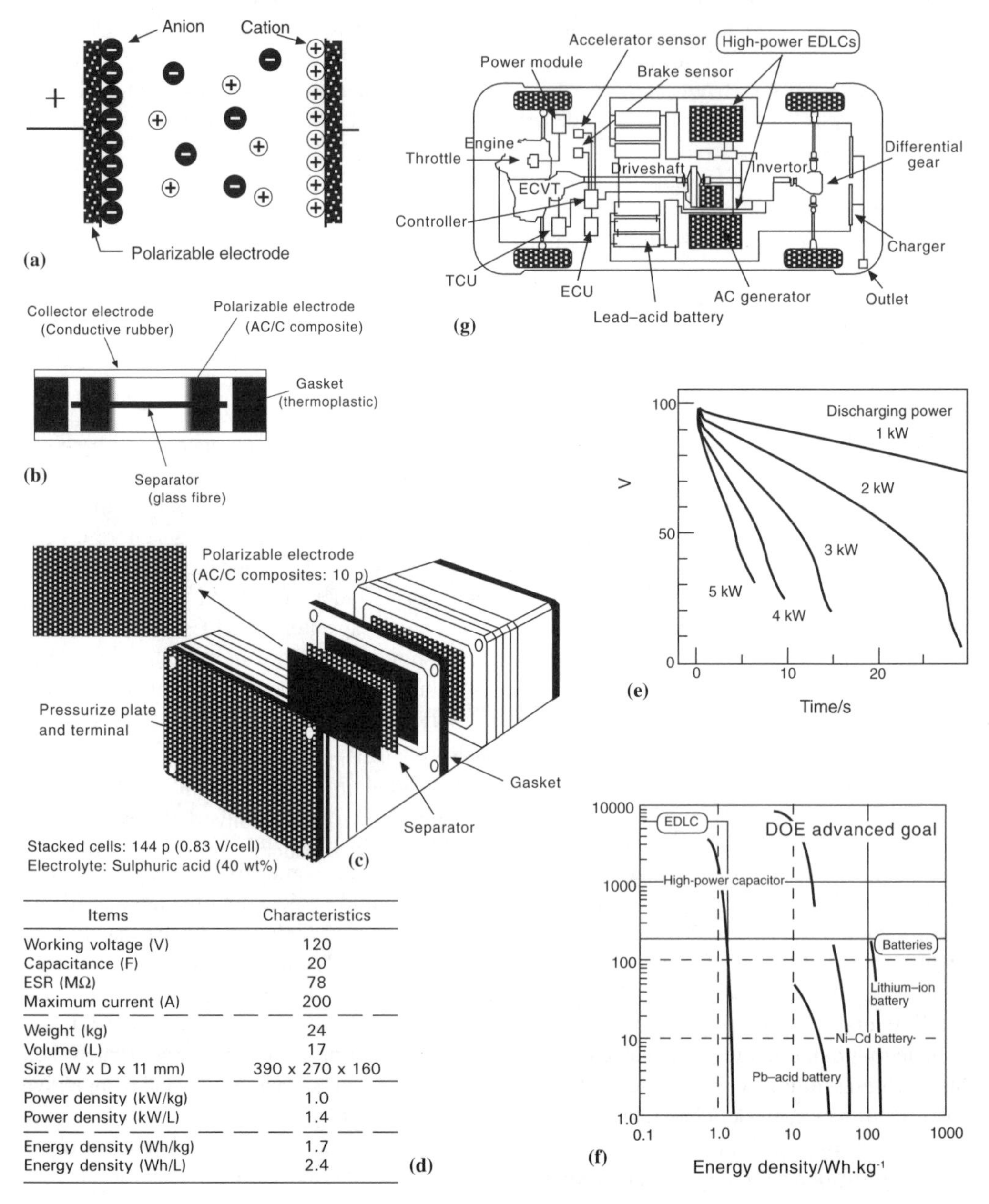

Items	Characteristics
Working voltage (V)	120
Capacitance (F)	20
ESR (MΩ)	78
Maximum current (A)	200
Weight (kg)	24
Volume (L)	17
Size (W x D x 11 mm)	390 x 270 x 160
Power density (kW/kg)	1.0
Power density (kW/L)	1.4
Energy density (Wh/kg)	1.7
Energy density (Wh/L)	2.4

(d)

Fig. 5.7 Supercapacitors: (a) EDLC model; (b) cell; (c) high-power EDLC schematic; (d) HP EDLC specification; (e) constant power discharge characteristics; (f) power density, *y*-axis in W/kg, vs energy density for high-power EDLC; (g) ELCAPA configuration.

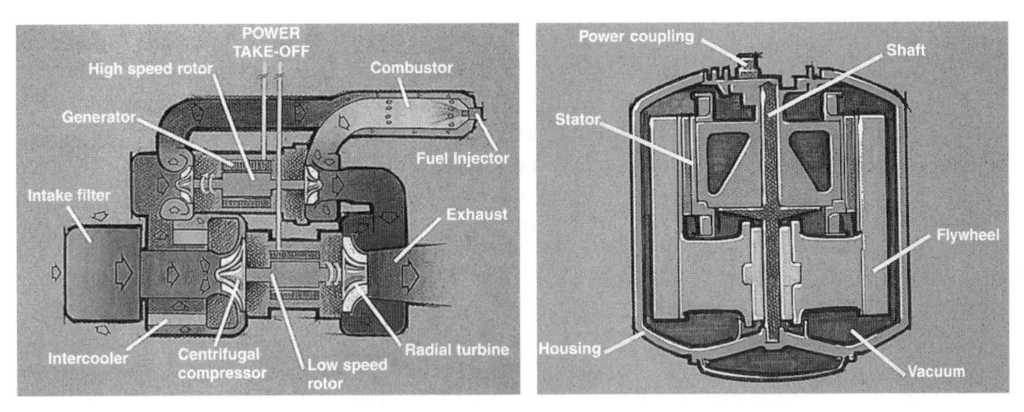

Fig. 5.8 Chrysler Patriot flywheel energy storage system: left, turbine; right, flywheel.

vacuum environment. The 57 kg unit rotates at 60 000 rpm and provides 4.3 kW of electrical energy. The flywheel is a gimbal-mounted carbon-fibre composite unit sitting in a carbo-fibre protective housing. In conjunction with its motor/generator it acts as a load leveller, taking in power in periods of low demand on the vehicle and contributing power for hill climbing or high acceleration performance demands.

European research work into flywheel storage systems includes that reported by Van der Graaf at the Technical University of Eindhoven[3]. Rather than using continuously variable transmission ratio between flywheel and driveline, a two-mode system is involved in this work. A slip coupling is used up to vehicle speeds of 13 km/h, when CVT comes in and upshifts when engine and flywheel speed fall simultaneously. At 55 km/h the drive is transferred from the first to the second sheave of the CVT variator, the engine simultaneously being linked to the first sheave. Thus a series hybrid drive exists at lower speeds and a parallel hybrid one at higher speeds. The 19 kg 390 mm diameter composite-fibre flywheel has energy content of 180 kW and rotates up to 19 000 rpm.

5.3 Battery car conversion technology

For OEM conversions of production petrol-engined vehicles the decades up to the 1970s, and up to the present day for aftermarket conversions, is typified by that used by many members of the UK Battery Vehicle Society and documented by Prigmore *et al*[4]. Such conversions rely on basic lead–acid batteries available at motor factors for replacement starter batteries. A ton of such batteries, at traction power loading of 10–15 kW/ton, stores little more than 20 kWh. Affordable motors and transmissions for this market sector have some 70% efficiency, to give only 14 kWh available at the wheels.

5.3.1 CONVERSION CASE STUDY

The level-ground range of the vehicle can be expressed in terms of an equivalent gradient *1: h*, representing rolling resistance, such that a resistance of 100 kgf/tonne is equivalent to a gradient of 1:10. If the fraction of the total vehicle weight contributed by the battery is f_b then range is given by $\{(14 \times 3600)/(9.81 \times 1000)\}f_b h$. Pessimistically h is about 30 at 50 km/h and if f_b is 0.4 then cruising range would be about 60–65 km. This of course is reduced by frequent acceleration and braking.

Series-wound DC motors, Fig. 5.9, have been chosen for low cost conversions because of their advantageous torque/speed characteristics, seen at (a), given relatively low expected road speeds.

Series winding of field and armature the same current is carried by both and as it increases in magnitude so does the magnetic flux and the torque increases more than proportionally with current. Rotation of the armature creates a back-EMF in opposition to the applied voltage because the wires at the edge of the armature are moving across the field flux. Motors are designed to equalize applied and back-EMFs at operational speed. This will be low when field current is high and vice versa. The speed/current curve can be made to move up the x-axis by reducing the field current to a fixed fraction of the armature current (0.5–0.7), with the help of the field diverter resistance

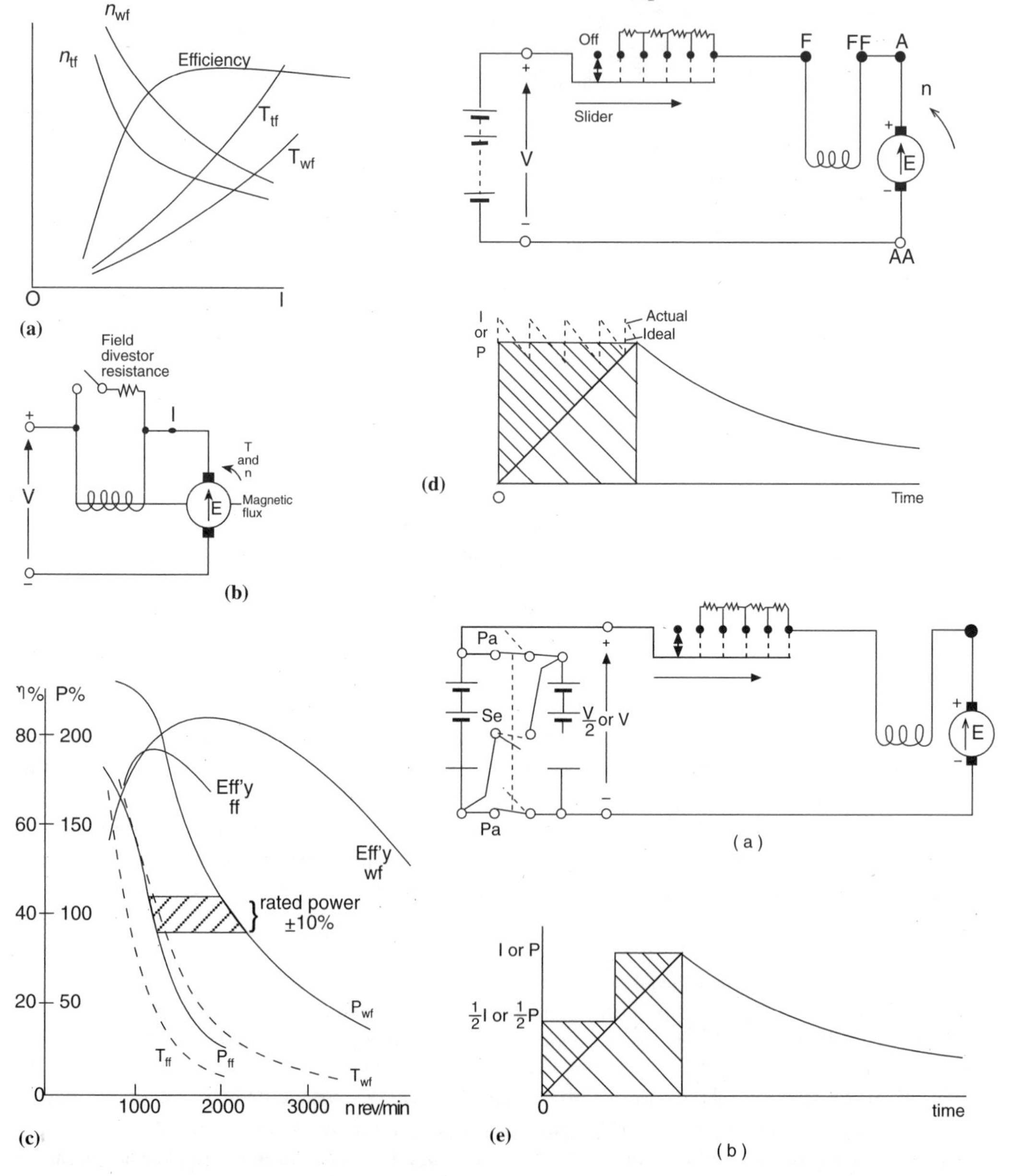

Fig. 5.9. Series-wound DC motor: (a) motor characteristics; (b) field diverter resistance; (c) speed-base motor characteristics; (d) rheostatic control; (e) parallel/series battery control.

shown but the torque for a given armature current is, of course, reduced, see (b). The efficiency of the motor is low at low speeds, in overcoming armature inertia, and again at high speeds as heating of the windings absorbs input power. Motors can thus be more highly rated by the provision of cooling fans. Average power in service should in general be arranged at 0.8 of the rated power and the transmission gear ratio be such that the motor is loaded to no more than its rated power for level-ground cruising. The motor characteristics shown at (c) are obtained by replotting the conventional characteristics on a speed base. The wide range of speeds available (up to 2:1) are around rated power and show how full field can be used for uphill running while weak field is used on the level enabling speed reduction to compensate for torque increase in limiting battery power requirements for negotiating gradients.

With little or no back-EMF to limit current at starting, resistance is added to keep the current down to a safe level, as at (d). The current is maintained at the required accelerating value, perhaps 2–4 times rated current. The starting resistance is reduced as the motor gains speed so as to keep the accelerating current constant to the point where the starting resistance is zero, at the 'full voltage point'. Thereafter a small increase in speed causes gradual reduction in current to the steady running value. As the current is supplied from the battery at constant voltage, the current curve can be rescaled as a power curve to a common time base, as at (e). The shaded area then gives energy taken during controlled acceleration with the heavily shaded portion showing the energy wasted in resistance. So rheostatic acceleration has an ideal efficiency of about 50% up to full voltage. This form of control is thus in order for vehicle operation involving, say, twice daily regular runs under cruise conditions but unwise for normal car applications.

5.3.2 MOTOR CONTROL ALTERNATIVES

Alternatives such as parallel/series (two-voltage) rheostatic control, or weak field control, can be better for certain applications, but the more elaborate thyristor, chopper, control of motor with respect to battery (Fig. 5.10) is preferred for maintaining efficiencies with drivers less used to electric drive, particularly in city-centre conditions. It involves repetitive on-off switching of the battery to the motor circuit and if the switch is on for a third of the time, the mean motor voltage is a third of the supply voltage (16 V for a 48 V battery), and so on, such that no starting resistance is needed. Effective chopper operation requires an inductive load and it may be necessary to add such load to the inherent field inductance. Because an inductive circuit opposes change in current then motor current rises relatively slowly during 'on' periods and similarly falls slowly during 'off' periods, provided it has a path through which to flow. The latter is provided by the 'flywheel' diode FD, a rectifier placed across the motor to oppose normal voltage. During chopper operation, current i_b flows in pulses from battery to motor while current i_m flows continuously through the motor. Electronic timing circuits control the switching of the thyristors, (a).

Single ratio drives from motor to driveline are not suitable for hilly terrain, despite the torque/speed characteristic, as the motor would have to be geared too low to avoid gradient overloading and thus be inefficient at cruise. A 5:1 CVT drive is preferred so that the motor can be kept at its rated power under different operating conditions. There is also a case for dispensing with the weight of a conventional final drive axle and differential gear by using two, say 3 kW, motors one for each driven wheel.

The behaviour of lead–acid batteries, (b), is such that in the discharged condition lead sulphate is the active material for both cell-plates which stand in dilute sulphuric acid at 1.1 specific gravity. During charging the positive plate material is converted to lead peroxide while that of the negative plate is converted into lead, as seen at (c). The sulphuric acid becomes more concentrated in the process and rises to SG = 1.5 when fully charged, the cells then developing over 2 volts. In discharge the acid is diluted by the reverse process. While thin plates with large surface area are

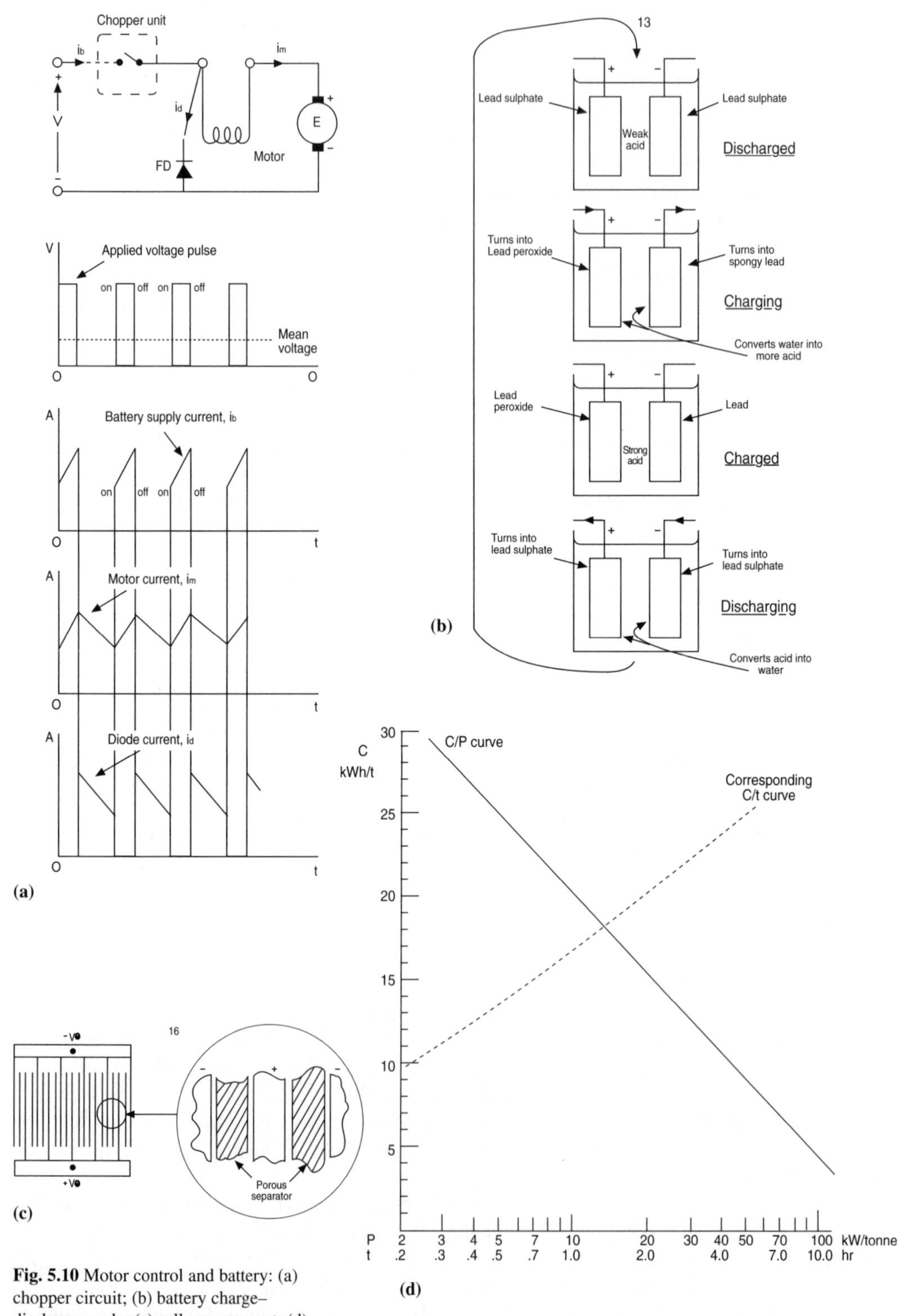

Fig. 5.10 Motor control and battery: (a) chopper circuit; (b) battery charge–discharge cycle; (c) cell arrangement; (d) battery time-of-discharge curves.

intended for batteries with high discharge rates, such as starter batteries, the expansion process of the active material increases in volume by three times during discharge and the active material of very thin plates becomes friable in numerous charge/discharge cycles, and a short life results. Normal cells, (b), comprise interleaved plates with porous plastic separators; there is one more negative than positive plates, reducing the tendency to buckle on rapid discharge. Expensive traction batteries have tubular plates in some cases with strong plastic tubes as separators to keep the active material in place. Discharge rates of less than half the nominal battery capacity in amp-hours are necessary to preserve the active material over a reasonable life-span, but short bursts at up to twice the nominal rate are allowable. The graphs at (d) permit more precise assessments of range than the simple formula at the beginning of the section which assumes heavy discharge causes battery capacity to be reduced by 70–80% of normal, 25 kWh becoming 20.

When charging the gassing of plates must be considered, caused by the rise in cell voltage which causes part of the current to electrolyse the water in the electrolyte to hydrogen. Gassing commences at about 75% full charge. At this point, after 3–4 hours' charging at 1/15th battery capacity, the rate should be decreased to 1/20th and carried on until 2.6 volts are shown at the cells. To ensure near-complete removal of the sulphate a periodic 'soak charge' should be provided for several hours until peak voltage remains steady at 2.6–2.8 V, with all cells gassing freely and with constant specific gravity. Such a charge should be followed by topping up with distilled water.

5.4 EV development history

According to pioneer UK EV developer and producer Geoffrey Harding[5], the Lucas programme was a major event in the renaissance of the electric vehicle. He set up a new Lucas Industries facility to develop battery EVs in 1974 because, as a major transport operator, he had asked Lucas to join him in an approach to a UK government department for some financial assistance to build a battery electric bus which would operate on a route between railway stations in Manchester. The reasons for his interest in this project were twofold. First, there was a major problem with the reliability of many of the diesel buses at that time and he wanted to find out whether electric buses would live up to the attributes of good reliability and minimal maintenance that had been afforded to EVs for many years. Second, a world shortage of oil at that time was causing an apparent continuous and alarming increase in the price.

Having subsequently joined Lucas and set up the new company, he was responsible for building the electric bus in question and providing technical support when it entered service. The bus – the performance of which was comparable with diesel buses, except for range – operated successfully for some years and was popular with both passengers and drivers. On the other hand, it was not popular with schedulers because its restricted range (about 70 km in city service) added yet more limitation to its uses, particularly at weekends. Nevertheless, much was learnt from the in-service operation of this vehicle which proved to be remarkably reliable. He then obtained agreement within Lucas that the battery EV most likely to succeed at that time was a 1-tonne payload van because it would be possible, with relatively minor changes to production vans, to modify the drive to battery electric without reducing either the payload volume or the weight, Fig. 5.11.

The converted Bedford vehicles underwent a significant testing programme on that company's test track, and were in fact built on the company's ICE van production line, interspersed between petrol- and diesel-powered versions of CF vans. This method of production was the first of its kind. Some hundreds of Bedford vans and a smaller number of Freight Rover vans were built and sold, all with a working range in excess of 80 km in city traffic, a payload of just under 1 tonne, an acceleration of 0–50 km in 13 s, a maximum speed of 85 km/h, and a battery design life of 4 years.

The vehicles had, for that time, sophisticated electronic controllers and DC/DC converters, as well as oilfired heating and demisting systems. Lucas designed and constructed the chargers and battery-watering systems. Some were sold in the USA as the GM Griffon, (a), and it was estimated that collectively their total service had exceeded 32 million km, and even today a few are still operating.

The Lucas Chloride converted Bedford CF van had two-pedal control and a simple selector for forward/reverse. Most of the vehicle's braking was regenerative and batteries were of the tubular cell lead–acid type. Thirty-six monobloc units of 6 volts were used – connected in series to give a

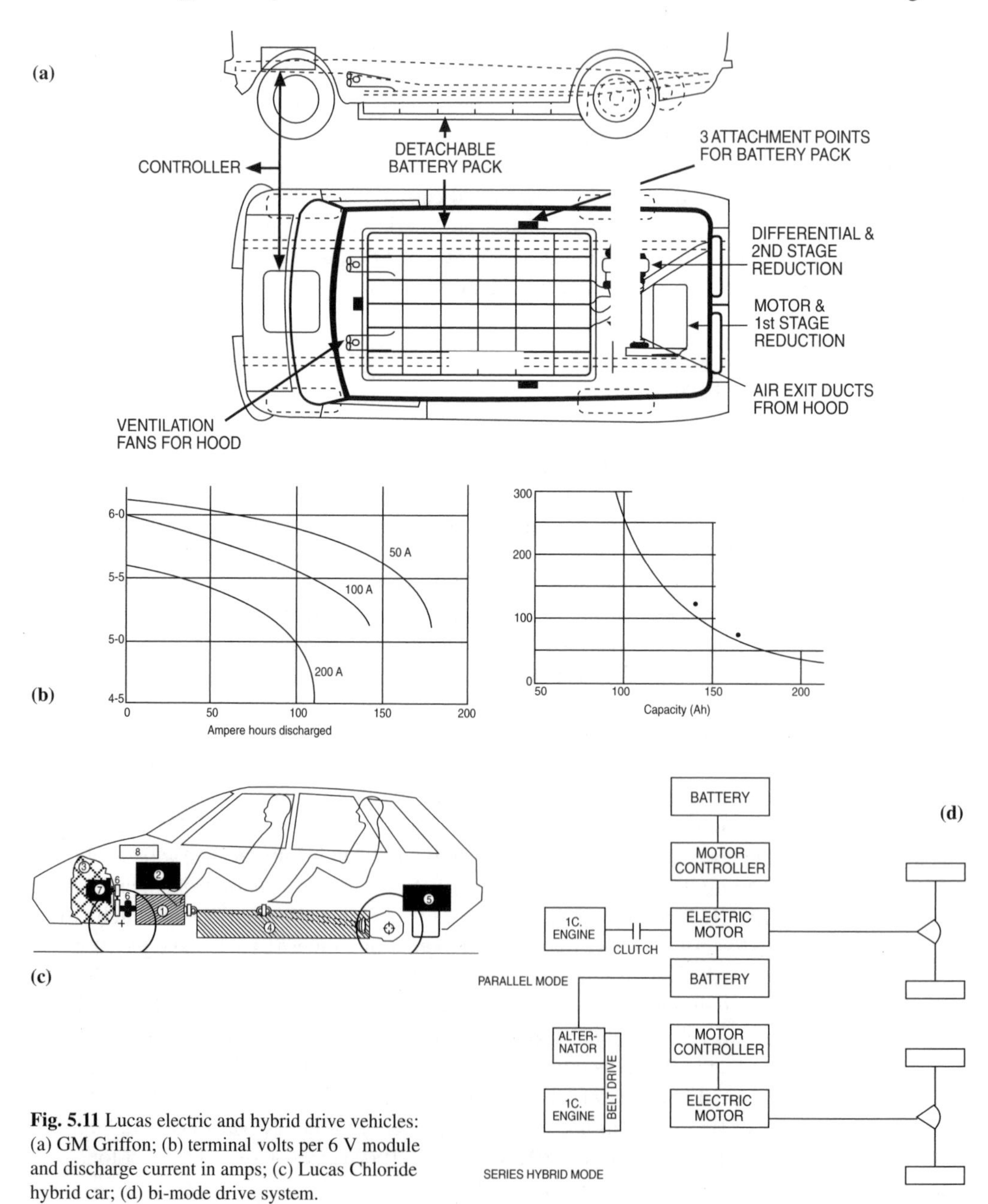

Fig. 5.11 Lucas electric and hybrid drive vehicles: (a) GM Griffon; (b) terminal volts per 6 V module and discharge current in amps; (c) Lucas Chloride hybrid car; (d) bi-mode drive system.

216 V, 188 Ah pack. The rear-mounted traction motor drove the wheels through a primary reduction unit coupled to a conventional rear axle, via a prop-shaft. Measured performance of the monobloc is shown at (b); an energy density of 34 Wh/kg was involved, at the 5 hour rate, and a 4 year service life was claimed. The motor used was a separately excited type in order to allow the electronics maximum flexibility in determining the power curve. It weighed 15 kg and had a controlled output of 40 kW; working speed was 6100 rpm corresponding to a vehicle speed of about 60 mph. The motor control system used an electronic bypass to leave the main thyristor uncommutated during field control. The latter uses power transistors which handle up to 25 A.

Within the Lucas development programme, which at one time employed close to 100 personnel, some work on HEVs was undertaken and one five-seat passenger car was designed and built. This utilized an electric Bedford drive system and could be operated either as a series hybrid or a parallel hybrid. The car had a maximum speed of 130 kph, and a pure-electric range of about 70 km. The Lucas Chloride hybrid, (c), has engine (3) driving through the motor (1) but midships positioning of the batteries (4) with on-board charger (5) at the rear. Clutches are shown at (6) while (7) and (8) are alternator and control unit. This used Reliant's 848 cc engine developing 30 kW alongside a 50 kW Lucas CAV traction motor. The 216 V battery set had capacity of 100 amp-hour on a 5 hour rate. Maximum speed in electric drive of 120 km/h rises to 137 km/h in combined mode, (d).

5.4.1 ELECTRIC VEHICLE DEVELOPMENT 1974–1998

In considering the changes which have taken place in the quarter century since the start of the Lucas project, Harding argues that the developments which have taken place in electric cars are not as great as had been hoped and expected. Some hybrids, he considers, are effectively ICEVs with an electric drive which assists when required. A major problem with HEVs has been their cost, which is exacerbated by having two drive systems in one vehicle. Fortunately, the automotive industry is so good at meeting challenges of this nature that who can say what can be achieved? However, it is claimed that micro-turbines together with their associated generators and accessories can be produced cheaply, mainly because they have a very low component count. These turbines are capable of operating on a wide variety of fuels and are considered to produce a very low level of pollutants, but with one or two exceptions such as Volvo and Chrysler, these claims have not been subjected to any extensive field testing. If what is claimed proves to be true, then such vehicles would be expected to play a large part in the transport scene in the new millennium.

At present, the great hope for the future, he believes, is the fuel cell. Hydrogen is the preferred fuel for fuel cells but its storage presents a problem. One of the ways of overcoming this problem is to convert a liquid fuel, such as methanol, into hydrogen. This was done in the 5 kW unit made by the Shell Oil Company as long ago as 1964. The unit was installed in the world's first fuel-cell powered car. Shell also produced a 300 W nett cell in 1965 which converted methanol directly into electricity, so it is not the case that this technology is new. The principal problem at the time this work was carried out was the cost of the unit. Although a number of fuel-cell powered cars

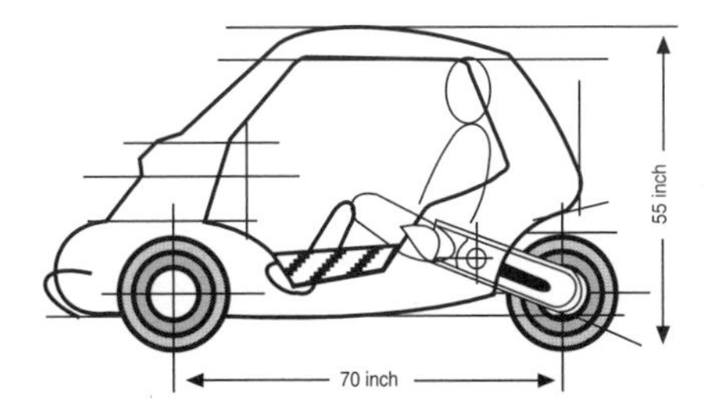

Fig. 5.12 Sinclair C10 proposal.

have been built recently by automobile manufacturers, the only vehicle so far offered for sale is the Zevco London taxi which was launched in London in July 1998. The propulsion system is a hybrid arrangement: a battery drives the vehicle and is recharged by a 5 kW fuel cell. The vehicle uses bottled hydrogen as fuel and has a service range of 145 km, and a performance similar to its diesel counterpart. This design works well because the stop-start nature of the traffic provides time for the low output of the fuel cell to replenish the energy drawn from the battery during previous spells of vehicle motion. At a later date, this type of taxi may be fitted with a cryogenic hydrogen-storage system, perhaps placed between the two layers of a sandwich-floor construction of the vehicle. With such an arrangement, it is expected that the fuel cell would be refuelled with very cold liquid hydrogen in minutes and, thereby, would extend the vehicle's range dramatically, but only in stop-start traffic.

Harding opines that what the world really needs are vehicles fitted with fast-response, high-output fuel cells together with on-board clean reformers which would enable a liquid fuel to be turned into hydrogen on vehicles. Initially, the most likely liquid fuel would seem to be methanol, but arranging for methanol to be widely available would necessitate some large changes in infrastructure. If all this is possible, then refuelling vehicles with liquid fuel would be, in principle, little or no different from today. The eventual aim is said, by those developing high-output fuel cells, to be the development of reformers which can produce hydrogen from gasoline. In this case, only the current gasoline infrastructure would be required. Interest and investment in fuel cells is increasing, and the joint arrangements between the Canadian fuel cell company Ballard and motor industry giants Mercedes and Ford would appear to be an almost irresistible force on a course aimed at solving some daunting problems. The Ballard unit is a proton exchange membrane (PEM) fuel cell and amongst early examples of road vehicles fitted with this are buses in the USA. Quite apart from the technical problems still to be resolved, the problem of cost is very great.

5.5 Contemporary electric car technology

According to Sir Clive Sinclair, whose abortive efforts to market an electric tricycle have led him to concentrate on economical bicycle conversions, peak efficiencies of 90% are available with EVs for converting electricity into tractive energy – and that attainable electrical generating efficiencies of over 50% meant a 45% fuel conversion efficiency could be obtained compared with 30% for the petrol engine. His C10 proposal shown in Fig. 5.12 must mean his faith in the future of the electric car is still maintained.

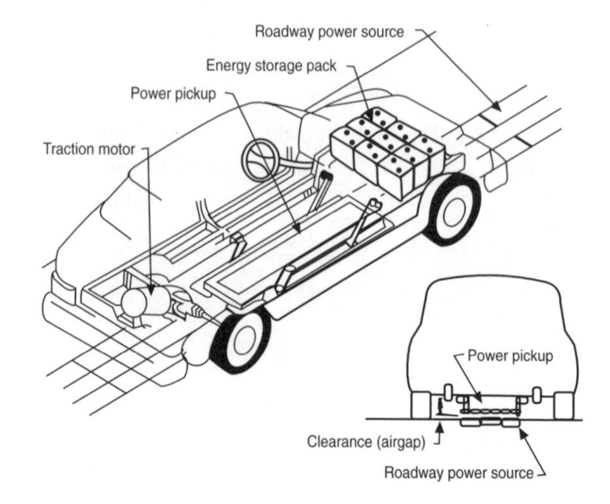

Fig. 5.13 Road-induced electricity.

There are other initiatives, too, such as the desire to make motorway driving under very high density peak traffic conditions less dangerous and less tiring. This is generating fresh interest in reserved lanes for vehicle guidance systems. Where these additionally provide roadway-induced powering, Fig. 5.13, as described by researchers from the Lawrence Livermore National Laboratory[6], a case for a car to suit relatively long-distant commuters can be made. The success of trials on GM's Impact electric car have so far pointed to the very considerable importance of light weight, good aerodynamics and low rolling resistance but the electrical breakthrough has come in the electronics technology of the DC/AC converter. Ford, too, have had very promising prototype results from their Ecostar l car-derived van, using a transistorized DC to AC inverter.

5.5.1 HONDA 'EV'

The state of the art in pilot-production electric cars is typified by Honda's nickel–metal hydride battery driven electric car, Fig. 5.14; it has been given the name 'EV' and claims twice the range obtainable with comparable lead–acid batteried cars. The car is not a conversion of an ICE model and has 95% new componentry. It is a 3-door, 4-seater with battery pack in a separated compartment between the floor. The pack comprises 24 × 12 V batteries and rests between virtually straight underframe longitudinal members running front to rear for maximum crash protection. The motor is a brushless DC type with rare earth high strength magnets and is said to give 96% efficiency. There is a fixed ratio transmission with parking lock. Maximum torque is 275 Nm, available from 0 to 1700 rpm, the speed at which a maximum power of 49 kW is developed and remains constant up to 8750 rpm. The under-bonnet power control unit comprises management and motor ECUs, power driver, junction board, 12 V DC/DC converter, air-conditioning inverter and 110/220 V on-board charger. Its aluminium container is liquid cooled in a system shared with motor and batteries. The controller uses IGBT switching devices in a PWM system. A phase control system involves both advanced angle control and field weakening to optimize operation in both urban and motorway conditions. A heat-pump climate control system has an inverter-controlled compressor with a remote-control facility to permit pre-cooling or pre-heating of the cabin prior to driving. Energy recovery is carried out in both braking and 'throttle-off' coasting modes. Low rolling-drag tyres are inflated to 300 kPa and are said to have just 57% the resistance of conventional tyres. A 'power-save' feature automatically reduces peak power when battery state-of-charge drops below 15%. An instrument display shows range and battery-charge state as a biaxial graph with clearly marked segments which even respond to throttle pedal depression. The car has a 125 mile urban range to the FUDS standard while top speed is 80 mph. Recharge time is 8 hours from 20% to fully charged.

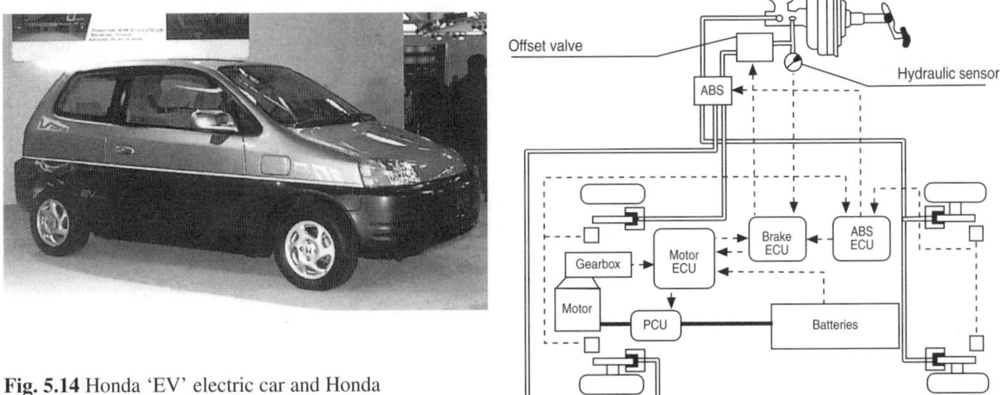

Fig. 5.14 Honda 'EV' electric car and Honda regenerative braking/coasting system.

5.5.2 GENERAL MOTORS 'EV1'

The latest generation GM EV1 (Fig. 5.15) is a purpose-built electric vehicle which offers two battery technologies: an advanced, high capacity lead–acid, and an optional nickel–metal hydride. The EV1 is currently available at selected GM Saturn retailers and is powered by a 137 (102 kW), 3 phase AC induction motor and uses a single speed dual reduction planetary gear set with a ratio of 10.946:1. The second generation propulsion system has an improved drive unit, battery pack, power electronics, 6.6 kW charger, and heating and thermal control module. Now, 26 valve-regulated, high capacity, lead–acid (PbA) batteries, 12 V each, are the standard for the EV1 battery pack and offer greater range and longer life. An optional nickel–metal hydride (NiMH) battery pack is also available for the Gen II model. This technology nearly doubles the range over the first generation battery and offers improved battery life as well. The EV1 with the high capacity lead–acid pack has an estimated real world driving range of 55 to 95 miles, depending on terrain, driving habits and temperature; range with the NiMH pack is even greater. Again, depending on terrain, driving habits, temperature and humidity, estimated real world driving range will vary from 75 to 130 miles, while only 10% of power is needed to maintain 100 km/h cruising speed, because of the low drag, now aided by Michelin 175/65R14 Proxima tyres mounted on squeeze-cast aluminium alloy wheels.

The 1990 Impact prototype from which the EV1 was developed had one or two more exotic features which could not be carried through to the production derivative but claimed an urban range of 125 miles on lead–acid batteries. In the Impact the 32 10 volt lead–acid batteries weighed 395 kg, some 30% of the car's kerb weight, housed into a central tunnel fared into the smooth underpanel and claimed to have a life of 18 500 miles. The Impact weighed 1 tonne and accelerated from 0 to 100 kph in 8 seconds, maximum power of the motor being 85 kW. The vehicle had 165/65R14 Goodyear low-drag tyres running at 4.5 bar. Two 3 phase induction motors were used, each of 42.5 kW at 6600 rpm; each can develop 64 Nm of constant torque from 0 to 6000 rpm, important in achieving 50–100 km/h acceleration in 4.6 seconds. Maximum current supply to each motor was 159 A, maximum voltage 400 V and frequency range 0–500 Hz. The battery charger was integrated into the regulator and charging current is 50 A for the 42.5 Ah lead–acid batteries, which could at 1990 prices be replaced for about £1000.

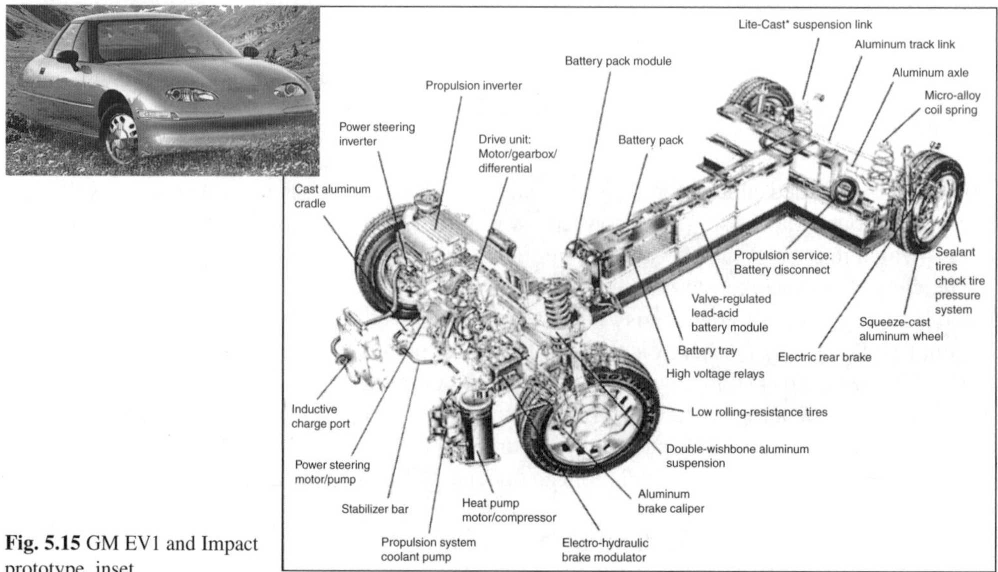

Fig. 5.15 GM EV1 and Impact prototype, inset.

The EV1 can be charged safely in all weather conditions with inductive charging. Using a 220 volt charger, charging from 0 to 100% for the new lead–acid pack takes up to 5.5–6 hours. Charging for the nickel–metal hydride pack, which stores more energy, is 6–8 hours. Braking is accomplished by using a blended combination of front hydraulic disk, and rear electrically applied drum brakes and the electric propulsion motor. During braking, the electric motor generates electricity (regenerative) which is then used to partially recharge the battery pack. The aluminium alloy structure weighs 290 pounds and is less than 10% of the total vehicle weight. The exterior composite body panels are dent and corrosion resistant and are made from SMC and RIM polymers. The EV1 is claimed to be the most aerodynamic production vehicle on the road today, with a 0.19 drag coefficient and 'tear drop' shape in plan view, the rear wheels being 9 inches closer together than the front wheels. The EV1 has an electronically regulated top speed of 80 mph. It comes with traction control, cruise control, anti-lock brakes, airbags, power windows, power door locks and power outside mirrors, AM/FM CD/cassette and also a tyre inflation monitor system.

5.5.3 AC DRIVES

An interesting variant on the AC-motored theme, Fig. 5.16, is the use of a two-speed transaxle gearbox which reduces the otherwise required weight of the high speed motor and its associated inverter. A system developed by Eaton Corporation is shown at (a) and has a 4 kW battery charger incorporated into the inverter. A 3 phase induction motor operates at 12 500 rpm – the speed being unconstrained by slip-ring commutator systems. A block diagram of the arrangement is at (b) and is based on an induction motor with 18.6 kW 1 hour rating – and base speed of 5640 rpm on a 192 V battery pack. The pulse width modulated inverter employs 100 A transistors. The view at (c) shows the controller drive system functions in association with the inverter. In an AC induction motor, current is applied to the stator windings and then induced into the windings of the rotor. Motor torque is developed by the interaction of rotor currents with the magnetic field in the air gap between rotor and stator. When the rotor is overdriven by coasting of the vehicle, say, it acts as a generator. Three phase winding of the stator armature suits motors of EV size; the rotor windings comprise conducting 'bars' short-circuited at either end to form a 'cage'. Rotation speed of the magnetic field in the air gap is known as the synchronous speed which is a function of the supply-current frequency and the number of stator poles. The running speed is related to synchronous speed by the 'slip'.

If two alternators were connected in parallel, and one was driven externally, the second would take current from the first and run as a 'synchronous motor' at a speed depending on the ratio of each machine's number of poles. While it is a high efficiency machine which runs at constant speed for all normal loads, it requires constant current for the rotor poles; it is not self-starting and will stop if overloaded enough for the rotor to slip too far behind the rotating stator-field. Normally, the synchronous motor is similar in construction to an induction motor but has no short-circuited rotor – which may be of the DC-excited, permanent-magnet or reluctance type.

The view at (d) shows a stator winding for a 2 pole 3 phase induction motor in diagrammatic form. If supply current frequency is f_s, then stator field speed is f_s/p for number of poles p. Rotor current frequency $f_r = sf_p$ where s is the slip. Power supplied to a 3 phase motor can be expressed as $EI(3)^{1/2}f_s n$ where n is efficiency. When a synchronous motor has no exciting voltage on the rotor it is termed a 'reluctance' motor which has very simple construction and, when used with power transistors, can be applied as a variable speed drive. Axial air-gap versions are possible as at (e); such an electronically commutated motor can operate with a DC source by periodic reversal of the rotor polarities.

The general form of control, with DC link inverter, for induction motors is shown diagrammatically at (f). The thyristors of the inverter are generally switched so as to route current

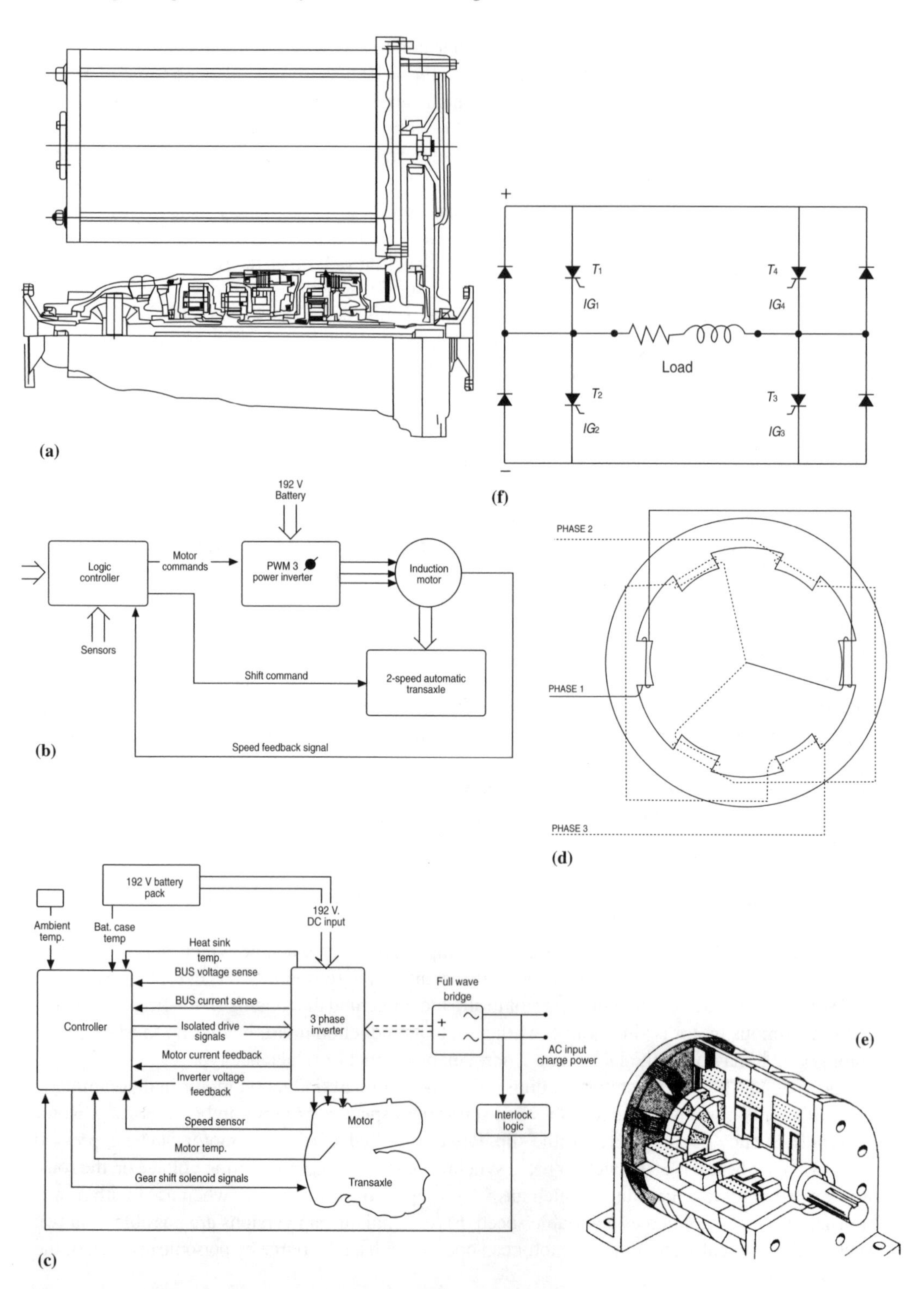

Fig. 5.16 AC drive systems: (a) transaxle motor; (b) controller arrangement; (c) controller drive system; (d) 3 phase induction motor stator; (e) air-gap type reluctance motor; (f) induction motor controller.

through the stator winding as though it were connected to a 3 phase AC. Frequency can be varied by timing of pulses to the thyristors and is typically 5–100 Hz, to give a speed range of almost twice synchronous speed. According to engineers from Chloride EV Systems Division[7], devices for building such an inverter are not yet available for current levels associated with EV traction. This situation may have already changed in the interim, however.

5.5.4 FORD E-KA: LITHIUM–ION BATTERY POWER

European Ford claim the first production-car use of lithium–ion batteries in a road vehicle by a major player in the industry. The high energy density and power-to-weight ratio of these storage units puts a prototype electric version of the small Ka hatchback, Fig. 5.17, on a par with the petrol-engined model in driving performance. Top speed is said to be 82 mph with acceleration to 62 mph in 12.7 seconds, although range between charges is still only 95 miles, but is extendible to 125 miles with a constant speed of 48 mph.

Until now Li–ion batteries have been used mainly in small consumer electronic products like notebook computers, cellular phones, baby monitors and smoke detectors. Output per cell of 3.6 volts is some three times that of nickel–cadmium and nickel–metal–hydride that they widely replace. They also retain full charge regardless of usage, can be recharged from zero to full capacity in 6 house with over 3000 repeated charge/discharge cycles, and are immune from the so-called 'memory effect' suffered by Ni–Cads. The basic Li–ion chemistry was initially redeveloped for automotive use by the French company SAFT SA, a leading battery manufacturer. Their advanced technology was then adapted to the e-Ka by Ford's Research Centre in Aachen, Germany, with financial assistance for the project from the German Ministry for Education and Research. The battery pack consists of 180 cells with 28 kWh rating and weighs only 280 kg (615 lb). This is 30% the weight of the power equivalent in lead–acid batteries, and substantially less than comparable Ni–Cads and Ni–MHs. In terms of volume the Ni-ion has approximately half the bulk of all the other three.

Batteries are divided into three individual sealed 'troughs', each with 30 modules containing six cells. One trough is located in the engine compartment, with the other two on either side of the back axle. Nominal output of 315 V DC is transformed by a solid-state inverter to 3 phase AC for the traction motor. Heat generated by the internal resistance of these second-generation Li–ion batteries is dissipated by one of two fluid cooling systems. A second independent system cools the drivetrain, with a 65 kW (88 bhp) asynchronous motor followed by a fixed-ratio transmission driving the front wheels. Torque rating is 190 Nm (140 lb ft).

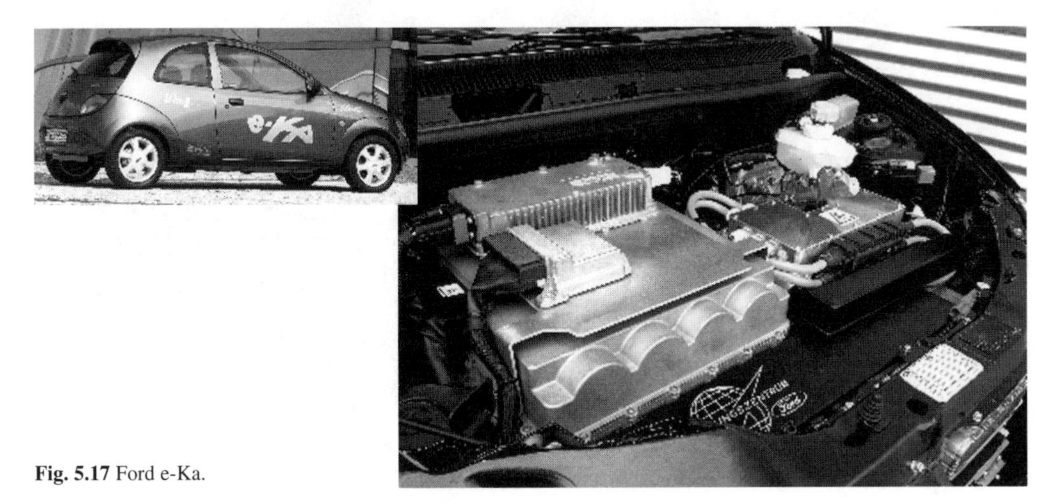

Fig. 5.17 Ford e-Ka.

Performance of the e-Ka is enhanced by a 45 kg (100 lb) weight reduction to counter the battery load. The roof and hood are of aluminium sandwich construction with a thermoplastic filling, while the front brake callipers with ceramics discs, rear drums, wheel rims and back axle are all in light alloy. Electric power steering supplied by Delphi provides further weight saving, where an electronic control module regulates the assistance needed to minimize battery demand. Brake servo and ABS system are also electric.

5.6 Electric van and truck design

5.6.1 GOODS VAN TO FLEET CAR CONVERSION

Europe's largest maker of EVs is Peugeot-Citroen whose Berlingo Dynavolt, Fig. 5.18, sets out to maximize the benefits of electric vehicles in a fleet car. It has a range extender in the form of an auxiliary generating system which does not quite make the vehicle a hybrid in the conventional sense. The generator feeds current into the traction motor rather than into the battery pack. The generator engine is a 16 ps, 500 cc Lombardini running on LPG which drives a Dynalto-style starter generator unit developing 8 kW at 3300 rpm, to supplement the supply from the 4 kW Saft Ni–Cad batteries. Company designed software controls the cut-in of the generator according to range requirements. The range is 80 km, which can be extended to 260 km with generator assistance. Series production was imminent as we went to press.

5.6.2 FORD EXT11

An early key US initiative in AC drives was the Ford EV project EXT11, Fig. 5.19, which has been exploring the use of an alternating current drive motor in the Aerostar Minivan, seen at (a). A sodium–sulphur battery was employed and a single-shaft propulsion system. A battery with the following specification was involved: nominal voltage 200 V; minimum voltage at 60 kW, 135 V; 50 kW capacity on FUDS cycling; 60 kW maximum power (20 seconds) and 35 kW continuous

Fig. 5.18 Citroen Berlingo Dynavolt: 1. electric motor and drive; 2. traction battery pack; 3. generator set; 4. motor controller; 5. generator controller; 6. drive programme selector; 7. LPG regulator; 8. LPG storage tank.

power (40 minutes). Much of the technology has since been carried over to the Ford Ecostar, described later. Dimensions of the battery were 1520 × 1065 × 460 mm and it was based on the use of small (10 Ah) cells connected in 4-cell series strings with parallel 8 V banks arranged to provide the required capacity. A voltage of 200 required 96 cells in series (each had a voltage of 2.076) and a parallel arrangement of 30 cells gave the 300 Ah capacity; overall 2880 cells are used and their discharge performance is shown at (b). Battery internal resistance was 30 milliohms with an appropriate thermal management system under development. The cells rested on a heater plate which incorporates a 710 W element used particularly in the initial warm-up. An associated air cooling system can dissipate 6 kW. Temperature must be maintained above 300° C to achieve an internal resistance value that will allow sensible current flow.

The US General Electric Company was also a partner in the programme, specifically on the AC drive system, at (c). This comprised a 50 bhp induction motor within a two-speed transaxle, a liquid-cooled transistor inverter being used to convert the 200 battery volts into variable-voltage 3 phase AC, (d). The induction motor was a 2 pole design with just under 13 cm stack length – within a stator diameter just under 23 cm. Calculated stall torque was 104 Nm and the transition between constant torque and constant power operation occurred at 3800 rpm. Design speed was 9000 rpm and absolute maximum 12 000 rpm. Full power was developed at a line current of 244.5 A. Control strategy relating inverter to motor is shown at (e).

Ford produced the transaxle on an in-house basis having coaxial motor and drive, (f). The motor rotor is hollow to allow one of the axle shafts to pass through it from the bevel-gear differential. The installation in a test saloon car achieved 30% gradient ability, 0–50 mph acceleration in less than 20 seconds, 60 mph top speed and 0.25 kWh/mile energy consumption. Gear ratios were 15.52 and 10.15:1, and rearward speed was obtained by electrically reversing the motor; in 'neutral' the motor was electrically disconnected.

5.6.3 UK EVA PRACTICE FOR CVS

In its manual of good practice for battery electric vehicles the Electric Vehicle Association lays down some useful ground rules for conceptual design of road-going electric trucks. Exploiting the obvious benefits of EV technology is the first consideration. Thus an ultra-low floor walk-through cab is a real possibility when batteries and motors can be mounted remotely. Lack of fuelling requirement, ease of start-up and getaway – also driving simplicity of two-pedal control without gearshifting – all these factors lend themselves to operations, such as busy city-centre deliveries where a substantial part of the driver's time is spent in off-loading and order-taking. Any aspect of vehicle design which minimizes the driving task thus maximizes his or her other workload duties. Successful builders of electric trucks are thus, say the EVA, specialists in assembling bought-in systems and components. Required expertise is in tailoring a motor/battery/speed-controller package to a given application. Principles to be followed in this process are keeping top speeds and motor power as low as possible consistent with fulfilling the task; using controllers which prevent any unnecessary acceleration once the vehicle has reached running speed in stop-start work; using generously rated motors rather than overspecifying battery capacity – the latter because large batteries cost more, displace payload and waste energy providing tractive force required for their extra weight.

Whereas over a decade ago the rule of thumb applied that 1 tonne of batteries plus one of vehicle and one of payload could be transported at 20 mph over a 20 mile range including 200 stop-starts per charge, now 40/40 to 50/50 mph/miles range is feasible by careful design. Belted radial tyres can now be run at 50% above normal inflation pressures. High voltage series motors of 72 V and above have working efficiencies of 85–90% over a 3:1 speed range and electronic controllers cut peak acceleration currents and avoid resistive losses – to extend ranges by 15–20%

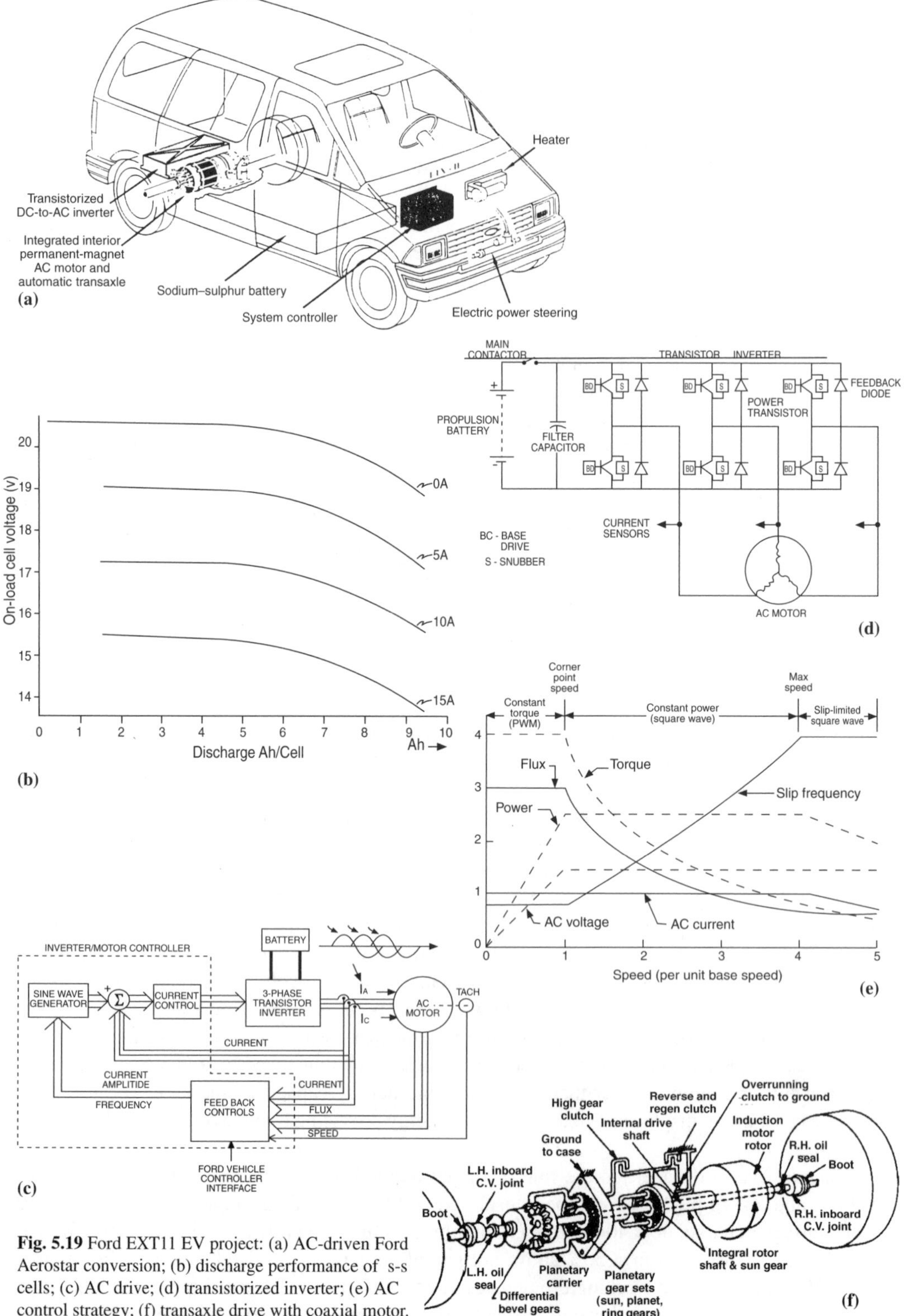

Fig. 5.19 Ford EXT11 EV project: (a) AC-driven Ford Aerostar conversion; (b) discharge performance of s-s cells; (c) AC drive; (d) transistorized inverter; (e) AC control strategy; (f) transaxle drive with coaxial motor.

over resistance controllers. A further 10% range increase is possible if high frequency controllers (15,000 Hz) are used – also much lower internal-resistance batteries reduce voltage fall-off and the use of DC/DC converters means that cells do not have to work so deeply to power auxiliaries, the EVA explain. The percentage of GVW allocated to the battery is a key parameter which should be kept below 35%. Lightweight lead–acid batteries giving 15–25% greater specific capacity than the BS 2550 standard traction cell should be considered, as they also have better discharge characteristics, but there is usually a trade-off in equivalent percentage life reductions. Separately-excited motors and regenerative braking give further bonuses.

5.6.4 THYRISTOR CONTROL

Thyristor control, Fig. 5.20, has been an important advance and units such as the controller by Sevcon are suitable for 24–72 V working with a current limit rating of 500 A for 5 minutes. Continuous rating at any frequency would be 160 A and a bank of eight 50 mF, 160 V commutation capacitors are employed. Overall dimensions are 430 × 301 × 147 mm and weight is 14 kg. Like other units of this type, it supplies the motor with fixed width pulses at a variable repetition frequency to provide stepless control over the average voltage supplied to the motor. The view at (a) shows the typical pulse width of 12 ms, giving a frequency range from 10 to 750 Hz and a good motor current form factor with minimal heating effect commensurate with some iron losses. Ratio of on-time to the total period would be 30–50 % while the commutation circuit which switches the main conducting thyristor operates every 2.5–3.5 milliseconds. The graph shows motor consumption at current limit; ripple is 170 while average motor current of 500 A compares with battery current of 165 A. At (b) is seen the energy being stored and reversed each cycle by the main commutation capacitor.

A typical DC chopper circuit for thyristor drive is shown at (c). Thyristor T1 is the 'on' one and T2 the 'off'; inductive load of a series motor is shown as Rl and Ll while D1 is a so-called 'freewheel' diode. When capacitor C is charged to a voltage *V*b in a direction opposing the battery

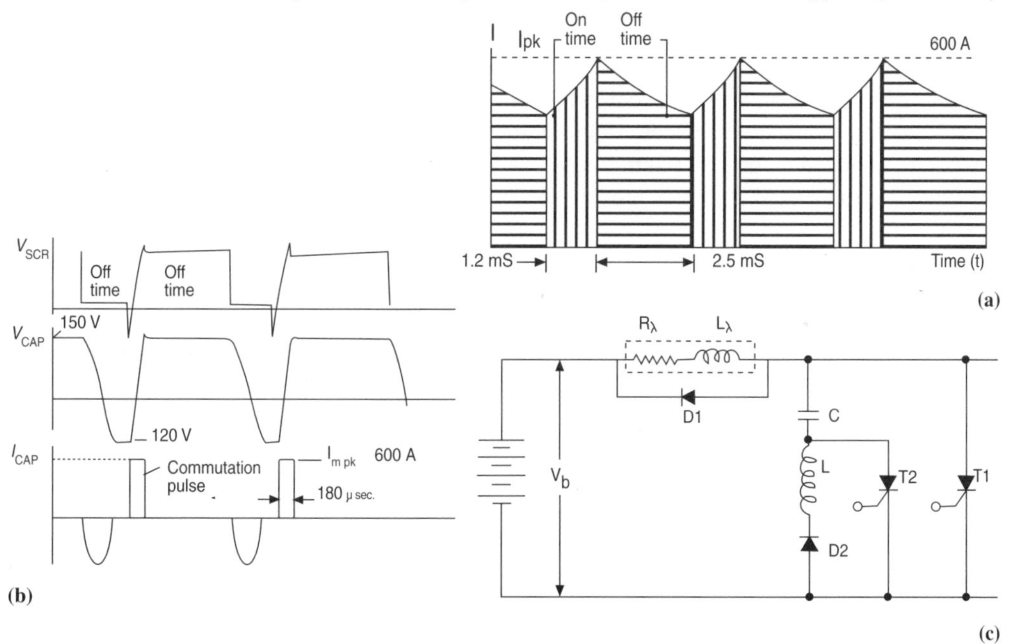

Fig. 5.20 Thyristor control: (a) controller pulses; (b) energy storage and reversal; (c) DC chopper circuit.

voltage, T1 and T2 are off and load current *Il* is flowing through Rl, Ll and D. A pulse is applied to the gate of T1 to turn it on; D1 becomes reverse-biased and load current *Il* begins to flow through Rl, Ll and T1. This causes a short-circuit, effectively, across CL and D2, creating a tuned circuit. Its resonance drives a current through T2, D2, L and C, sinusoidally rising to a maximum then decaying to zero, at which point C is effectively charged to V2 in the reverse direction.

When the current attempts the second half of its cycle, D2 prevents any reversal and the reverse charge on C is caught by D2, the 'catching' diode. At time t_{on} a pulse applied to the gate of T2 turns it on, applying reverse voltage on C across T1 – which rapidly turns off and diverts load current through Rl, Ll, C, L and T2. C is thus charged at a linear rate by *Il* passing through it. When capacitor voltage again approaches *V*b, opposing battery voltage, load current begins to flow through Rl, Ll and D1 again, current through T2 dropping below the holding value so turning off T2 – completing the cycle ready for reinitiation by pulsing the gate of T2 again. The circuit thus represents a rapid on/off switch which enables effective DC output voltage to be variable according to:

$$[t_{on}/(t_{on} + t_{off})]Vb$$

and output from the device is a train of rectangular voltage pulses. The steady current related to such a train is obtained by connecting D1 across the motor load. Output voltage can thus be controlled by altering t_{off}/t_{on} while holding $t_{on} + t_{off}$ constant – a constant frequency variable mark/space chopper – or altering $1/(t_{on} + t_{off})$ while holding t_{on} constant – a variable frequency constant mark chopper.

5.6.5 FORD ECOSTAR

An electric vehicle with equivalent performance to an IC-engined one, and yet based on a conventional Escort van platform, Fig. 5.23, has been achieved by Ford with the use of a sodium–sulphur battery. The battery is enclosed in a hermetically sealed metal casing, which contains sodium and sulphur on each side of a porous ceramic separator. In order to react on a molecular level, the sodium and sulphur must be free to flow through the separator in liquid form. This requires maintaining an operating temperature of between 290 and 350° C. The Ecostar's battery pack is formed of a double-skinned, stainless steel case, with a glass fibre insulated cavity which has been evacuated to form a vacuum flask. Inside the case, 480 individual sodium–sulphur cells are embedded in a sand matrix to prevent movement and contain the elements in the event of a severe impact.

Each cell is encased in a corrosion-resistant aluminium cylinder and produces about 2 volts. They are linked to provide a 330 volt power supply and a steady-load current of 80 amps. Power of up to 50 kW is available for short periods. The battery pack is mounted under the vehicle floor, ahead of the rear axle. The vehicle is fitted with an intelligent on-board charging system that automatically adjusts to the supply voltage from the mains. It can be used for overnight recharging and temperature control of the sodium–sulphur battery. A full recharge takes between 5 and 7 hours. In the event of electrical malfunction, the battery pack is isolated by replacing external 200 amp fuses at the battery connection box. The batteries are also equipped with internal fuses that isolate the electrical supply from the vehicle if there is any kind of internal overload.

The Ecostar, Fig. 5.21, is powered by a high speed, 3 phase alternating current (AC) motor, rated at 56 kW (75 bhp). It is directly coupled to the transaxle by a single-speed planetary reduction gear set. The motor operates at 330 volts, provided by the sodium–sulphur battery pack. Compared to equivalent DC systems, the AC motor works at a higher rate of efficiency, is smaller, lighter, lower in cost, easier to cool, more reliable and generally more durable. Maximum speed is 13 500 rpm and maximum torque 193 Nm at zero rpm. Regenerative braking comes from the motor acting as a generator, so recharging the battery.

The supply from the sodium–sulphur battery is fed along special heavy-duty cables to the power electronics centre (PEC), which is housed in the engine compartment. The candy-striped, red-on-black cables are strikingly marked to avoid confusion with any other wiring in the van. Encased in aluminium, the PEC incorporates the battery charging electronics and inverters that convert the 330 volt DC supply to AC power. It also includes a transformer which enables a 12 volt auxiliary battery to be recharged from the high voltage traction batteries. The electrical supply to the PEC is connected and isolated by power relays inside a contactor box, controlled by the ignition key and the electronic modules.

On-board microprocessors are linked by a multiplex database that allows synchronous, highspeed communication between all the vehicle's systems. The vehicle system controller (VSC) acts as the user/vehicle interface and is operated by electrical signals from the accelerator pedal. This 'drive-by-wire' system has no mechanical connection to the speed controller. It is supplemented by a battery controller that monitors the sodium–sulphur operating temperature, the state of charge and recharging. The control system also incorporates a diagnostic data recorder which stores information from all the on-board electronic systems. This locates any operational malfunctions quickly and precisely.

A fault detection system (known as the power protection centre (PPC)), has also been built into the battery controller to monitor continuously the main electrical functions. Every 4 seconds it checks for internal and external leakage between the high-voltage system, the vehicle chassis and the battery case. Should a leak be detected in any wire, a warning light tells the driver that service action should be taken. The vehicle can then be driven safely for a short distance so that repairs can be made. If leakage is detected from both battery leads, the vehicle system cuts off the power to the motor and illuminates a red warning light. The auxiliary power supply is maintained to operate the battery cooling system. An inertia switch is also fitted, which is activated in the event of a vehicle collision and isolates power from the main battery pack. Auxiliary power to the battery coolant pump is also cut off to reduce the risk of hot fluids escaping. The vehicle incorporates a small amount of 'creep' whereby slight brake pressure is required to prevent it from moving forwards (or backwards when in reverse gear). This results in easier manoeuvring and smoother transmission of power. Auxiliary vehicle systems are powered by a standard automotive 12 volt lead–acid battery, the exception being the electrically driven cabin air conditioning system, which is powered directly from the main sodium–sulphur traction batteries, via a special AC inverter in the PEC module.

The climate control unit handles both the air conditioning and a highly efficient 4.5 kW ceramic element PTC heater. The heater elements are made from barium titanate with a multi-layer metallic coating on each side, impregnated with special chemical additives. The material has low resistance at low temperature for a very fast warm-up, while at higher temperatures the power supply is automatically regulated to save electrical energy. A strip of solar panels across the top of the windscreen supply power to a supplementary extractor fan that ventilates the cabin when the vehicle is parked in direct sunlight. This relieves the load on the air conditioning system when the journey is resumed. Some lightweight materials have been used in the Ecostar to offset the 350 kg weight of the battery pack.

Elimination of the clutch, torque converter and additional gearing is supplemented by a magnesium transmission casing, aluminium alloy wheels, air conditioning compressor and power electronics housing. Plastic composite materials have been used for the rear suspension springs, load floor and rear bulkhead. Use of these materials has helped keep the Ecostar's kerb weight to between 1338 and 1452 kg, which is 25% heavier than a standard diesel-powered van. The vehicle also has a useful load carrying capacity of up to 463 kg and retains similar load space dimensions to the standard Escort van. The Ecostar has been developed for optimum performance in urban conditions, where it is expected to be driven most frequently. Its top speed is restricted to 70 mph, whilst 0 to 50 mph takes approximately 12 seconds. Because of the high torque at low speed, acceleration from standing to 30 mph is quicker than diesel and petrol driven vehicles of the same size. The average vehicle range between charges to date has been 94 miles, with a maximum recorded range of 155 miles.

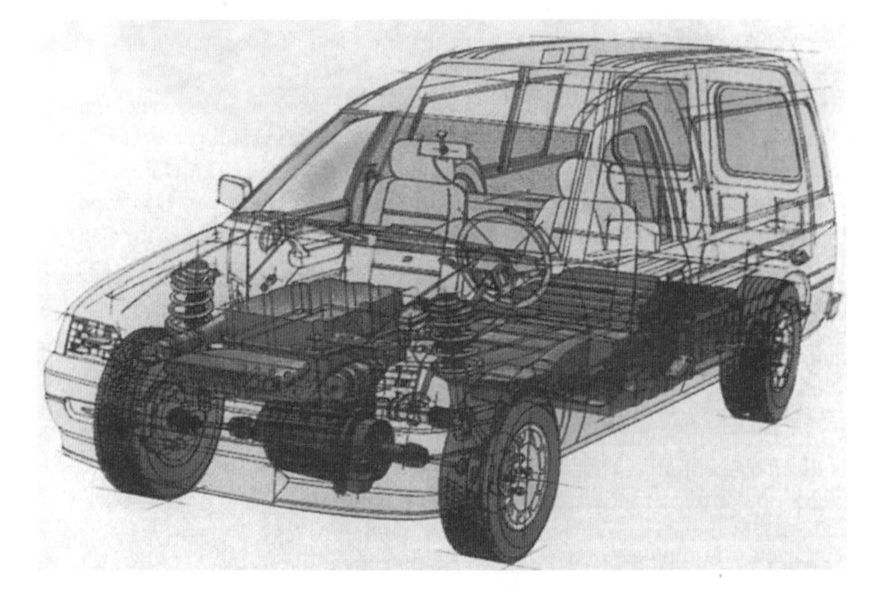

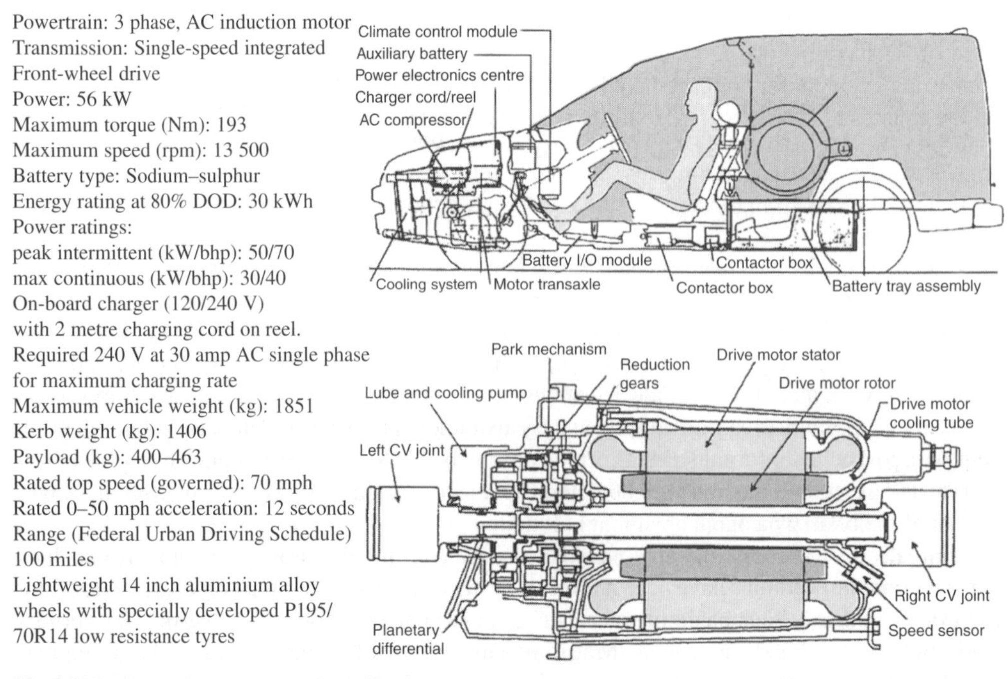

Powertrain: 3 phase, AC induction motor
Transmission: Single-speed integrated
Front-wheel drive
Power: 56 kW
Maximum torque (Nm): 193
Maximum speed (rpm): 13 500
Battery type: Sodium–sulphur
Energy rating at 80% DOD: 30 kWh
Power ratings:
peak intermittent (kW/bhp): 50/70
max continuous (kW/bhp): 30/40
On-board charger (120/240 V)
with 2 metre charging cord on reel.
Required 240 V at 30 amp AC single phase
for maximum charging rate
Maximum vehicle weight (kg): 1851
Kerb weight (kg): 1406
Payload (kg): 400–463
Rated top speed (governed): 70 mph
Rated 0–50 mph acceleration: 12 seconds
Range (Federal Urban Driving Schedule)
100 miles
Lightweight 14 inch aluminium alloy
wheels with specially developed P195/
70R14 low resistance tyres

Fig. 5.21 Ecostar package, motor and specification.

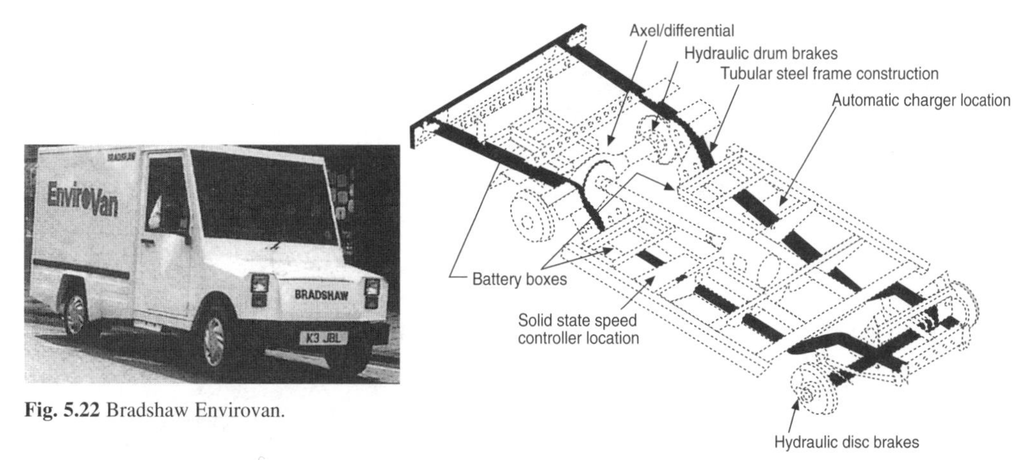

Fig. 5.22 Bradshaw Envirovan.

5.6.6 BRADSHAW ENVIROVAN

DC drive is used on the higher payload capacity purpose-built Bradshaw Envirovan. Figure 5.22 shows the Envirovan built in conjunction with US collaborator Taylor-Dunn. This can carry 1500 lb on a 3.55 square metre platform at speeds up to 32.5 mph and is aimed specifically at city deliveries. The vehicle relies on 12 6 volt deep-cycle, rechargeable lead–acid batteries for a total of 72 V. All accessories, such as internal lights, windscreen wipers, and gauges, run off the 72 V system through a DC/DC converter, which steps the power down to 12 V, so that all batteries discharge equally. This distributes power requirements evenly across all 12 batteries and prevents one or two of the batteries from draining prematurely. A battery warning indicator shows the current percentage of battery power available, with a visual warning when battery charge is below 20%. An on-board battery charger, featured on the Envirovan, can be used to recharge the battery packs simply by plugging it into any standard 240 V AC power socket. The entire 72 V system requires approximately 9.5 hours to fully charge the batteries from an 80% state of discharge (20% remaining charge). The battery pack provides approximately 1000 recharging cycles before replacement is required. Battery packs are available for less than £1000 which equates to less than 2p per mile. Recharging costs add an additional 1p per mile giving a total cost per mile of 3p. A 20 bhp General Electric motor has been designed for the Envirovan. A range of 8 hours/50 miles is available for the vehicle which measures 4.21 metres long × 1.65 wide. It can accelerate to 25 mph in 6 seconds and its controller can generate up to 28 bhp for quick response.

5.7 Fuel-cell powered vehicles

5.7.1 GENERAL MOTORS ZAFIRA PROJECTS

GM and its Opel subsidiary are aiming at a compact fuel-cell driven vehicle by 2004, Fig. 5.23. By 2010, up to 10% of total sales are expected to be taken by this category. The efficiency of cells tested by the company is over 60% and CO_2 emissions, produced during the reformation of methanol to obtain hydrogen, are about half that of an equivalent powered IC engine. Fuel cells have already been successfully exploited in power generation, at Westervoort in the Netherlands, and experimental versions have been shown to successfully power lap-top computers. According to GM, in principle four basic fuels are suitable: sulphur-free modified gasoline, a synthetic fuel, methanol or pure hydrogen. Modified gasoline is preferred because of the existing distribution infrastructure but CO_2 emission in reforming is higher than with methanol. Synthetic

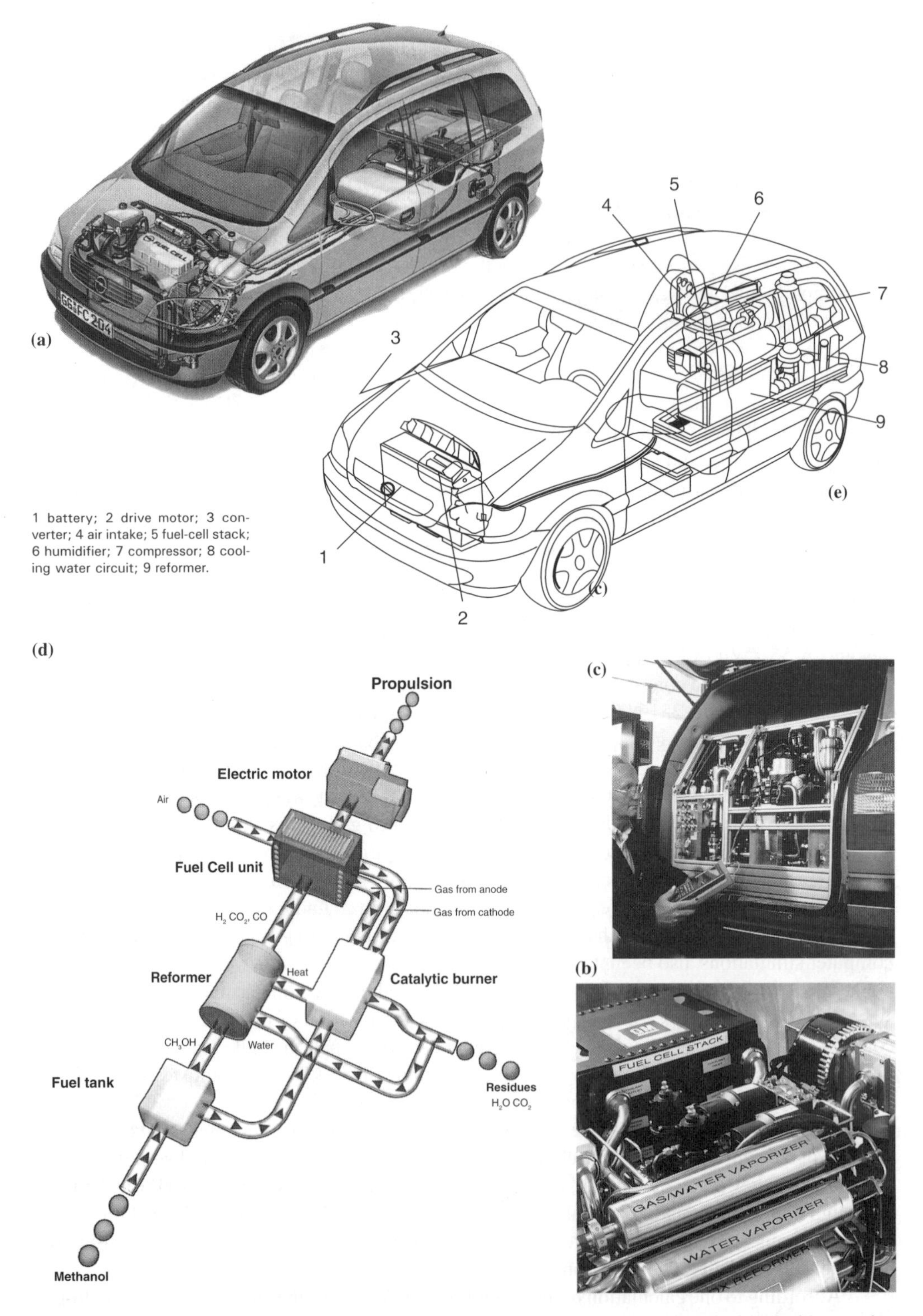

Fig. 5.23 GM fuel-cell developments: (a) Zafira conversion package; (b) under-bonnet power-pack; (c) reformer and cells; (d) flow diagram; (e) latest package with on-board hydrogen storage.

fuel and methanol can be obtained from some primary energy sources including natural gas. Transportation and storage of hydrogen is still at the development stage for commercial viability, Liquefying by low temperature and/or pressure being seen as the only means of on-vehicle storage.

Currently GM engineers are working on a fuel-cell drive version of the Zafira van (a) in which electric motor, battery and controller are accommodated in the former engine compartment (b). The 'cold combustion' of the fuel-cell reaction, hydrogen combining with oxygen to form water, takes place at 80–90°C and a single cell develops 0.6–0.8 V. Sufficient cells are combined to power a 50 kW asynchronous motor driving the front wheels through a fixed gear reduction. The cell comprises fuel anode, electrolyte and oxygen cathode. Protons migrate through the electrolyte towards the cathode, to form water, and in doing so produce electric current. Prospects for operating efficiencies above 60% are in view, pending successful waste heat utilization and optimization of gas paths within the system. The reforming process involved in producing hydrogen from the fuel involves no special safety measures for handling methanol and the long-term goal is to produce no more than 90 g/km of CO_2. In the final version it is hoped to miniaturize the reformer, which now takes up most of the load space, (c), and part of the passenger area, so that it also fits within the former engine compartment. Rate of production of hydrogen in the reformer, and rate of current production in the fuel cell, both have to be accelerated to obtain acceptable throttle response times – the flow diagram is seen at (d). The 20 second start-up time also has to be reduced to 2 seconds, while tolerating outside temperatures of –30°C.

Currently GM Opel are reportedly working in the jointly operated Global Alternative Propulsion Centre (GAPC) on a version of their fuel-celled MPV which is now seen as close to a production design. A 55 kW (75 hp) 3 phase synchronous traction motor drives the front wheels through fixed gearing, with the complete electromechanical package weighing only 68 kg (150 lb). With a maximum torque of 251 Nm (181 lb ft) at all times it accelerates the Zafira to 100 km/h (62 mph) in 16 seconds, and gives a top speed of 140 km/h (85 mph). Range is about 400 km (240 miles).

In contrast to the earlier vehicle fuelled by a chemical hydride system for on-board hydrogen storage, this car uses liquid hydrogen. Up to 75 litres (20 gallons) is stored at a temperature of –253°C, just short of absolute zero, in a stainless steel cylinder 1 metre (39 in) long and 400 mm (15.7 in) in diameter. This cryostat is lined with special fibre glass matting said to provide insulating properties equal to several metres of polystyrene. It is stowed under the elevated rear passenger seat, and has been shown to withstand an impact force of up to 30 g. Crash behaviour in several computer simulations also been tested.

Fuel cells as well as the drive motor are in the normal engine compartment. In the 6 months since mid-2000 the 'stack' generating electricity by the reaction of hydrogen and oxygen now consists of a block of 195 single fuel cells, a reduction to just half the bulk. Running at a process temperature of about 80°C, it has a maximum output of 80 kW. Cold-start tests at ambient temperatures down to –40°C have been successfully conducted.

GAPC has created strong alliances with several major petroleum companies to investigate the creation of the national infrastructures needed to support a reasonable number of hydrogen-fuelled vehicles once they reach the market, possibly in 5 years' time. Fuel cost is another critical factor. Although hydrogen is readily available on a commercial basis from various industrial processes, its cost in terms of energy density presents a real problem for the many auto-makers who research both fuel cells and direct combustion.

According to one calculation based on current market prices, the energy content of hydrogen generated by electrolysis using solar radiation with photovoltaic cells equals gasoline at roughly $10 a gallon.

5.7.2 FORD P2000

Mounting most of the fuel-cell installation beneath the vehicle floor has been achieved on Ford's FC5, seen as a static display in 1999, with the result of space for five passengers in a medium-sized package. Their aim is to achieve an efficiency twice that of an IC engine. The company point out that very little alteration is required to a petrol-distributing infrastructure to distribute methanol which can also be obtained from a variety of biomass sources. Oxygen is supplied in the form of compressed air and fed to the Ballard fuel-cell stack alongside reformed hydrogen. Ford use an AC drive motor, requiring conversion of the fuel cell's DC output. Even the boot is accessible on the 5-door hatchback so much miniaturization has already been done to the propulsion system. The vehicle also uses an advanced lighting system involving HID headlamps, with fibre-optic transmission of light in low beam, and tail-lights using high efficiency LED blade manifold optics. The company's running P2000 demonstrator, Fig. 5.24, uses fuel in the form of pure gaseous hydrogen in a system developed with Proton Energy Systems.

5.7.3 LIQUID HYDROGEN OR FUEL REFORMATION, FIG. 5.25

Renault and five European partners have produced a Laguna conversion with a 250 mile range using fuel-cell propulsion. The 135 cell stack produces 30 kW at a voltage of 90 V, which is transformed up to 250 V for powering the synchronous electric motor, at a 92% transformer efficiency and 90–92% motor efficiency. Nickel–metal hydride batteries are used to start up the fuel cell auxiliary systems and for braking energy regeneration. Some 8 kg of liquid hydrogen is stored in an on-board cryogenic container, (a), at –253°C to achieve the excellent range. Renault insist that an on-board reformer would emit only 15% less CO_2 than an IC engine against the 50% reduction they obtain by on-board liquid hydrogen storage.

According to Arthur D. Little consultants, who have developed a petrol reforming system, a fuel-cell vehicle thus fitted can realize 80 mpg fuel economy with near zero exhaust emissions. The Cambridge subsidiary Epyx is developing the system which can also reform methanol and ethanol. It uses hybrid partial oxidation and carbon monoxide clean-up technologies to give it a claimed advantage over existing reformers. The view at (b) shows how the fuel is first vaporized (1) using waste energy from the fuel cell and vaporized fuel is burnt with a small amount of air in a partial oxidation reactor (2) which produces CO and O_2. Sulphur compounds are removed from

Fig. 5.24 Ford P2000 fuel cell platform with two 35 kW Ballard stacks.

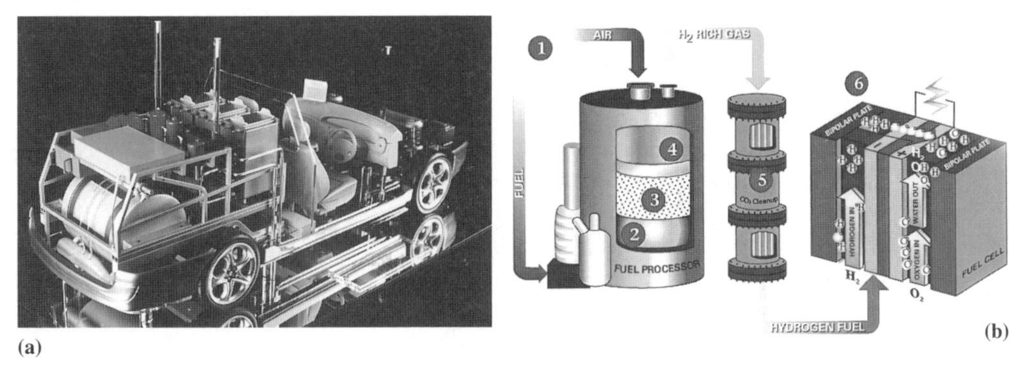

Fig. 5.25 Liquid hydrogen or reformed fuel: (a) Renault cryogenic storage; (b) Arthur D. Little reformer.

the fuel (3) and a catalytic reactor (4) is used with steam to turn the CO into H_2 and CO_2. The remaining CO is burnt over the catalyst (5) to reduce CO_2 concentration down to 10 ppm before passing to the fuel cell (6).

5.7.4 PROTOTYPE FUEL-CELL CAR

Daimler-Chrysler's Necar IV, Fig. 5.26, is based on the Mercedes-Benz A-class car and exploits that vehicle's duplex floor construction to mount key propulsion systems. The fuel cell is a Ballard proton exchange membrane type, 400 in the stack, developing 55 kW at the wheels to give a top speed of 145 km/h and a range of 450 km. Fuel consumption is equivalent to 88 mpg and torque response to throttle movement is virtually instantaneous. While the first prototype weighs 1580 kg, the target weight is 1320 kg, just 150 kg above the standard A-class. Tank to wheel efficiency is quoted as 40% now, with 88% in prospect for a vehicle with a reformer instead of compressed hydrogen. The American Methanol Institute is predicting 2 million thus-fitted cars on the road by 2010 and 35 million by 2020.

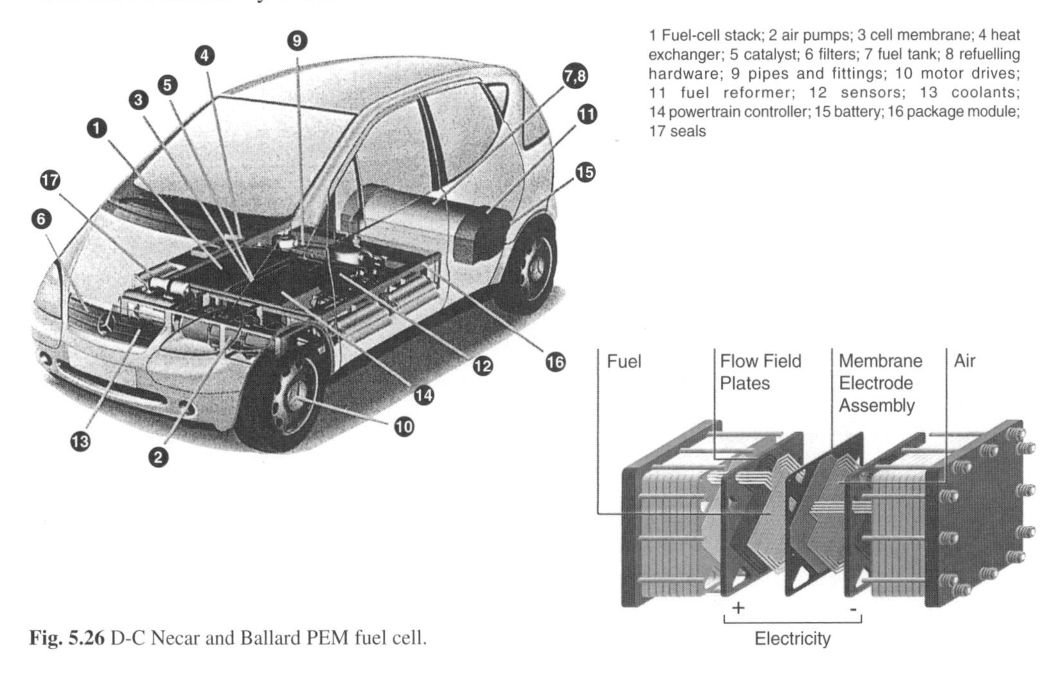

Fig. 5.26 D-C Necar and Ballard PEM fuel cell.

In a summer 1999 interview Ballard chief Firoz Rasul put the cost of electricity produced by fuel cells as $500/kW so that car power plants between 50 and 200 kW amount to $25–100 000. PEM cells operate at 80°C and employ just a thin plastic sheet as their electrolyte. The sheet can tolerate modest pressure differentials across it, which can increase power density. Ballard's breakthrough in power density came in 1995 with the design of a stack which produced 1000 watts/litre, ten times the 1990 state of the art. Cell energy conversion efficiency, from chemical energy to electricity, is about 50% and the cell does not 'discharge' in the manner of a conventional storage battery. Electrodes are made from porous carbon separated by the porous ion-conduction electrolyte membrane. It is both an electron insulator and proton conductor and is impermeable to gas. A catalyst is integrated between each electrode and the membrane while flow field plates are placed on each side of the membrane/electrode assembly. These have channels formed in their surface through which the reactants flow. The plates are bi-polar in a stack, forming the anode of one cell and the cathode of the adjacent one. The catalyst causes the hydrogen atoms to dissociate into protons and electrons. The protons are carried through to the cathode and the free electrons conducted as a usable current.

References

1. Origuchi *et al.*, Development of a lithium–ion battery system for EVs, SAE paper 970238
2. Saito *et al.*, Super capacitor for energy recycling hybrid vehicle, Convergence 96 proceedings
3. Van der Graaf, R., EAEC paper 87031
4. Prigmore *et al., Battery car conversions,* Battery Vehicle Society, 1978
5. Harding, G., Electric vehicles in the next millennium, *Journal of Power Sources,* 3335, 1999
6. Huettl *et al., Transport Technology USA*, 1996
7. SAE paper 900578, 1990

Further reading

Smith & Alley, *Electrical circuits, an introduction,* Cambridge, 1992

Copus, A., DC traction motors for electric vehicles, *Electric Vehicles for Europe,* EVA conference report, 1991

EVA manual, Electric Vehicle Association of GB Ltd

Unnewehr and Nasar, *Electric vehicle technology,* Wiley, 1982

Huettl *et al., Transport Technology USA*, 1996

Argonne National Laboratory authors, SAE publication: *Alternative Transportation Problems,* 1996

Strategies in electric and hybrid vehicle design, SAE publication SP-1156, 1996

(ed.) Dorgham, M., *Electric and hybrid vehicles,* Interscience Enterprises, 1982

Electric vehicle technology, MIRA seminar report, 1992

Battery electric and hybrid vehicles, IMechE seminar report, 1992

(ed.) Lovering, D., *Fuel cells,* Elsevier, 1989

The urban transport industries report, Campden, 1993

The MIRA electric vehicle forecast, 1992

Niewenhuis *et al., The green car guide,* Merlin, 1992

Combustion engines and hybrid vehicles, IMechE, 1998

6

Hybrid vehicle design

6.1 Introduction

The hybrid-drive concept appears in many forms depending on the mix of energy sources and propulsion systems used on the vehicle. The term can be used for drives taking energy from two separate energy sources, for series or parallel drive configurations or any combination of these. Here the layout and development of systems for cars and buses is described in terms of drive configuration and package-design case studies of recent-year introductions.

6.1.1 THE HYBRID VEHICLE

This solution is considered by coauthor Ron Hodkinson to be a short-term remedy to the pollution problem. It has two forms, parallel and series hybrid which he illustrates in Fig. 6.1. Conventionally, parallel hybrids are used in lower power electric vehicles where both drives can be operated in parallel to enhance high power performance. Series hybrids are used in high power systems. Typically, a gas turbine drives a turbo-alternator to feed electricity into the electric drive. It is this

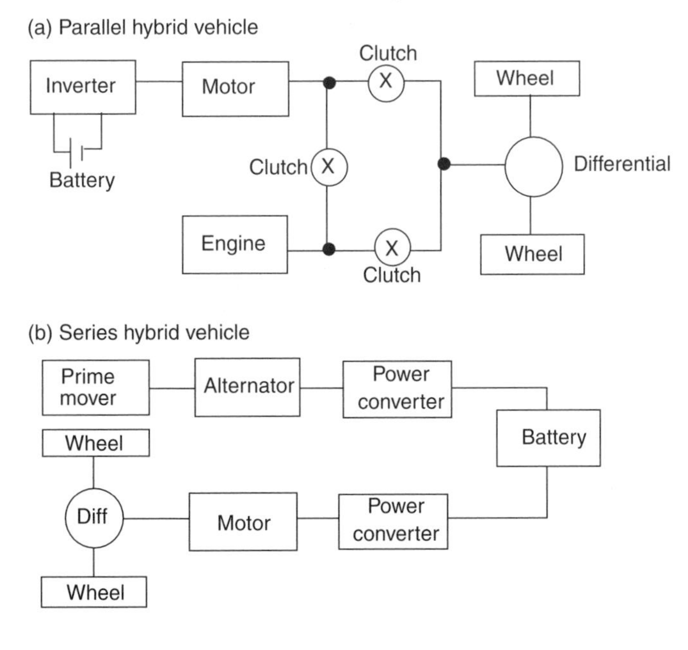

Fig. 6.1 Types of hybrid drive.

type of drive that would be used on trucks between 150 kW and 1000 kW. In pollution and fuel economy terms, hybrid technology should be able to deliver two-thirds fuel consumption and one-third noxious emission levels of IC engined vehicles. This technology would just about maintain the overall emissions status quo in 10 years overall. If hybrid vehicles were used on battery only in cities, this would have a major impact on local pollution levels.

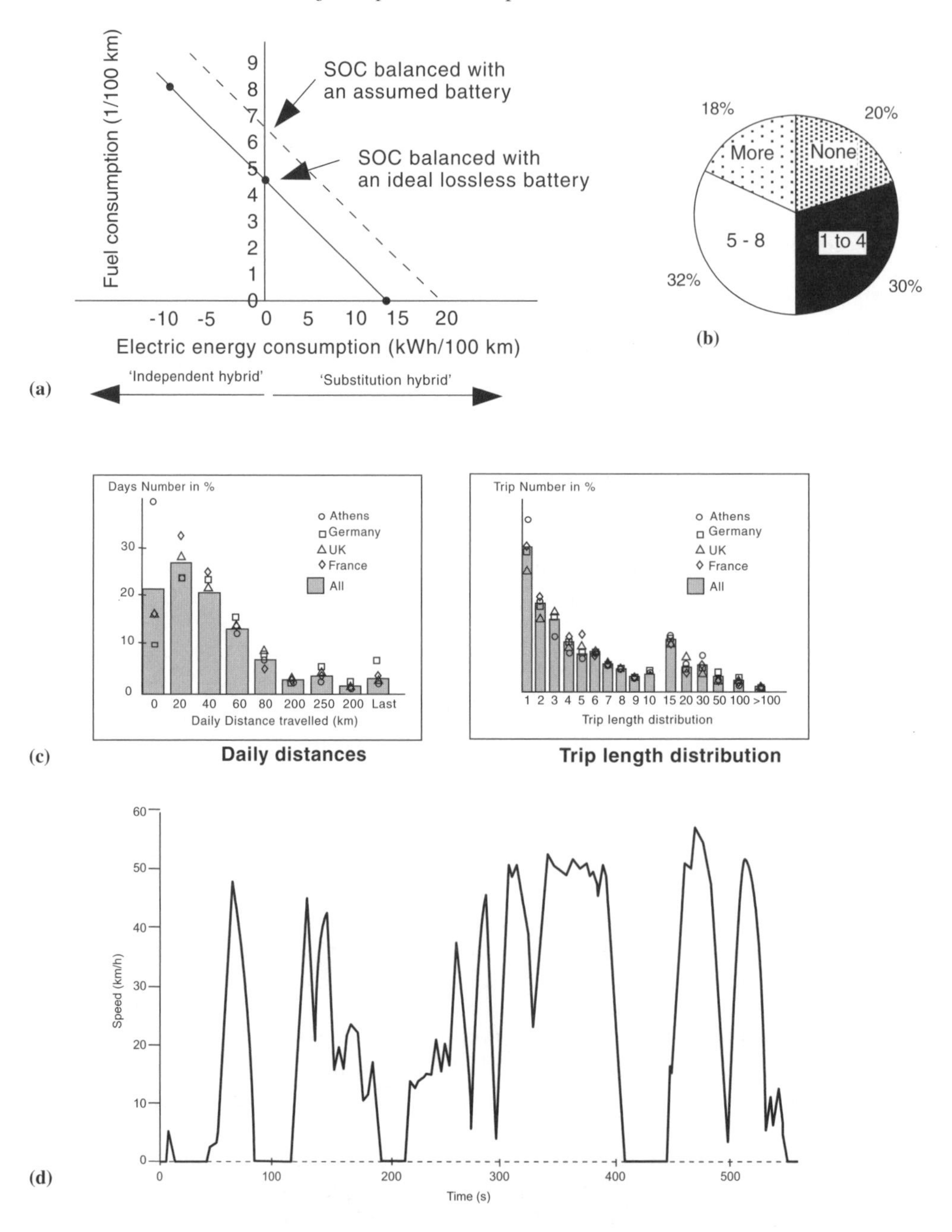

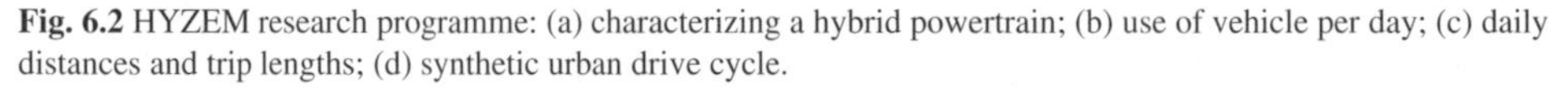

Fig. 6.2 HYZEM research programme: (a) characterizing a hybrid powertrain; (b) use of vehicle per day; (c) daily distances and trip lengths; (d) synthetic urban drive cycle.

6.2 Hybrid-drive prospects

A neat description of the problems of hybrid-drive vehicles has come out of the results of the 3 year HYZEM research programme undertaken by European manufacturers, Fig. 6.2. According to Rover participants[1], controlled comparisons of different hybrid-drive configurations, using verified simulation tools, are able to highlight the profitable fields of development needed to arrive at a fully competitive hybrid-drive vehicle and demonstrate, in quantitative terms, the trade-off between emissions, electrical energy and fuel consumption. Only two standard test points are required to describe the almost linear relationship: fuel consumption at point of no overall change in battery state-of-charge (SOC) and point of electrical consumption over the same cycle in pure electric mode. A linear characteristic representing an ideal lossless battery can also be added to the graph, to show the potential for battery development, as at (a).

Confirmation was also given to such empirical assessments that parallel hybrids give particularly good fuel economy because of the inherent efficiency of transferring energy direct to the wheels as against the series hybrids' relatively inefficient energy conversion from mechanical to electrical drive. The need for a battery which can cope with much more frequent charge/discharge cycles than one for a pure electric-drive vehicle was also confirmed. Although electric energy capability requirement is less stringent, a need to reduce weight is paramount in overcoming the problem of the redundant drive in hybrid designs.

A useful analysis of over 10 000 car journeys throughout Europe was undertaken for a better understanding of 'mission profile' for the driving cycles involved. Cars were found to be used typically between one and eight times per day, as at (b), and total daily distances travelled were mostly less than 55 km. Some 13% of trips, (c), were less than 500 metres, showing that we are in danger of becoming like the Americans who drive even to visit their next door neighbours! Even more useful velocity and acceleration profiles were obtained, by data recoding at 1 Hz frequency, so that valuable synthetic drive cycles were obtained such as the urban driving one shown at (d).

6.2.1 MAP-CONTROLLED DRIVE MANAGEMENT

BMW researchers[2] have shown the possibility of challenging the fuel consumption levels of conventional cars with parallel hybrid levels, by using map-controlled drive management, Fig. 6.3. The two-shaft system used by the company, seen at (a), uses a rod-shaped asynchronous motor, by Siemens, fitted parallel to the crankshaft beneath the intake manifold of the 4-cylinder engine, driving the tooth-belt drive system as seen at (b); overall specification compared with the 518i production car from which it is derived is shown at (c). The vehicle still has top speed of 180 kph (100 kph in electric mode) and a range of 500 km; relative performance of the battery options is shown at (d). Electric servo pumps for steering and braking systems are specified for the hybrid vehicle and a cooling system for the electric motor is incorporated. The motor is energized by the battery via a 13.8 V/50 A DC/DC converter. The key electronic control unit links with the main systems of the vehicle as seen at (e).

To implement the driving modes of either hybrid, electric or IC engine the operating strategy is broken down into tasks processed parallel to one another by the CPU, to control and monitor engine, motor, battery and electric clutch. The mode task determines which traction condition is appropriate, balancing the inputs from the power sources; the performance/output task controls power flow within the total system; the battery task controls battery charging. According to accelerator/braking pedal inputs, the monitoring unit transfers the power target required by the driver to the CPU where the optimal operating point for both drive units is calculated in a continuous, iterative process. The graphs at (f) give an example of three iterations for charge efficiency, also determined by the CPU, based on current charge level of the battery.

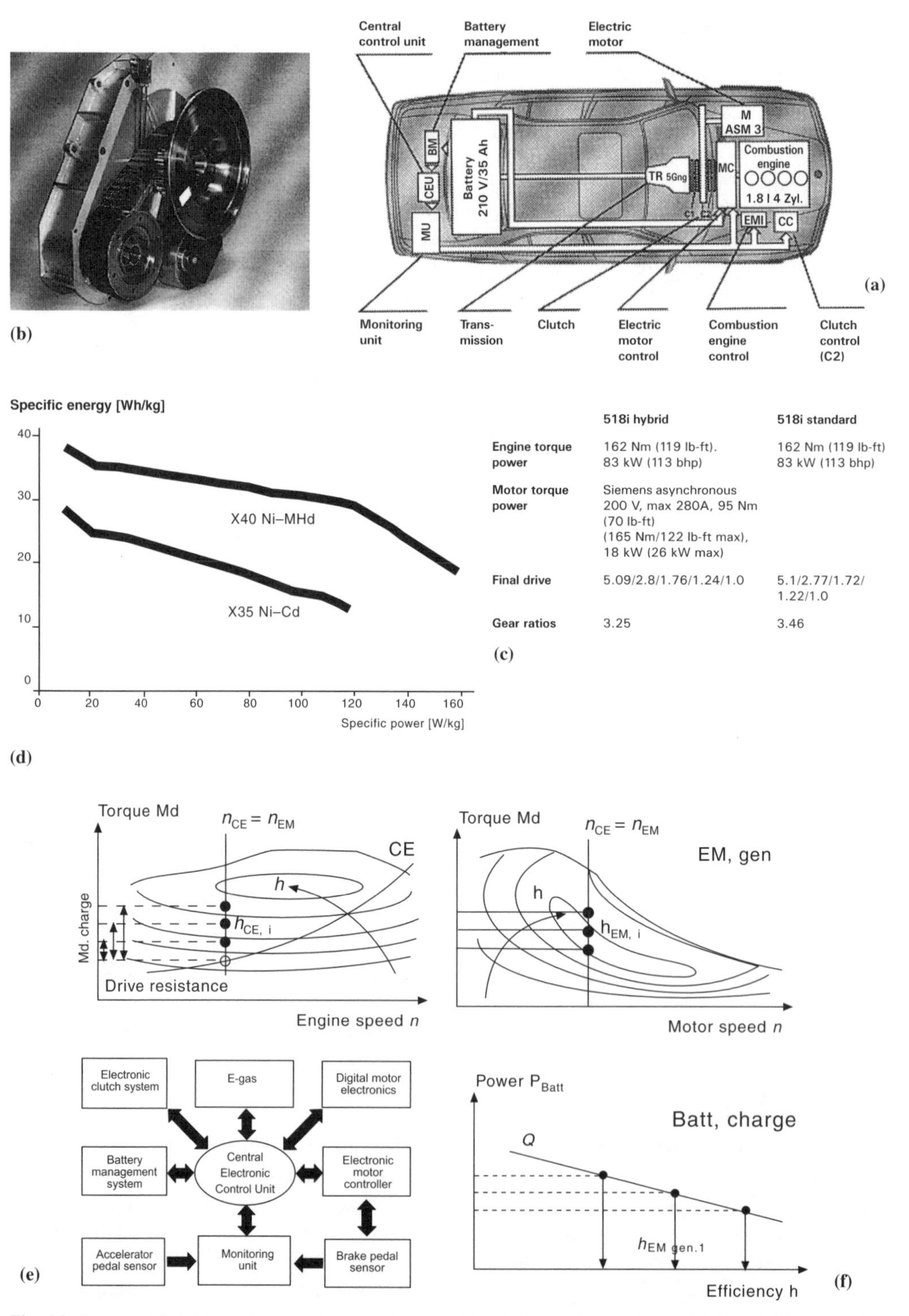

	518i hybrid	518i standard
Engine torque power	162 Nm (119 lb-ft). 83 kW (113 bhp)	162 Nm (119 lb-ft) 83 kW (113 bhp)
Motor torque power	Siemens asynchronous 200 V, max 280A, 95 Nm (70 lb-ft) (165 Nm/122 lb-ft max), 18 kW (26 kW max)	
Final drive	5.09/2.8/1.76/1.24/1.0	5.1/2.77/1.72/ 1.22/1.0
Gear ratios	3.25	3.46

(c)

Fig. 6.3 Map-controlled drive management: (a) BMW parallel hybrid drive; (b) parallel hybrid drive mechanism; (c) vehicle specification; (d) ragone diagram for the two battery systems; (e) vehicle management; (f) optimized recharge strategy.

6.2.2 JUSTIFYING HYBRID DRIVE, FIG. 6.4

Studies carried out at the General Research Corporation in California, where legislation on zero emission vehicles is hotly contested, have shown that the 160 km range electric car could electrify some 80% of urban travel based on the average range requirements of city households, (a). It is unlikely, however, that a driver would take trips such that the full range of electric cars could be totally used before switching to the IC engine car for the remainder of the day's travel. This does not arise with a hybrid car whose entire electric range could be utilized before switching and it has been estimated that with similar electric range such a vehicle would cover 96% of urban travel requirements. In two or more car households, the second (and more) car could meet 100% of urban demand, if of the hybrid drive type.

Because of the system complexities of hybrid-drive vehicles, computer techniques have been developed to optimize the operating strategies. Ford researchers[3], as well as studying series and parallel systems, have also examined the combined series/parallel one shown at (b). The complexity of the analysis is shown by the fact that in one system, having four clutches, there are 16 possible configurations depending on state of engagement. They also differentiated between types with and without wall-plug re-energization of the batteries between trips.

6.2.3 MIXED HYBRID-DRIVE CONFIGURATIONS

Coauthor Ron Hodkinson argues that while initially parallel and series hybrid-drive configurations were seen as possible contenders (parallel for small vehicles and series for larger ones) it has been found in building 'real world' vehicles that a mixture of the two is needed. For cars a mainly parallel layout is required with a small series element. The latter is required in case the vehicle becomes stationary for a long time in a traffic jam to make sure the traction battery always remains charged to sustain the 'hotel loads' (air conditioning etc.) on the vehicle's electrical system. Cars like the Toyota Prius have 3–4 kW series capability but detail configuration of the system as a

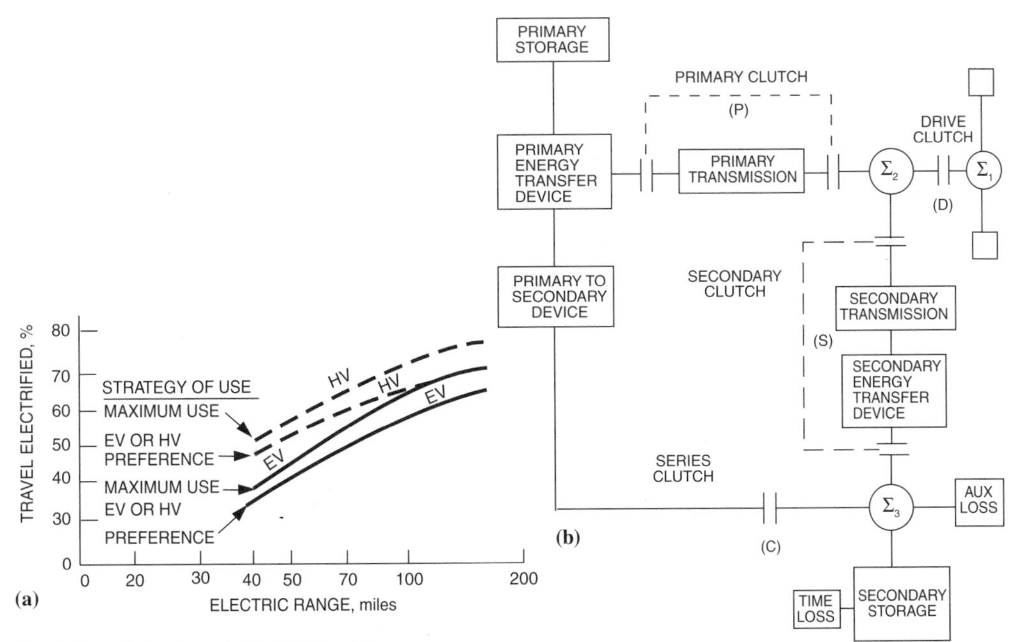

Fig. 6.4 Justifying the hybrid: (a) EV traffic potential; (b) combined series–parallel mode.

whole is just a matter of cost vs performance. Generally the most economical solution for passenger cars is with front wheel drive and a conventional differential/final-drive gearbox driven by a single electric motor. No change-speed gearbox is required, where the motor can give constant power over a 4:1 speed range, but reduction gearing is required to match 13 500 rpm typical motor speed with some 800 rpm road-wheel speed. This is usually in the form of a two-stage reduction by epicyclic gear trains, the first down to 4000 rpm, and the final drive gearing providing the second stage – typically two stages of 3–4:1 are involved. A change-speed gearbox only becomes necessary in simple lightweight vehicles using brushed DC motors and Curtis controllers. Weight can be saved by using a motor of one-quarter the normal torque capacity and multiplying up the torque via the gearbox.

6.3 Hybrid technology case studies

6.3.1 THE HYBRID ELECTRIC SOLUTION FOR SMALL CARS

Ron Hodkinson[4] points out that US President Bill Clinton's initiative for the American family car sets the target that, in 2003, cars will run for 100 miles on one US gallon of unleaded gasoline. The objectives are reduced fuel consumption, reduced imported oil dependency, and reduced pollution to improve air quality. Can it be done? The answer is yes. Work carried out on GM's ultra-lightweight car programme (see Chapter 7) involved composite structure to achieve a body weight of 450 kg, drag coefficient $C_d < 0.2$ by means of streamlining the underside of the car, reduced frontal area (<1.5 m^2), conventional drive train with a 30 bhp two-stroke orbital engine and low-rolling-resistance tyres. Overall, this leads to a vehicle weight of 750 kg, a 400 kg payload at a top speed of 80 mph and an acceleration of 0–60 mph in 20 seconds. This illustrates the dilemma. Reduce the engine size to improve the fuel consumption, and the acceleration performance is sacrificed.

6.3.2 HYBRID POWER PACK, A BETTER SOLUTION

In the long term we may use electric vehicles using flywheel storage or fuel cells. Until these systems are available the best answer is to use a hybrid drive line consisting of a small battery, a 45 kW electric drive, and a 22.5 kW engine. This solution would increase the vehicle weight from

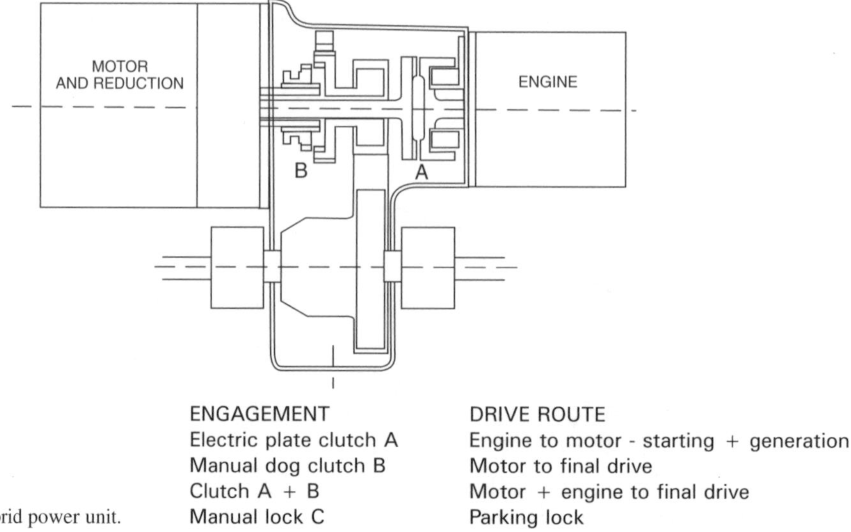

Fig. 6.5 The hybrid power unit.

750 kg to 860 kg, but it would now accelerate from 0 to 60 mph in 8 seconds. In addition, the vehicle will have automatic transmission with regenerative braking, could operate in electric-only mode with a 30 mile range for use in zero emission mode in city centres, and could be recharged from a wall socket or charging point if desired.

The Polaron subsidiary, Nelco, worked with Wychwood Engineering and Midwest Aero Engines on a parallel hybrid replacement for a front-wheel drive train in family cars and delivery vans, Fig. 6.5.

6.3.3 ROTARY ENGINE WITH PM MOTOR, THE MECHANICAL OUTLINES

The drive line is a marriage of two techniques: a permanent-magnet brushless DC motor and a Wankel two-stroke engine. The electric motor provides instant acceleration with 45 kW of power available from 1500 to 6000 rpm, on this design. A permanent-magnet design is used because it is lightweight, highly efficient, and results in an economical inverter. The concept is to exploit the machine characteristics using vector control. At low speeds, the permanent magnets provide the motor field. At high speeds, the field is weakened by introducing a reactive *Id* component at right angles to the torque-producing component *Iq*. The control objective is to maintain the terminal voltage of the motor constant in the high speed region, Fig. 6.6.

The system efficiency is achieved by using a dual-mode control system for the inverter: (i) at low speeds the inverter operates in the current-source mode; the current in the motor windings is pulse-width modulated; (ii) when the rectified motor voltage exceeds that of the DC link, the inverter changes to voltage-source control. Since the system operates with field weakening, the machine has a leading power factor, consequently there is virtually no switching loss in the inverter transistors if square-wave excitation is used. Since the motor has 30% impedance at full load, the impedance at the 5th/7th harmonic is 150/210%, and consequently there are few harmonics in the current. As a result, low-saturation IGBTs (insulation gate bipolar transistors) can be used, and they switch at less than 2 kHz with high efficiency in the cruise mode and low RF interference.

A key benefit of the PM brushless DC motor is the wide band constant-power curve. Consequently there is no need for gear changing for high-torque operation, and the motor gives high efficiency and low rotor heating both on the flat and in hilly terrain. The motor inverter and battery are oil cooled to ensure compact dimensions. The battery uses lead/tinfoil plates to achieve low internal resistance and is thermally managed to ensure charge equalization in the cells. A chopper is used to give a stabilized 300 V DC link. The outstanding feature of the lead–acid battery is the peak power capability of 50 kW for 2 minutes in a weight of 170 kg, Fig. 6.7(a).

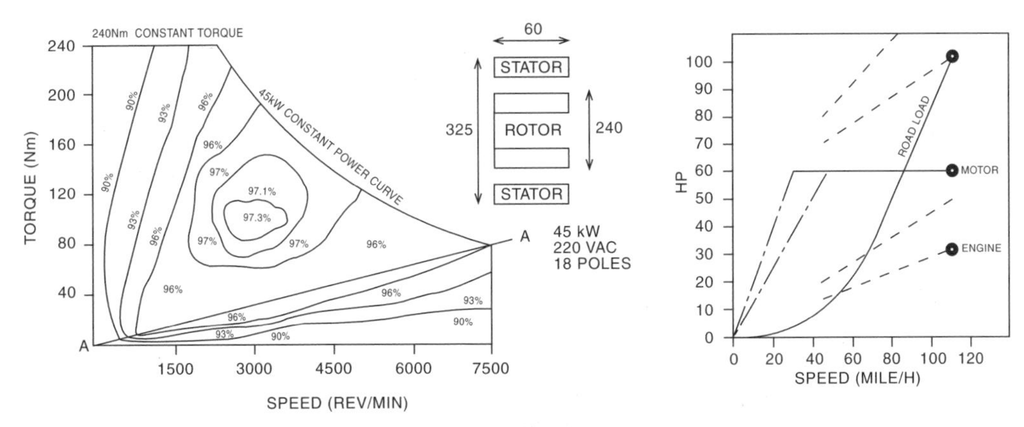

Fig. 6.6 Efficiency map, torque/speed curves and load matching.

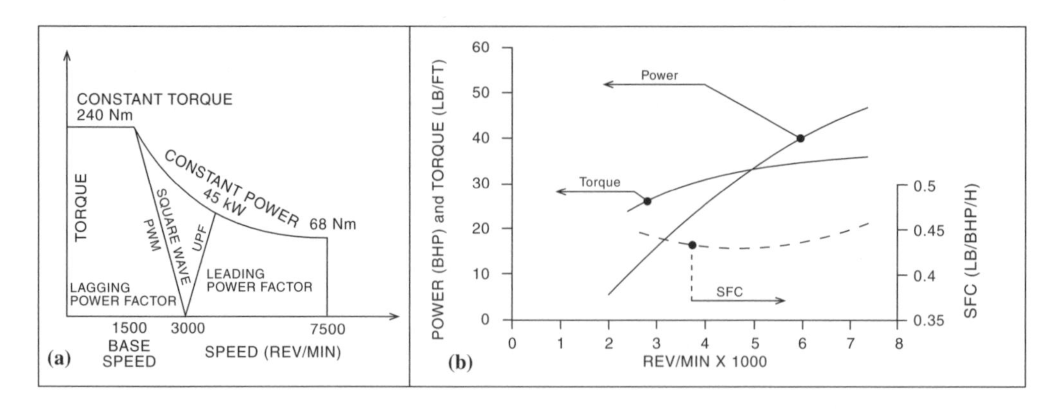

Fig. 6.7 System operation states (a), with performance curves of 300 cc fuel-injection Wankel engine (b).

6.3.4 WANKEL ROTARY ENGINE

The Wankel engine is well proven at 300 cc size, having full Air Registration Board certification for use in drones and microlights. The main benefits of the Wankel engine are: light weight and '6 cylinder' smoothness; flat torque/speed curve and good fuel economy with fuel injection very low emissions; on natural gas it is possible to comply without a catalytic converter. On unleaded petrol, two electrically preheated catalytic converters are used. Another advantage is multi-fuel capability. In the control scheme, the electric drive runs continuously. The Wankel engine switches on and off. At speeds above 60 mph, the engine will run continuously. The objective is to avoid discharging the battery by more than 30%. In this way we can obtain 11 000 cycles or 100 000 miles on a small battery. Consequently the economics of this scheme make sense.

In mass production (>100 000 systems) the additional cost per car of this system would be £2500. We believe this cost could be recouped, in fuel savings and reduced maintenance, by a 3 year first user, Fig. 6.7(b). The problem is to break through the dichotomy of the electric-vehicle market. Market logic says start with luxury vehicles and work down. The environment requires an impact on the mass market, that is the 'repmobile', as soon as possible. At least we now have a system that can meet the environmental demands at a reasonable cost, and which the market will be prepared to buy.

6.3.5 HYBRID PASSENGER CARS

A recent method of construction for permanent magnet motors (Fig. 6.8), by Fichtel & Sachs and Magnet Motor GmbH, exploits the relationship that specific motor torque, on the basis of weight and bulk, is proportional to the product of magnetic field in the air gap, radius of the air gap squared and the axial length of the motor. The requirement for maximum air gap focuses on the construction of the outside of the rotor. On the inside tangentially magnetized permanent magnets are fitted to the circumference. Shown inset in (a), trapezoidal iron conductors are seen between the rare-earth element magnets which are also trapezoidal in section. These collect magnetic field and divert it to the stator. The laminations of the stator are arranged radially and wound in individual coils which are connected individually or in groups, series or parallel, to the single phase power electronic DC/AC converters.

The latter are supplied with power from a DC link circuit and commutate the coil current at the amplitude required by the rotor angle as detected by remote sensors. The converters only supply the section of the motor assigned to them and thus work independently. They are made as 4 quadrant controllers and have IGBT switching. These are described as Multiple Electronic Permanent (MEP) magnet motors and are made by the Magnet Motor Co. With liquid cooling, the specific performance of these MEP machines is significantly increased over conventional EV motors as seen by the table at (b) and the corresponding characteristic curves at (c). Both companies are jointly engaged in further development for volume production with target performances seen in the tables at (d).

A test vehicle, Fig. 6.9, based on an Audi 100 Quattro of 100 kW has been constructed and the drive configuration chosen is seen at (a). The four MEP motors have a nominal performance of 25 kW each and are direct connected to each road wheel without the need for mechanical reduction. The tandem configuration of the motors has the advantages of an electrical torque apportioning differential; the motor is part of the vehicle sprung mass and relatively long drive shafts can be used. The generator is also of MEP construction and direct-flange-connected to the IC engine. Power electronics are used to provide power through a DC link circuit to the four motors.

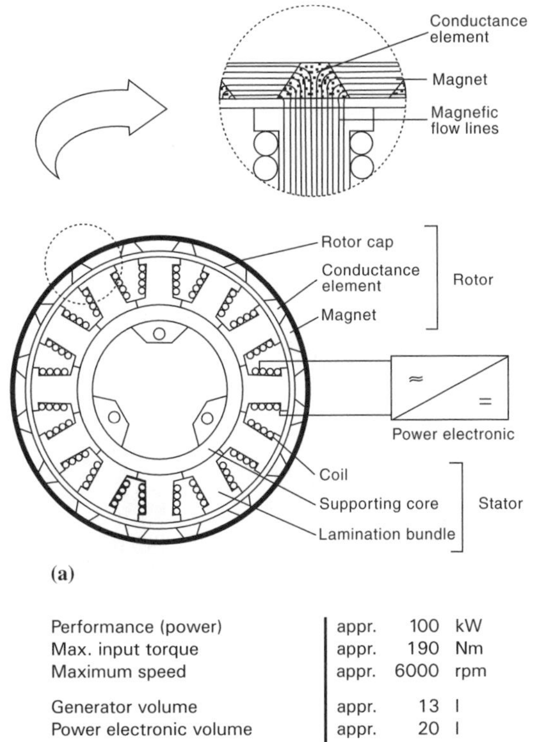

(a)

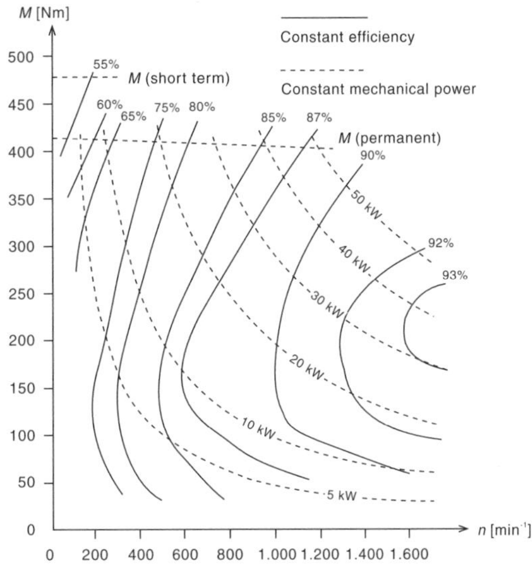

(c)

Performance (power)	appr.	100	kW
Max. input torque	appr.	190	Nm
Maximum speed	appr.	6000	rpm
Generator volume	appr.	13	l
Power electronic volume	appr.	20	l
Generator mass	≤	25	kg
Power-electronics mass	≤	13	kg
Efficiency at rated performance	≥	0.93	
Joint cooling method		liquid cooling	

(b)

Coolant temperature: 50°C		
Maximum torque (short term)	482	Nm
(permanent)	410	Nm
Maximum mechanical performance	49	kW
Mass, including power electronic	30	kg
Maximum efficiency	0.935	
Operating voltage	600	V DC

Performance	appr.	2 x 55	kW
Max. output torque	appr.	2 x 550	Nm
Maximum speed	appr.	1800	rpm
Tandem motor volume	appr.	17	litres
Power electronic volume	appr.	20	litres
Tandem motor mass	≤	0	kg
Power electronic mass	≤	13	kg
Efficiency at rated performance	≥	0.93	
Joint cooling method		liquid cooling	

(d)

Fig. 6.8 Advanced PM motor system: (a) trapezoidal iron conductors; (b) motor performance; (c) motor curves; (d) target performances.

A drive-by-wire arrangement is involved in the IC-engine throttle control, an ECU matching engine speed to generator output. The view at (b) shows the speed/torque map of the engine set to a low SFC value. The ECU also controls the commutation of the MEP machines. The drive configuration allows the engine to operate at constant speed and the wheel speed reductions are also controlled by software in the ECU and thus different driving programs can be instituted. Handling control is also affected by the software in that different torque distribution to the road wheel can be programmed. The MEP generator also acts as a starter motor for the IC engine.

Test results and simulated performance are seen at (c). The simulation has also been used to test the effect of further developments such as the removal of the IC engine flywheel, redesign of the Audi platform to better exploit electric propulsion and the achievement of the target MEP machine performances previously listed. These estimated performances are seen at (d). A 10% fuel consumption is achieved for road performance equivalent to the standard car and work is under way to design ULEV versions incorporating on-board storage batteries.

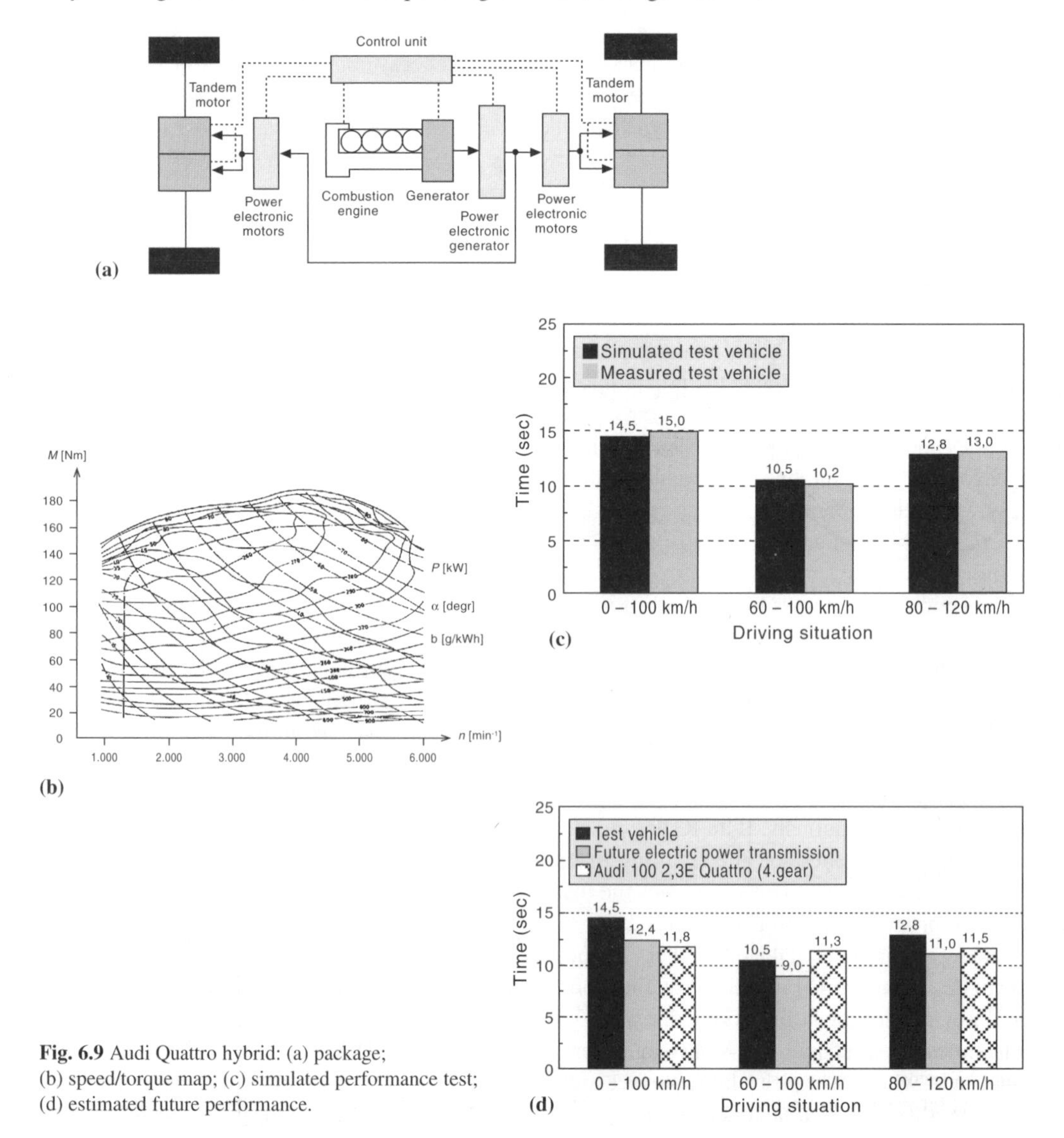

Fig. 6.9 Audi Quattro hybrid: (a) package; (b) speed/torque map; (c) simulated performance test; (d) estimated future performance.

6.3.6 TAXI HYBRID DRIVE

Based on a Range Rover, an EC project involving Rover, gas turbine makers OPRA, Athens Technical University, Hawker Batteries and Renault was carried out at Imperial College so as to provide a hybrid powertrain for a European taxi, Fig. 6.10, and in particular the development of a turbogenerator. A system involving a gas turbine and high speed generator together with a small battery pack for power storage and zero-emission city driving in a series configuration has been proposed by the EC and the turbogenerator work is reported by Pullen and Etemad[5], (a).

Its proponents point to the unpleasant side-effects of traffic congestion with conventional vehicles in that reduced vehicle speeds cause the engine to run more than necessary for each journey; the average operating point for the engine is further away from its design point and greater energy wastage is involved in braking. In urban usage, the average power consumption of the vehicle is quite low since average vehicle speeds are low. Typical power consumption figures are shown at (b) with only 5 kW being sufficient to drive the vehicle at a speed of 60 km/h, generally above the average vehicle speed for European cities. However, large, short duration power demands must be met to provide sufficient acceleration. Since the rate of energy consumption is low in the urban regime, this energy can be provided by means of a lead–acid battery and give an acceptable range of 50–70 km. Such a battery must, however, be capable of meeting the peak power demands for acceleration without overheating. High specific energy 'advanced' lead–acid batteries are hence required.

While the vehicle will operate at zero emissions in the above mode of operation, if the journey length is greater than the range, the gas turbine engine must be operated to allow the batteries to be recharged. Again, 'advanced' lead–acid batteries are required to allow the full power of the gas turbine to be absorbed, hence preventing it from operating in a part load condition. Although the vehicle will not be ZEV in this mode, the composite emissions will still meet the ultra-low emissions requirements (ULEV) due to the inherent low emissions characteristics of the gas turbine engine. Combustion in the gas turbine is steady state as compared to the intermittent combustion in IC engines and is always lean burn. The potential for further reduction of small gas turbine emissions is also very good and has been demonstrated in much larger machines in the industrial sector. In highway usage, operation requires a much greater rate of energy consumption due to the increased average vehicle speeds. Typically 30–50 kW is required for passenger vehicles operating at motorway speeds.

It is currently not possible to store sufficient energy in a battery or any other accumulator such as a flywheel to give anywhere near an acceptable range at such a discharge rate. The gas turbine must hence operate continuously although not necessarily at constant power. Generally, the power output of the gas turbine will be matched to the vehicle power demand although some form of smoothing of the power demand will be needed to avoid rapid cycling of the gas turbine engine. Any excess power generated or power deficit will be handled by absorbing this or taking this from the battery. Since the power demand is high and continuous, the engine must be able to produce such power continuously. If an IC engine generator set was to be used instead of the gas turbine and high speed generator, the resulting machinery would be too voluminous and heavy for the series hybrid concept. This eliminates the feasibility of a small 'battery charger' IC engine which is often proposed for hybrid vehicles.

A key requirement of the researchers' proposal is to produce a gas turbine at a price corresponding to an IC engine while providing low fuel consumption over the medium to full power range. Tests are now under way at Imperial College on gas turbines of 30 and 50 kW; these are simple cycle machines but a recuperator is envisaged for the larger engine. The generator envisaged is a disc alternator which provides a large surface area for cooling from a relatively low volume machine, (c). The flux path is axial and is returned at the rotor ends by high strength steel keeper discs, the retainment ring being of carbon fibre. A key factor is that the rotor magnet retaining ring is not between the magnets and the stator; the flux within the stator is consequently higher and the diameter of the rotor can be greater. This results in a compact shaft length suitable for high speeds. An induction motor used with the system is considered to be the most reliable of AC types, which are generally lighter than DC ones, and 94% overall efficiency could be expected, the researchers maintain. It is suggested that one motor per drive wheel be used with differential speed action being geared to the position of the steering wheel and an anti-spin system is envisaged.

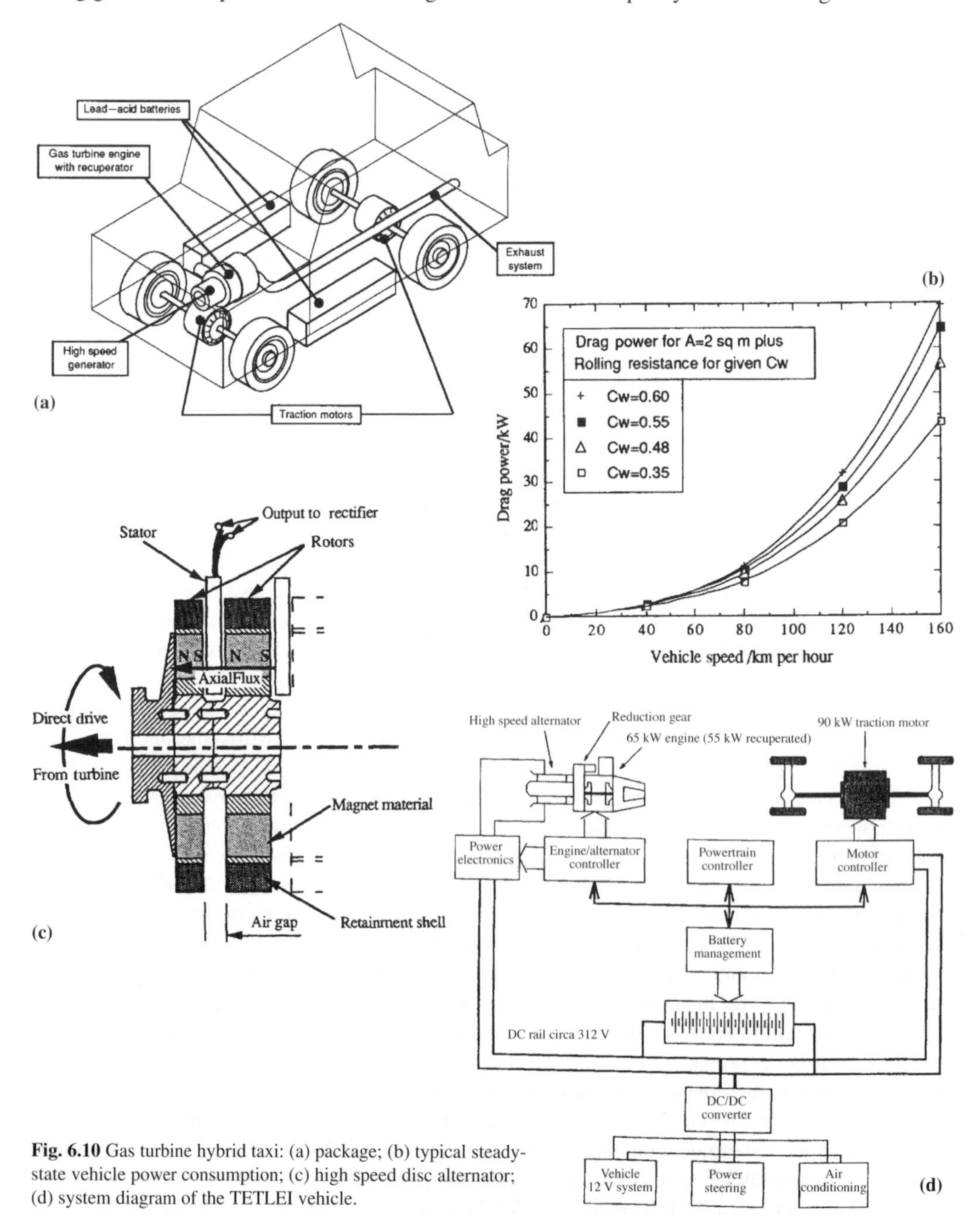

Fig. 6.10 Gas turbine hybrid taxi: (a) package; (b) typical steady-state vehicle power consumption; (c) high speed disc alternator; (d) system diagram of the TETLEI vehicle.

The Rover Group have been involved in the TETLEI Euro-taxi project and K. Lillie, with Warwick University co-workers[6], have also gone into print on the gas turbine series hybrid concept. The taxi will be based on the latest Range Rover and the series hybrid mode allows the gas turbine to be decoupled from the wheels so as to operate at its optimum speed and load and avoid the classic limitation of this type of power unit, poor fuel economy at light loads, poor dynamic response and a high rotational speed required, over 60 000 rpm. A sophisticated control system is required to run the turbine in on-off mode according to power demand. One of the vehicle schematics under consideration is seen at (d) which will be computer modelled to assess its effectiveness. The researchers argue that the development of a valid simulation requires that a number of factors are fully considered. As all of the data stems from an initial calculation of the battery current, it is important that this value is accurate. Small variations in the internal resistance of the battery can cause large variations in the rail voltage (square of variation). It is important to have an accurate model of the battery which considers both the variation of internal resistance and open-circuit voltage of the battery at different states of charge. Modelling the drive cycle at the two extremes of battery operation (80% and 20% discharge) gives a good indication of the range of values over which the voltage and currents in the system may vary for specification purposes. The next step is to introduce a more realistic battery model and practical limits on the power sourced from regenerative braking. During electrical regenerative braking, there is a practical upper limit to the voltage permitted across a battery in order to avoid 'gassing'. This restriction may be overcome by the use of an additional form of power sink if the alternator is to continue operating at a fixed load point.

6.3.7 DUAL HYBRID SYSTEM

Japanese researchers[7] from Equos Research have described the dual system of hybrid drive which differs from the more familiar series–parallel drives and their combinations. It allows free control of the IC engine while keeping mechanical connection between it and the drive wheels; a compact transaxle design integrates the two electric drive motors, to simplify the conversion of conventional vehicles, and use of the generator as a motor in combination permits flexible adaptation to driving conditions. Essentially, the 'split' drive system divides the output from the engine using a planetary gear, Fig. 6.11.

Instead of using a switching system, between series and parallel drive, (a), the split system acts as a series and parallel system at all times, the planetary gear dividing the drive between the series path of engine to generator and parallel path of engine to drive wheels. As parallel-path engine speed increases in proportion to vehicle speed, output energy from the engine also increases with vehicle speed, as is normally required. At high speeds most of the engine output is supplied by the parallel path and a smaller generator for the series system can therefore be employed.

The dual system, (b), is an optimized arrangement of the split system and thus far has been applied to a Toyota Corolla with an all-up weight of 1345 kg, involving a 660 cc engine adapted to drive-by-wire throttle control and giving 90–100 kph cruising speed. In the Toyota the dual system engine is mounted, for front wheel drive, onto a transaxle (c) of just 359 mm overall length, which is shorter than the production automatic transmission installation and 30 kg lighter than the engine/transaxle assembly of the standard model. The transaxle is of four shaft configuration, with compactness achieved by mounting motor and engine on separate shafts, each having optimized gear reduction, of 4.19 overall for the engine and 7.99 for the motor. The planetary splitter gear has a carrier connected to the engine and ring gear to the output shaft; it also acts as a speed increasing and torque reducing device for the motor/generator, with a 3.21 reduction ratio.

The motor/generator is a brushless 8 pole DC machine with 6 kW output, having generator brake and planetary gear installed within the coil ends for compactness. The generator functions

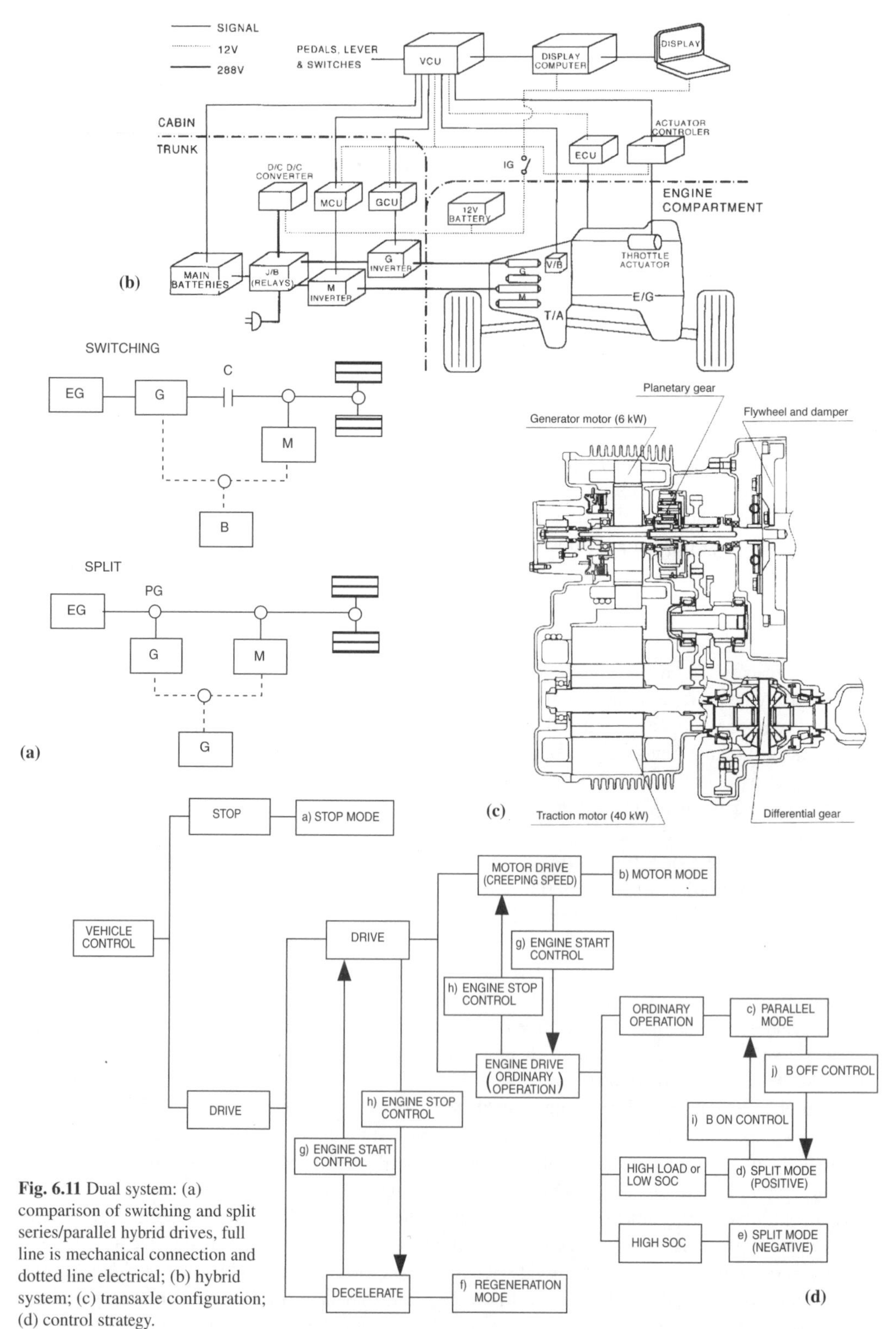

Fig. 6.11 Dual system: (a) comparison of switching and split series/parallel hybrid drives, full line is mechanical connection and dotted line electrical; (b) hybrid system; (c) transaxle configuration; (d) control strategy.

as a starting device, clutch and form of CVT. The 40 kW traction motor is a 4 pole brushless motor which functions as a torque levelling device of the parallel hybrid system. Under low-load cruising conditions, the system uses parallel hybrid mode with the brake engaged, preventing the motor/generator from causing energy conversion losses. Brake cooling oil is also used for cooling the motor coils. Twenty-four lead–acid batteries are used, of 25 Ah capacity each, to give a total voltage of 288 V, the type being Cyclon-25C VRLA. Overall control strategy is seen at (d).

6.3.8 *FLYWHEEL ADDITION TO HYBRID DRIVE*

According to Thoolen[8], the problem of providing peak power for acceleration, and recuperation of braking energy, in an efficient hybrid-drive vehicle can be overcome with an electromechanical accumulator. Such systems also admirably suit multiple stop-start vehicles such as city buses by using flywheel and electric power transmission.

In the Emafer concept, Fig. 6.12, a flywheel motor/generator unit is controlled by a power electronic converter. The flywheel (a) is of advanced composites construction and the motor/generator of the synchronous permanent-magnet type while the converter uses high frequency power switches. The flywheel is comprised of four discs of tangentially wound prestressed fibre composite, designed to achieve a modularity of energy capacity as well as improved failure protection under the 100 000 *g* loading. The motor is of the exterior type with rotor outside the stator. High speed operation is possible as the rotor is merely a steel cylinder with permanent magnets on its inside. All windings are contained in the stator, having hollow journals at its ends

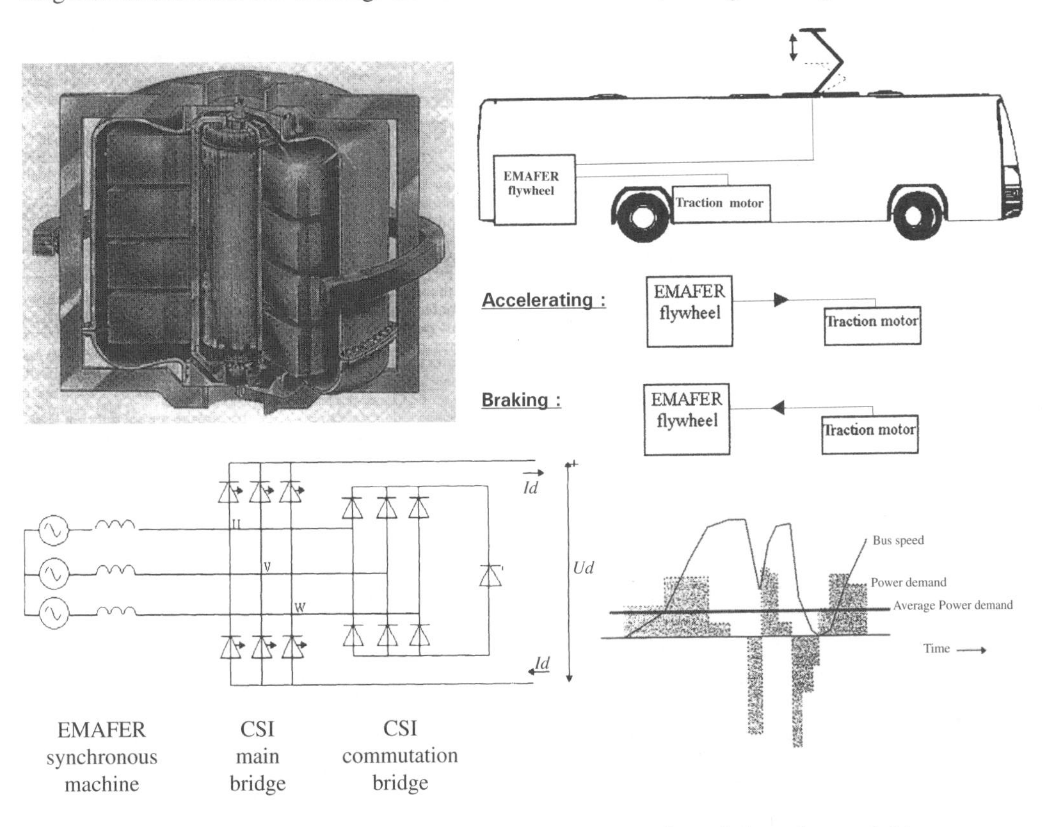

Fig. 6.12 Flywheel motor/generator: (a) Emafer flywheel/generator construction and electronic control; (b) power converter: shaded areas show power the Emafer has to supply/extract for accelerating or braking.

for feeding electrical energy, cooling and lubrication fluids. Carefully designed supports for the high speed ball bearings allows the rotor to run 'overcritical' without serious vibration modes. Bearings, rotor and stator are vacuum-enclosed for reducing windage losses and for safety reasons. The containment is cardanically suspended to avoid gyroscopic effects.

The power converter, (b), controls exchange of electrical power between the 3 phase terminals of the motor/generator and the DC load. Of the current source inverter (CSI) type, it comprises a full bridge with six semiconductor switches and GTO thyristors. The latter are driven by measurements by the CSI of rotor position, DC voltage and current. When used as a sole driving source the Emafer is charged at bus stops by overhead supply contacts. In a hybrid drive-line, an IC engine on board, with generator, supplies the average power demand, with the Emafer taking care of fluctuations about the average, the flywheel extracting or applying power according to braking or accelerating mode.

6.4 Series-production hybrid-drive cars

During the early stages of introducing hybrid vehicles into the urban scene, state or local authorities may well offer direct and indirect financial inducements to get these 'clean vehicles' into areas that suffer from atmospheric pollution by motor transport. Now Toyota are manufacturing their Prius, Fig. 6.13, at the rate of 1000 a month and these cars are selling well in Japan. The Japanese Government, in a deliberate effort to curb urban pollution in Japan, is subsidizing the manufacture and sales drive by a variety of tax concessions, including one that directly benefits the user/operator of the hybrid saloon. The deal with the Japanese Government has enabled Toyota to offer these cars as a competitive package, when taking these taxation inducements into account. Toyota have found a technical solution, which, in engineering terms, is both ingenious and realistic. The company have made use of various new technologies to reduce the weight of the vehicle and its major components and systems. For example, the rolling resistance of the tyres has also been minimized, which reduces power demand by about 5 to 8%.

6.4.1 TOYOTA PRIUS SYSTEMS, FIG. 6.14

The Toyota Hybrid System (THS), (a), has two motive power sources, which are selectively engaged, depending on driving conditions: (1) A 1.5 litre petrol engine, developing 42.5 kW at 4000 rpm and a peak torque of 102 Nm at 4000 rpm; (2) A battery-powered permanent magnet synchronous electric motor with a maximum output of 30 kW over the speed range of 940–2000 rpm and peak torque of 305 Nm from standstill to 940 rpm. The petrol engine is the hybrid's main power source. It is a 1.5 litre DOHC 16 valve, 13.5:1 compression ratio, engine with VVT-I (Variable Valve Timing: Intelligent, a continuously variable valve mechanism) and electronic fuel

(a)

(b)

Fig. 6.13 Toyota Prius and its THS inverter.

injection, using the highly heat-efficient Miller cycle, that, in turn, is a further development of the high expansion Atkinson cycle. In this cycle the expansion continues for longer than in the conventional 4-stroke engine, thereby extracting more of the thermal energy of the burning gases than can be achieved in either a 2-stroke or 4-stroke engine of conventional design. Engine revolutions are restricted to 4000 rpm maximum, and the engine is electronically controlled to run always within a relatively narrow band of engine speed and load, corresponding to optimum fuel efficiency. Toyota claim that this sacrifice of a wide span of engine speed is more than made up for by the greater flexibility of the epicyclic drive system and the power split between the two motive-power units.

THS functions as a continuously variable transmission and combines power from the petrol engine and the electric motor, to give smooth power delivery with little lag between the driver depressing the accelerator pedal and vehicle response. The innovative features of the Prius are in the design details of the power sources and the power split device in the hybrid transmission that allocates power from the petrol engine either directly to the vehicle's front wheels or to the electric generator. The power-split device, (b), employs a planetary gear system, which can steplessly effect the optimum power flow to suit the driving conditions encountered at any one moment. One of the output shafts of the power-split device is linked to the electric generator, while the other is linked to the electric motor and road wheels. The complex transmission system (c), which also includes a reduction gear, is electronically controlled, with the power flow allocation constantly being reviewed by the special control unit. This means that the information, which has been gathered by a number of key sensors, is compared with the target values encoded in the ECU, the system's brain. This ECU ensures that the appropriate elements in the epicyclic transmission are being braked or released, so that the respective speed of the petrol engine, the electric generator, and the electric motor are held within the optimum performance band. This power flow allocation split will depend on whether the car is being driven at a steady rate, accelerated or slowing down. The distribution of the petrol engine's power, which is so regulated that it will generally operate mainly in its optimum fuel efficiency band, the high torque zone, is determined by such factors as throttle opening, vehicle speed, and state of battery charge. The portion that is used to turn the wheels is balanced against that which is used to generate electric power. Electric power created by the largish generator may then be used to power the electric motor, to help drive the vehicle. There are a number of systems operating conditions.

In 'normal driving' the engine power is divided into two power-flow paths by the power-split device, one route will directly power the road wheels, and the other will drive the electric generator. Electric current from the generator may be used to power the electric motor, to assist in driving the road wheels. Electric current may also flow into the traction battery pack, to top up its charge. The power-split electronic control system determines the ratio of power flow to these outlets in such a manner that optimum fuel efficiency and responsive driveability are maintained at all times. The battery pack is made up of 40 individual nickel–metal hydride batteries and has a relatively small capacity of only 6.5 Ah, which would not give it much of a range in driving the vehicle in an all-electric mode.

During 'full throttle acceleration' drive mode, (d), power is also supplied from the battery to augment the drive power supplied by the petrol engine. Such a power boost, though adequate for overtaking and short bursts of speed, can generally not be maintained for extended periods of high speed motoring. The vehicle is being promoted as a car that produces only half the amount of CO_2 of conventionally powered compact-size cars and only one-tenth of the amount of HC, CO, and NO_x permitted under current Japanese emission regulations. Despite the vehicle having a kerb weight of 1.5 tonne, Toyota claim that Prius will accelerate from standstill to 400 m in 19.4 seconds and reach a top speed of 160 km/h.

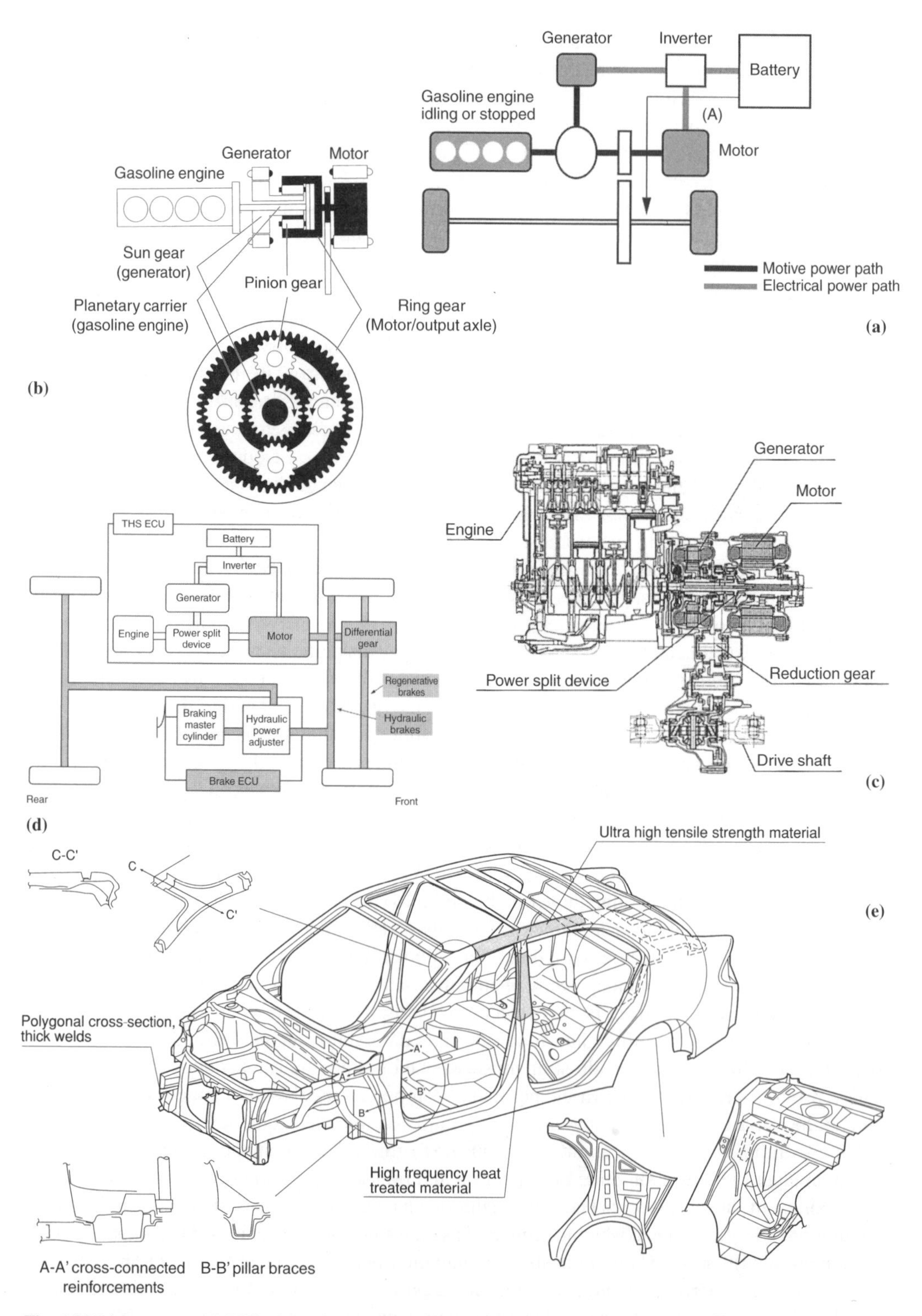

Fig. 6.14 Prius systems: (a) THS schematic; (b) power-split device; (c) engine and hybrid transmission; (d) hybrid system in full-throttle acceleration; (e) lightweight structure detail.

In 'starting from rest and light load' mode (moving at low speed or descending a slight gradient) the electric motor drives the vehicle and the petrol engine is stopped. The high torque characteristic of the electric motor helps to get the car moving and will sustain it during low load demand slow speed progress in urban centres. Should additional power be required from the petrol engine, the computer control system will ensure that the engine will play its part, either by charging the traction battery pack or by some direct contribution to driving the road wheels. But when coming to rest at traffic lights, the fuel supply to the engine is cut off and the engine is automatically stopped.

During 'deceleration and braking' mode, the kinetic energy of the moving mass of the vehicle passes from the road wheels through the epicyclic transmission gearing of the power-split device to the electric motor. This then acts as an electric generator, delivering this energy as a charging current to top up the traction battery pack. This feature of regenerative braking comes into play, regardless of whether the operator applies the foot brake or relies on engine braking to slow down the car. A complex but compact full power inverter and control unit ensures that the traction battery pack is being maintained at a constant charge. When the charge is low, the electric generator routes power to the battery. In most instances, this energy will come from the internal combustion engine rather than the energy recovered during braking. The system has been so designed that the batteries do not require external charging, which means that there is no practical restriction to the operating range of the vehicle.

Toyota have compensated for the dual drive and battery weight penalty with a number of ingenious measures: since the petrol engine has been restricted to a maximum of 4000 rpm, key components have been pared down to save weight. Compared with an engine of comparable size but 5600 rpm maximum speed, the internal dynamic loadings on many of the moving parts are halved. Consequently there is scope for reducing the dimensions of, for instance, crankshaft journals and also pistons, which have remarkably short skirts; the overall effect of paring down of individual components is reduced weight of the built-up assembly. Overall length is only 4.28 m, but the car provides an interior space equal to that of many medium-class cars, by having a relatively long wheelbase of 2.55 m, a 1.7 m wide body, and short overhangs front and rear. The Prius boot has a reasonable capacity, thanks to the newly developed rear suspension which has no internal protrusions into the luggage compartment. The slanted, short bonnet covers the transversely mounted and very compact THS power train assembly. With a height of 1.49 m, the car stands taller than others in its class. The blending of the three box layout into a good aerodynamic shape has resulted in a drag factor of $C_d = 0.30$. Considerable weight saving, (e), without any sacrifice in passive safety, has been achieved in the body-in-white. The platform is based on Toyota's GOA (Global Outstanding Assessment) concept study of an impact-absorbing body and high integrity occupant cabin design, developed to meet 1999 US safety standards. Ribs made of energy-absorbing materials are embedded inside the pillars and roof side rails. GOA also features strong cross members, several produced in higher-tensile-strength sheet steel, linking the various body frame elements. These provide strength and stiffness, particularly in potential collision damage zones, and also spread the impact loading, thereby minimizing intrusion into the cabin, the occupant safety cell.

Air conditioning and power-assisted steering are featured. The automatic air conditioner creates a double layer of air, recirculating only internal air around the leg areas, even when the fresh air intake mode has been selected. The glass in the side and rear windows is of a type which inhibits heating up of the cabin space, by blocking most of the sun's ultraviolet rays. Insulating materials in the roof and floor panels also contribute to maintaining a comfortable cabin atmosphere. They also offer good sound insulation. Steering has a power-assist system using an electric motor, which consumes power only during steering operations. The front suspension has MacPherson struts with L-arms for locating their lower ends. In the semi-trailing-arm rear suspension, the combined coil spring and hydraulic damper units are much shorter. Their lower attachment is to a trailing arm each, which, in turn, is attached to an innovative type of torsion beam, of an inverted channel section. It incorporates toe-control links, to improve handling stability and the double-layer anti-vibration mounts joining the suspension to the chassis suppress much of the road noise. Passenger comfort is appropriate to a car which retails at around 27 to 30% above a comparable, but conventionally propelled model.

For the power-split device, Fig. 6.15, which is a key part of the system, company engineers[9] have provided the diagram at (a) to show how the engine, generator and motor operate under different conditions. At A level with the vehicle at rest, the engine, generator and motor are also at rest; on engine start-up the generator produces electricity acting as a starter to start the engine as well as operating the motor causing the vehicle to move off as at B. For normal driving the engine supplies enough power and there is no need for the generation of electricity, C. As the vehicle accelerates from the cruise condition, generator output increases and the motor sends extra power to the drive shaft for assisting acceleration, D. The system can change engine speed by controlling generator speed; some of the engine output goes to the motor via the generator as extra acceleration

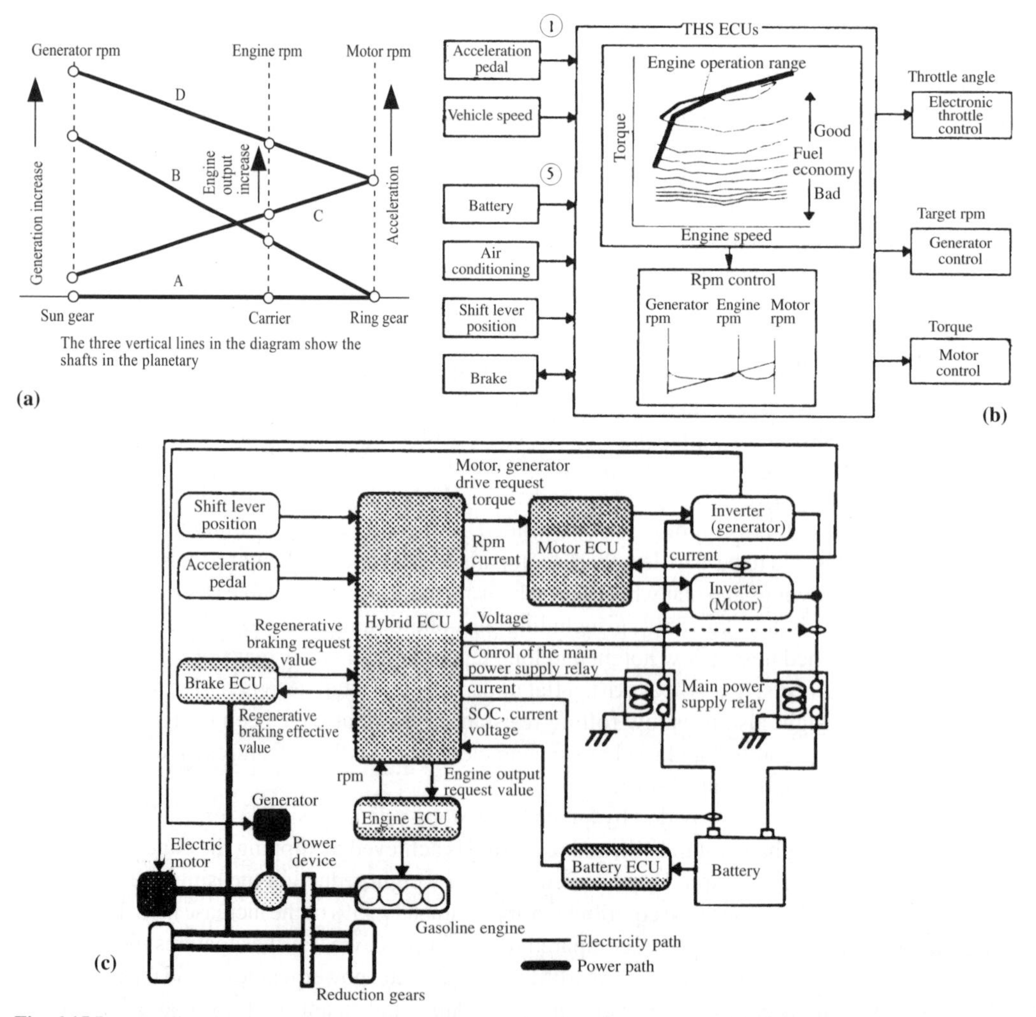

Fig. 6.15 Power-split control: (a) power interaction diagram; (b) THS control system; (c) ECU schematic.

power and there is no need for a conventional transmission. The control system schematic for the vehicle is at (b), the THS (Toyota Hybrid System) calculates desired and existing operating conditions and controls the vehicle systems accordingly, in real time.

The ECU keeps the engine operating in a predetermined high torque to maximize fuel economy. The corresponding schematic for the ECU is at (c). It is made up of five separate ECUs for the major vehicle systems. The hybrid ECU controls overall drive force by calculating engine output, motor torque and generator drive torque, based on accelerator and shift position. Request values sent out are received by other ECUs; the motor one controls the generator inverters to output a 3 phase DC current for desired torque; the engine ECU controls the electronic throttle in accordance with requested output; the braking ECU coordinates braking effort of motor regeneration and mechanical brakes; the battery ECU controls charge rate.

Toyota claim that Prius has achieved a remarkably low fuel consumption of 28 km/litre (79.5 mpg or 3.57 litre/100 km) on the 10/15 mode standard Japanese driving cycle.

6.4.2 RECENT ADDITION TO PRODUCTION HYBRID VEHICLES

Honda's Insight hybrid-drive car, Fig. 6.16, uses the company's Integrated Motor Assist (IMA) hybrid system, comprising high efficiency petrol engine, electric motor and lightweight 5-speed manual transmission, in combination with a lightweight and aerodynamic aluminium body, seen at (a), to provide acceleration of 0 to 62 mph in 12 seconds and a top speed of 112 mph, without compromising fuel economy of 83 mpg (3.41/100 km) and 80 g/km CO_2 (EUDC) emission. The car is claimed to have the world's lightest 1.0 litre, 3-cylinder petrol engine, which uses lean burn technology, low friction characteristics and lightweight materials in combination with a new lean burn compatible NO_x catalyst.

The electrical drive consists of an ultra-thin (60 mm) brushless motor, (b), directly connected to the crankshaft, (c), 144 V nickel–metal hydride (Ni–MH) batteries (weighing just 20 kg) and an electronic Power Control Unit (PCU). The electric motor draws power from the batteries during acceleration (so-called motor assist) to boost engine performance to the level of a 1.5 litre petrol engine as well as acting as a generator during deceleration to recharge the batteries. As a result engine output is increased from a high 50 to 56 kW with motor assist, but it is low speed torque that mainly benefits, boosting a non-assist 91 Nm at 4800 rpm to 113 Nm at 1500 rpm.

A new type of lightweight aluminium body, (d), offers a high level of rigidity and advanced safety performance. It is a combination of extruded, stamped and die cast aluminium components and body weight is said to be 40% less than a comparable steel body. All outer panels are aluminium except for the front wings and rear wheel skirt which are made from recyclable abs/nylon composite. Total kerb weight is 835 kg (850 kg including air conditioning). Aerodynamic characteristics include a streamlined nose, a low height and long tapered roof, narrow rear track, low drag grille, aluminium aero wheels, rear wheel skirt, a flat underside, and a tail designed to reduce the area of air separation. Insight also uses low rolling resistance tyres that have been designed to provide good handling, ride comfort and road noise characteristics. All these features give the Insight an aerodynamic drag coefficient of 0.25.

Further fuel savings are provided by an auto idle stop system. In simple terms, the engine cuts out as the car is brought to a standstill, and restarting is achieved by dipping the clutch and placing the car in gear. In combination, Honda calculates that weight reduction measures, aerodynamics and reduction of rolling resistance contribute to approximately 35% of the increase in fuel efficiency, and the IMA system a further 65% compared to a 1.5 litre Civic. Further features include ABS, electric power steering, dual air bags, AM/FM stereo cassette, power windows and mirrors, power door locks with keyless entry, automatic air conditioning and an anti-theft immobilizer.

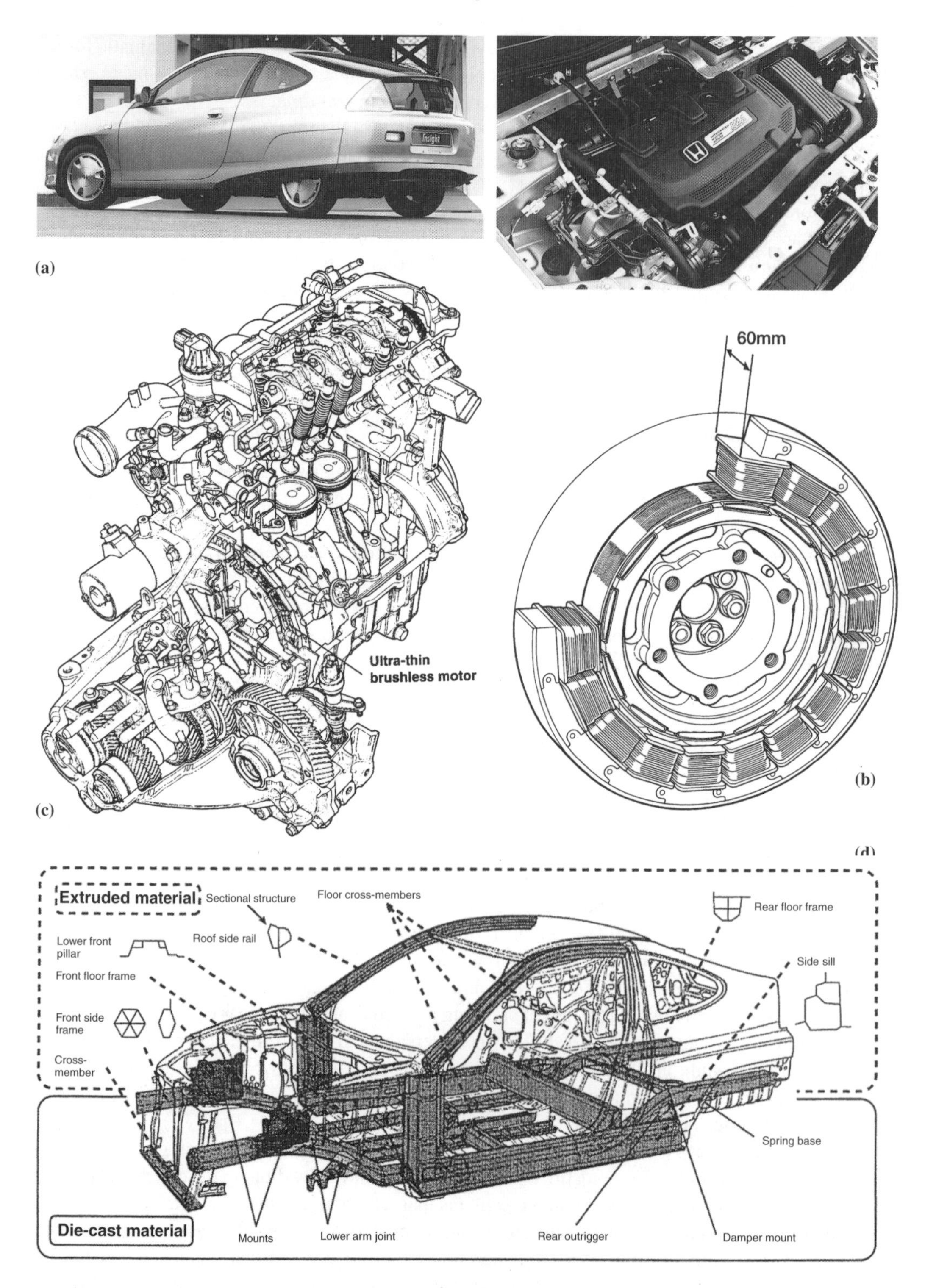

Fig. 6.16 Honda Insight hybrid: (a) aerodynamic tailed body and underbonnet power unit; (b) motor; (c) motor installation; (d) body structure.

The battery system is designed to avoid overcharging or complete emptying and in the unlikely event of motor failure, the Insight will run on the petrol engine alone. At the front, the suspension consists of struts, with an aluminium forged knuckle and lower arm, anti-roll bar linked to the dampers, and light aluminium cast wheels; while at the rear, a light and compact suspension features a twist beam with variable cross-section, and trailing arms with bushes having a toe-control function. Electric power steering, optimized for feel and feedback, has been used to make further fuel savings. It features a centre takeoff and aluminium forged tie-rod.

The company argue that in conventional petrol/electric hybrid systems, the vehicle is powered by the electric motor alone at low speeds. At higher vehicle speeds, or when recharging is required, engine torque is directed to the driven wheels or used to drive a generator. Such systems require complex control mechanisms, large capacity batteries, as well as a separate motor and generator. Honda chose instead a system in which the motor is linked directly to the engine, assisting it during acceleration for a reduction in consumption and acting as a generator during deceleration. When cruising, there is no assistance and lean burn keeps fuel consumption to a minimum. A very wide, flat torque curve is achieved through the benefits of VTEC technology at high engine speeds and the substantial boost provided by the electric motor at low and mid-range engine speeds. This approach allows for superior fuel efficiency and excellent driving performance over a wide range of driving situations.

The key to the engine operating at exceptionally low air/fuel ratios is rapid combustion of the mixture, since combustion time increases as the mixture becomes leaner. By adopting a new swirl port to enhance the turbulence of the mixture in the cylinder, a compact combustion chamber and a high compression ratio are achieved. The design is an evolution of the conventional VTEC-E mechanism where swirls are generated by almost closing one of the pair of inlet valves. In the new design, the inlet ports are set up in a more vertical direction to generate more powerful swirls flowing into the cylinder. This has been made possible with a new VTEC mechanism. Rather than inlet and exhaust rocker arms carried on separate rocker shafts, the Insight features just one rocker shaft with the included angle of the valves narrowed from 46 to 30°, allowing the high swirl port shape and the compact combustion chamber to be realized. Conventional lean burn engines, with their oxygen rich exhaust gases, mean reducing NO_x emissions is technologically difficult. The Insight's improvement in combustion efficiency goes some way towards solving the problem. However, a newly developed catalytic converter containing additives able to absorb NO_x, provides an elegant solution to the problem. During lean-burn driving, NO_x is directly absorbed; it is later reduced to harmless nitrogen in stoichiometric driving conditions. The system also helps to boost fuel efficiency, since it allows a widening of the lean burn range and therefore improved efficiency. Emissions performance is further improved by an exhaust manifold-integrated cylinder head. Rather than a conventional arrangement of an independent exhaust port for each cylinder, the ports are combined into one in the cylinder head structure. Considerable weight reduction is the result, but just as important, the small radiation area minimizes heat loss, enabling quick activation of the catalytic converter.

New technologies have reduced the overall friction of the engine by 38% compared with a conventional 1.5 litre engine. Among the measures adopted are roller type rocker arms, adapted to the single cam VTEC mechanism, providing a 70% reduction in friction losses. A special 'micro dimple' surface treatment of the piston skirt improves the retention of the oil film between the piston and the cylinder reducing friction by approximately 30%, in conjunction with offset cylinders and low tensile piston rings. By using case hardening for significantly increased strength, slimmer connecting rods have been adopted, achieving a reduction in weight of 30%. A newly developed magnesium alloy, with a high degree of heat resistance, has been used for the engine sump in place of aluminium alloy, giving a 35% weight reduction. Other weight saving technology includes: a thin sleeve block, the new VTEC cylinder head, bracketless ancillary equipment, a magnesium PCU case, and an increase in plastic parts (intake manifold, cylinder head cover, water pump pulley).

The ultra-thin brushless motor of 10 kW output sandwiched between the engine and transmission has a central rotor manufactured using the lost wax method, to give a precise shape and high strength, which achieves a 20% weight reduction. For the rotor magnet, improvements to the neodymium sintered magnet used in the Honda EV Plus mean an improvement in the magnetic flux density or torque ratio by 8%, while improved heat resistance has made a cooling system unnecessary. In order to create a thin motor, a split stator with compact salient-pole field winding and centralized bus ring forms a very simple structure allowing a width of 60 mm, 40% thinner than if conventional technologies were used. The Ni–MH battery pack installed at the rear of the car is held in a compact cylindrical pack. A series connection of 120 cells each with 1.2 V provides a voltage of 144 V. Ni–MH batteries are said by the company to offer stable output characteristics regardless of the charging condition, as well as excellent durability. The Power Control Unit (PCU), mounted alongside the battery pack, provides precision control of the motor assist and battery regeneration functions, as well as the supply of electricity to the standard 12 V battery through a DC/DC converter. The inverter which drives the motor, and is the most important element in the PCU, consists of a compact 3 phase integrated type switching module.

The weight target was a body-in-white of 150 kg, or half that of the Civic 3 door, the closest comparable sized Honda model; compared with the Civic it was reduced to 47%, yet torsional rigidity is up by 38%, and bending rigidity up by 13%. Hexagonal extruded aluminium cross-sections are used for the front side frame, bringing a weight saving of 37% while also attaining high energy absorbing characteristics compared to a conventional steel frame. The side sill and roof side rail which contribute considerably to the overall body rigidity, although simpler in cross-section, achieve 47% and 53% weight reductions respectively. A new manufacturing technique, 'three-dimensional bending forming', provides a degree of freedom in design, and a reduction in the number of parts required has been adopted, for example to produce the roof side rails. Widely different sections and the need for high rigidity called for the special jointing method involving die cast aluminium, which permits a high degree of shaping and flexibility in joining different shaped sections. However, in the case of the rear outrigger, where structural frames meet from three directions and which serves as the installation area for the suspension frame, an alternative was required. Its deep box-like shape means that if it were formed with the conventional die cast method its wall thickness would become too thick and too heavy. So the thixo-cast method was used, said to be a first in body frame construction. This involves pouring aluminium in a half solidified rather than molten state to create a uniform and fine metal structure allowing a 22% thinner wall thickness, 20% higher strength, and a 20% weight reduction. In comparison with the NSX, the Insight uses 15% fewer body parts and 24% fewer welding spots to give weight and productivity savings.

6.5 Hybrid passenger and goods vehicles

6.5.1 HYBRID-DRIVE BUSES

Passenger service vehicles have been the first to use hybrid drives on a commercial scale, usually employing a series layout. In a series hybrid configuration, part of the traction energy is converted into electrical energy, and then into mechanical energy, and part flows to the wheels directly via a mechanical transmission. It is argued that this configuration can potentially offer higher overall efficiency. In a series layout, all the IC engine energy is converted into electrical energy and then into mechanical energy. Such a configuration could offer advantages where the electric motor is designed as a very high efficiency unit regardless of the load upon it. Furthermore, the ability to use pure electrical transmission allows for flexibility, in system management, in optimizing engine operating conditions and reducing noise output. It also allows greater freedom in mechanical packaging, including electric motors direct driving the wheels, and for the incorporation of future fuel-cell technology.

This was the thinking behind the choice of a series system by Fiat in their pioneering hybrid bus, Fig. 6.17, a flow diagram for the system being shown at (a). The DC compound motor used, with separately excited field, had interpoles and compensating windings to expand the load range of maximum efficiency operation both as a motor and generator. Armature current was varied, by means of a thyristor chopper, from zero to base speed, above which field current was varied by an independent chopper. The reversible chopper allowed regenerative braking, down to zero speed. The motor had characteristics as seen at (b) and was direct coupled to the rear-axle driving head. Its nominal voltage was 600 V, continuous power 90 kW, maximum power 180 kW and field control ratio at continuous power 1150/3200 rpm. Efficiency at continuous power in field control was over 90%. The 600 V generating unit comprised a diesel generator set with power rating of 56 kW and maximum power 78 kW. Battery storage comprised 50 12 V lead–acid cells of 135 Ah capacity at a 5 hour discharge rate, their weight of 1930 kg corresponding to 12% of the GVW. Vehicle layout was as seen at (c); in tests the vehicle recorded diesel consumption of 32.3 kg/100 km compared with 37.8 for a conventional vehicle, with battery state of charge found to be the same at the beginning and end of the tests. Range in purely electric drive was 30 km of city driving from 100% to 20% state of charge of the batteries.

During the design stages Fiat examined a typical town route between two termini as seen at (d), on a time base of seconds. It was also established that the maximum acceptable acceleration for standing passengers was 1.5 m/sec^2 on level ground and 0.27 at a gradient. The subtended area in the power diagram gave energy required between stops, the negative portion representing energy flowing back to the batteries having taken the various system efficiencies into account.

For E_s the total energy required at the wheels, E_m the engine energy and E_r the mains electrical energy, then

$$E_s = (\eta_g E_m + E_r)\eta_b \eta_t \eta_o$$

where efficiencies subscripted g, b, t and o refer to generator, battery, motors/controllers and transmission respectively, with their product the overall efficiency *h*. Then for total duration of daily service T_s, terminus turnaround time T_c and number of daily runs between termini *N*, the power required from the engine is:

$$P_m = (E_s - \eta E_r)/[\eta_g \eta \{T_s - (N-1)T_c\}]$$

The total daily energy is calculated for a typical route of length L_p, divided into N_t segments of length L_t. The heaviest cycle is shown at (e), in which Ea = 584.6 Wh; Ed = 442 Wh. $Ead = \eta_t{}'\eta_s \eta Ed$ = 337.6 Wh and energy spent per run $Ep = Ead(Lp/Lad)$ = 12 250 Wh.

6.5.2 CNG-ELECTRIC HYBRID

Smaller buses have been built with pure electric and alternative forms of hybrid drive. An interesting project by Unique Mobility in North America put a CNG-electric hybrid system into a 25 ft (7.62 m), 24 passenger vehicle (Fig. 6.18). Here the compactness and locational flexibility of the hybrid-

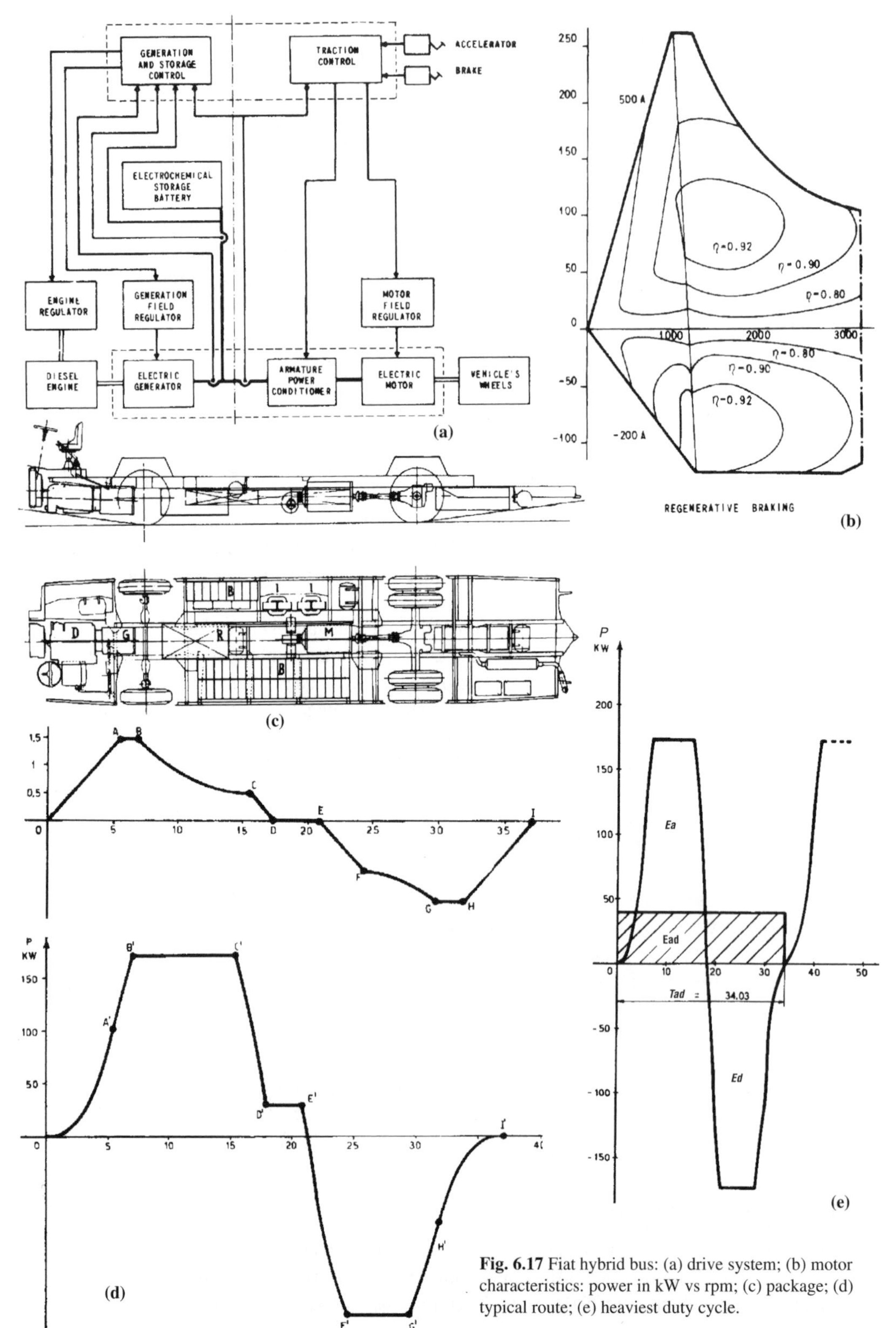

Fig. 6.17 Fiat hybrid bus: (a) drive system; (b) motor characteristics: power in kW vs rpm; (c) package; (d) typical route; (e) heaviest duty cycle.

drive elements meant that considerable gains could be obtained in packaging the vehicle occupants. Using high power-density permanent-magnet motors driving the rear wheels allowed a particularly low floor of 12 in (305 mm) from the ground.

The CNG-engine generator provided steady state power and was augmented by storage batteries to supply the power required above that base level while recharging of the batteries would take

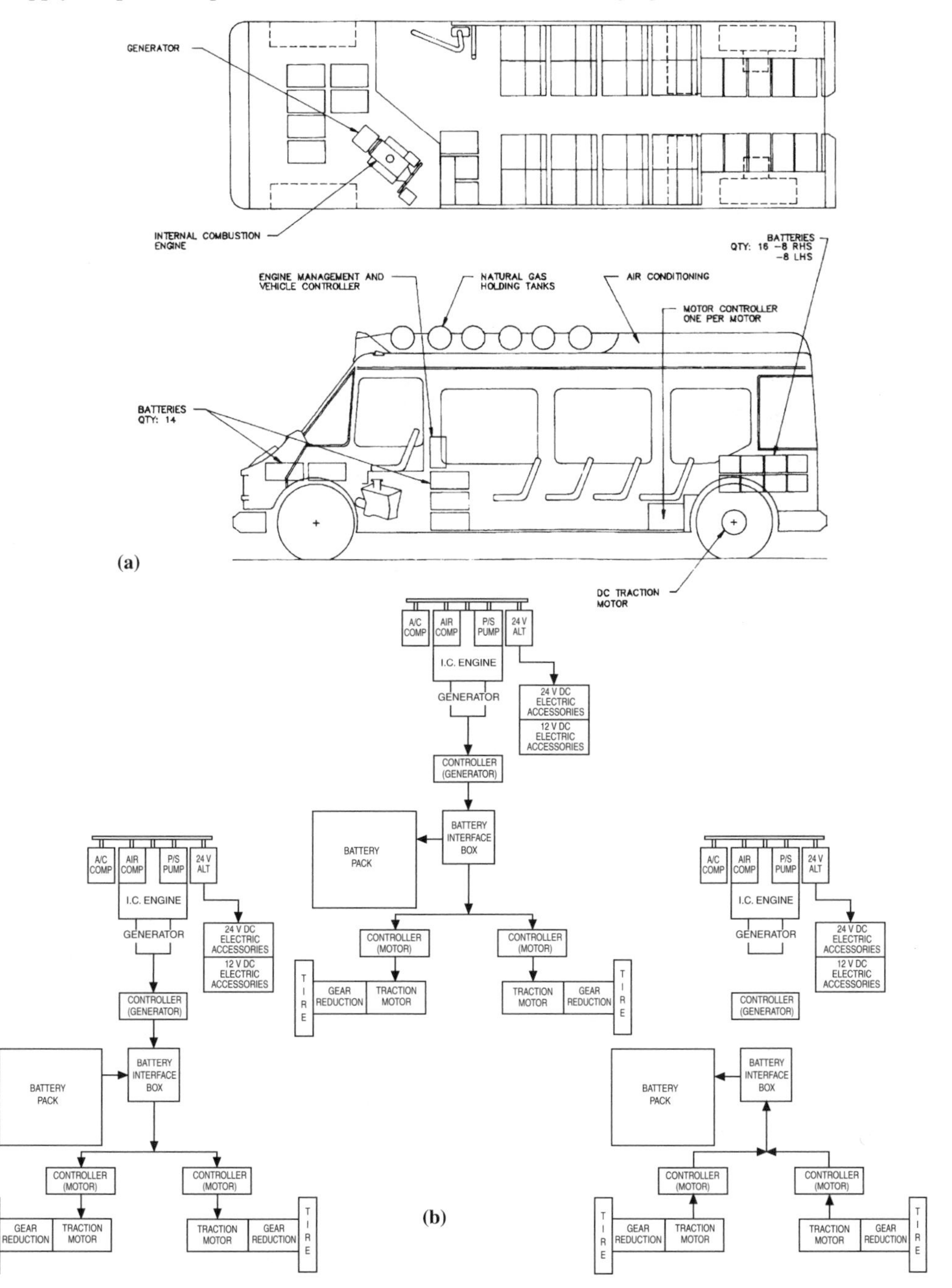

Fig. 6.18 Unique Mobility small hybrid bus: (a) Unique Mobility midibus; (b) power flow charts.

place when the power requirement fell below the base level. CNG tanks were roof mounted while batteries were positioned over the rear wheel wells and inside the engine compartment. The 11 tonne GVW bus is seen at (a).

The 90 bhp gas engine drove the generator through a flywheel-positioned step-up planetary gear set and an engine management system allowed engine speed and power to vary with load conditions. The rate at which speed was increased was minimized by the controller in order to avoid poor fuel economy and high emissions associated with transients. The two 70 kW traction motors were provided with a single planetary reduction gear of 2.77:1, directly coupled to the drive wheels through a secondary set of 5.2:1 included in the wheel hub to give an attainable speed of 55 mph. Rear suspension was an independent trailing arm system with traction motors direct mounted to the arms, so as to maximize floor area between the wheels. Motor differential speeds for cornering are electronically controlled with reference to steering wheel angle and road wheel speed.

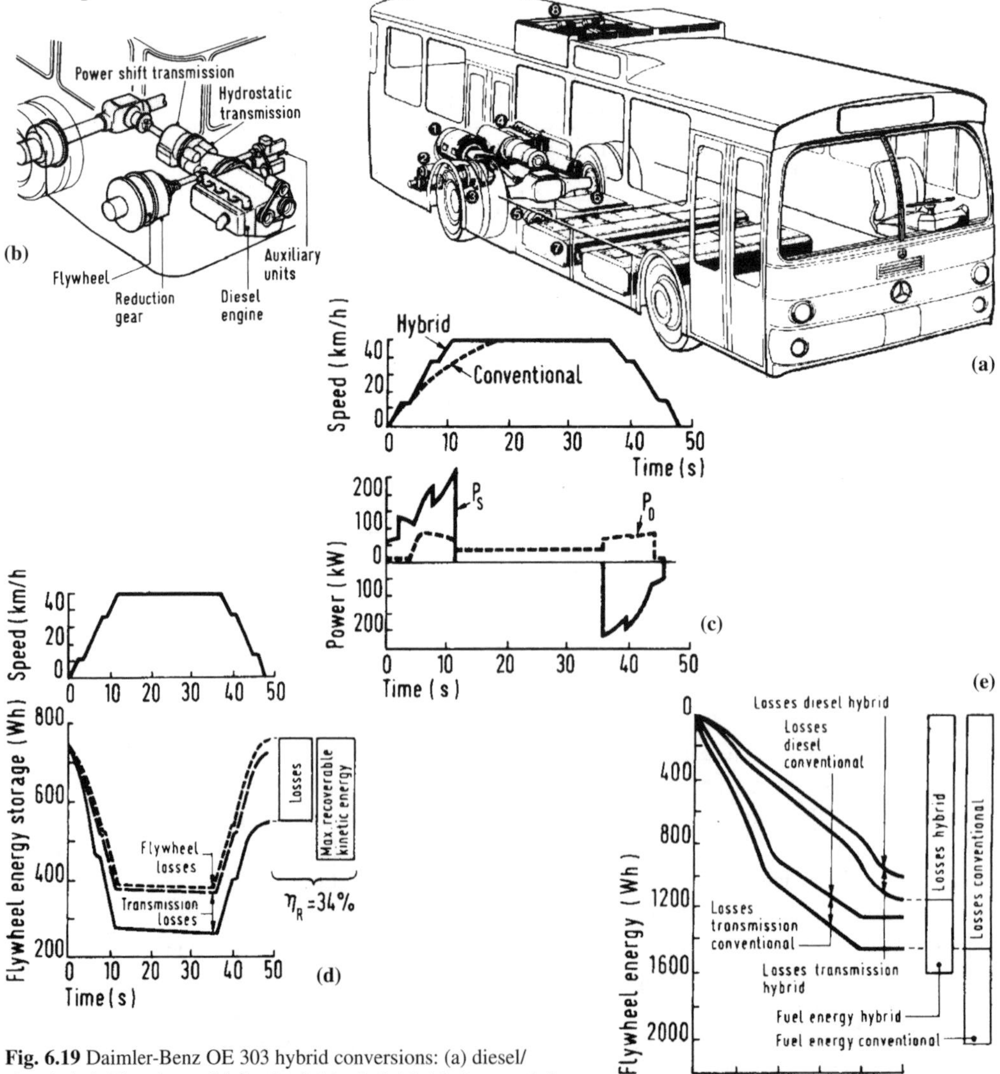

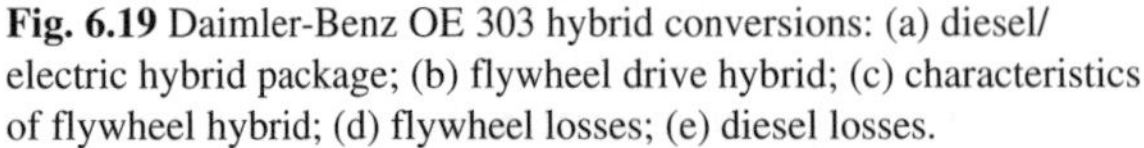

Fig. 6.19 Daimler-Benz OE 303 hybrid conversions: (a) diesel/electric hybrid package; (b) flywheel drive hybrid; (c) characteristics of flywheel hybrid; (d) flywheel losses; (e) diesel losses.

The view at (b) shows the power flow charts for different modes of operation. In the first, on IC engine power only, a speed of 37 mph was achieved. On IC engine and battery power, higher speeds were possible and a reserve was available for gradients and acceleration; in the final mode of regenerative braking with the IC engine operating, the engine provided power only to the accessories and that from braking was fed into the storage batteries. The latter were used primarily for supplying accelerative power and were 12 V units with 160 Ah capacity. Two series strings of 15 batteries were connected in parallel to yield 180 V and 320 Ah total capacity.

Another pioneering series of hybrid buses has been the Daimler-Benz OE 305 city bus conversions (Fig. 6.19), some 20 of the first type were evaluated in trials in German cities in the early 1980s. Electric drive in the city centre and diesel drive in the suburbs was the mode of operation. Seen at (a), the set-up was electric motor (1), air compressor and power-steering pump (2), motor fan (3), diesel engine and generator (4), battery-fan (5), electronic controller (6), traction batteries (7) and battery cooling unit (8). Range was 30–45 miles on batteries alone and 190 miles as a hybrid diesel combination. The 100 passenger vehicle had a maximum speed of 43 mph and the 19 tonne GVW vehicle had batteries weighing 3.5 tonnes. The motor could develop up to 200 bhp while the diesel engine was rated at 100 bhp. The five 275 Ah batteries operated at 360 V.

At the same period a Daimler-Benz 305 was also converted to flywheel hybrid operation in a study which involved MAN and Berlin University, too. The team calculated the flywheel storage energy requirements of a city bus to be 750 kW for absorbing the kinetic energy of the vehicle at top speed and a 220 lb flywheel was chosen, with a 1500 kW total energy to allow a reserve, which suffered 2 kW power loss at 12 000 rpm. For a typical urban operating cycle, characteristics were plotted, from the starting point of the bus stationary with diesel engine at idle and flywheel charged up from previous operation, for the configuration shown at (b).

During acceleration the mechanical gear stage is automatically shifted as the vari-drive hydraulic transmission changes speed. This results in the almost constant slope (full line) in the curves shown at (c) compared with a conventional bus (dotted line). Initially the flywheel alone is used, then the diesel is brought in during deceleration, as seen in the bottom half of the figure. Flywheel power P_s is 260 kW immediately before the constant-power cruising phase, during which the diesel drives. During deceleration the flywheel is recharged and its power is 200 kW, corresponding diesel power being shown by P_d. Flywheel energy content and losses are seen at (d); the inertia losses are replaced by diesel energy. The diesel losses are seen at (e) together with the transmission losses between engine and drive wheels, compared with those for a conventional vehicle.

6.5.3 ADVANCED HYBRID BUS

A joint venture between MAN and Voith has resulted in the NL 202 DE low floor concept city bus, Fig. 6.20, designed to carry 98 passengers at a maximum speed of 70 kph, (a). No steps are involved at any of the entrances which lead directly to a completely level deck height of between 317 to 340 mm. The rear-mounted horizontally positioned diesel engine allows fitment of a bench seat at the rear of the bus; it drives a generator with only electrical connection to the Voith transverse-flux wheel motors which drive the wheels through two-stage hub-reduction gearsets, (b). The diesel is rated at 127 kW and the generator at 135 kW; the controller, (c), is of the IGBT converter type and also developed by Voith. It provides a differential action to the wheel motors on cornering. Permanent-magnet synchronous wheel motors are rated at 57 kW and have a maximum speed of 2500 rev/min; see table at (d). The bus is 12 metres long and has water cooling for its generator, converters and wheel motors. As well as providing virtually jerk-free acceleration, the drive system is seen by MAN as providing the possibility of four-wheel drive on future articulated buses to improve traction and stability in slippery road conditions. The term transverse flux motor refers to the means used to guide the magnetic flux in the stator; this is new to inverter-supplied PM types and involves a novel collector configuration. Double-sided magnetic force generation is also new and involves a patented double air gap construction having high idling inductances and force densities up to 120 kNm/m^2, with relatively low losses. A new control process permits operation of the motor in a field-weakening type mode, in spite of PM excitation. The generator is almost identical in concept but involves no field weakening. Each has concentric construction of permanent magnets, rotor/stator soft-iron elements and stator winding; see below. Armature elements are U-shaped cut strip-wound core sections, embedded in the ring-shaped supporting structures of inner and outer stators. Each core surrounds the windings and forms a stator pole with its cut surfaces facing the rotor. The latter is pot-shaped and positioned between poles of the outer and inner stators. In the stator pole region it comprises magnet and soft-iron element while in the winding region a ring of GRP serves as the connecting element. The inverters supply the motors with sinusoidal currents and voltages until the nominal operating point is reached; operating frequency is 10 kHz. In field-weakening mode the induced voltage exceeds intermediate circuit voltage and only square wave voltages are supplied to the motor. Power output then remains constant and the operating frequency equals the fundamental motor frequency. A large speed ratio, 1.5:1, is thus possible.

(a)

	TFM wheel motor	TFM generator
Power	57 kW	135 kW
Rated speed	735 rpm	1750 rpm
Approx. max. speed	2500 rpm	200 rpm
Max. fundamental frequency of stator	1350 Hz	—
Rated torque	740 Nm	740 Nm
Approx. max. torque	1050 Nm	740 Nm
Approx. torque conversion	1:3	
Power/weight ratio	1.8 kg/kW	0.9 kg/kW

(d)

(b)

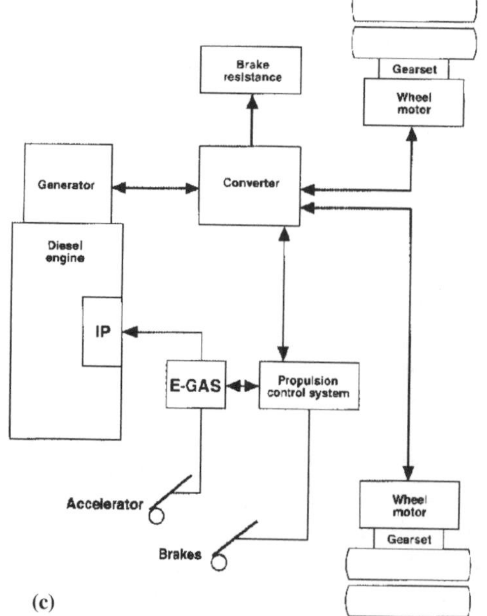

(c)

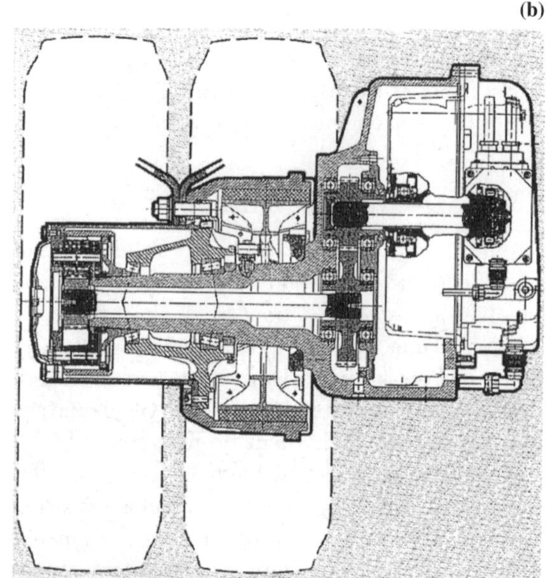

Fig. 6.20 MAN/Voith concept city bus: (a) low floor package; (b) wheel motor; (c) controller; (d) drive characteristics.

6.5.4 ADVANCED HYBRID TRUCK

Mitsubishi have been prominent in hybrid truck manufacture and have recently developed a heavier, municipal, version of the light hybrid truck launched in 1995. Because added cost limited market acceptance of the lighter, Canter-based, hybrid the decision to build a heavier municipal version, Fig. 6.21, was taken on the grounds of low noise, and greatly reduced emissions, which made the vehicle attractive for city-centre operation, a lift-platform version being particularly popular. The hydraulic pump for operating auxiliaries such as a lift platform is electric motor driven, with the benefit of near silent operation.

Series hybrid mode, (a) was chosen first, because the engine is used solely for power generation and so can be operated in a peak efficiency speed band and secondly, since the engine is isolated from the drive system, it results in a simpler and more flexible drive-system layout with greater freedom for hydraulic equipment mounting. Two electric motors are involved. Shown at (b) are typical operational modes of the truck: when the battery has high state-of-charge (SOC) the vehicle operates exclusively in battery mode. At less than 65% SOC the power-generating engine starts and hybrid mode is invoked; when 70% SOC is achieved again the vehicle reverts to battery operation. Provision is also made to inhibit hybrid operation until 30% SOC is reached so silent and zero-emission night-time, or in-tunnel, operation is made possible. In hybrid mode SOC is maintained at 65–70%, at which point the generated power, from the generator, and the regenerative power, from the motor, provide sufficient charging.

The overall package layout is shown at (c); dimensions are 5.78 m long × 1.88 m wide × 3.35 m high, with a wheelbase of 2.5 m. Gross vehicle weight is 6.965 tonnes and tyre size 205/85R16. While an elevating platform vehicle normally requires counterweighting, in this case the mass of the dual drive suffices with just modest additions. The central positioning of the battery above the chassis frame was chosen to optimize weight distribution and avoid the weight of cantilever frames.

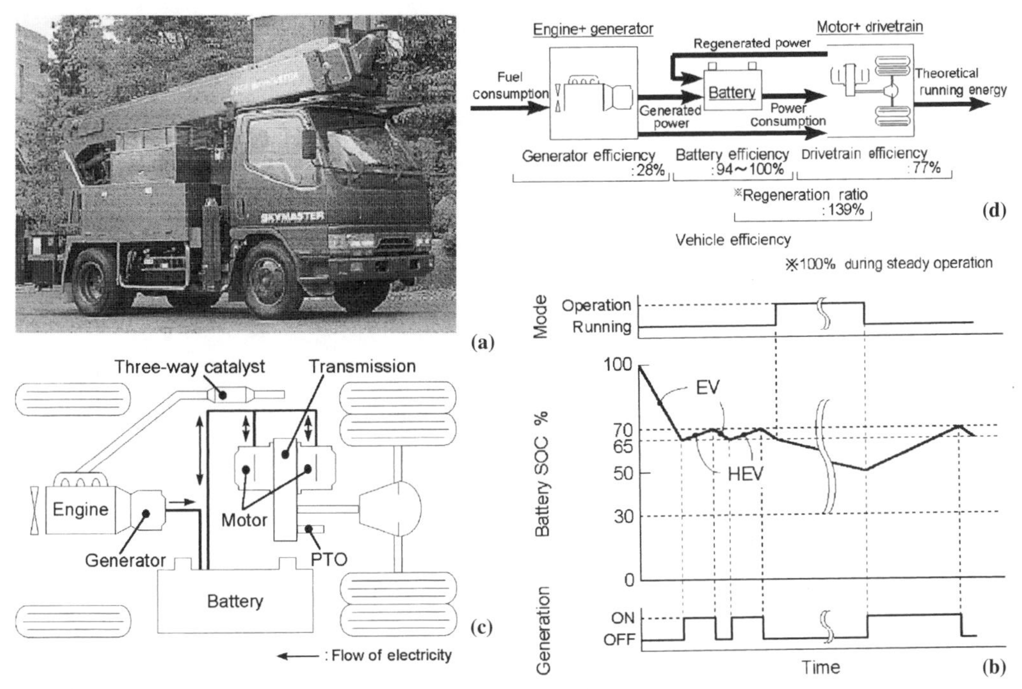

Fig. 6.21 Mitsubishi Canter-based hybrid municipal truck: (a) Complete package; (b) hybrid drive; (c) operating modes; (d) unit efficiencies.

The motors, of 55 kW, are of the induction type and each develop 150 Nm at 3500 rpm, rated voltage being 288 V. The simple two speed transmission has a PTO for driving the hydraulic pump and reverse motion is achieved by altering the rotation of the electric motor. The generator has a maximum output of 30 kW at 3500 rpm; it operates at 220–360 V and weighs 70 kg.

The petrol engine is a 16 valve unit of 1834 cc which has a 1.935:drive gear to the generator. Lead–acid traction batteries are employed, 24 units each weighing 25 kg and having 65 Ah capacity at a 5 hour rate. The company's estimations of unit efficiency are shown at (d).

References

1. Jones *et al.,* HYZEM – a joint approach towards understanding hybrid vehicle introduction into Europe, Proceedings of the IMechE combustion engines and hybrid vehicles conference, 1998
2. Friedmann *et al.,* Development and application of map-controlled drive management for a BMW parallel hybrid vehicle, SAE Special Publication SP1331, 1998
3. SAE paper 830350, 1983
4. Hodkinson, R., The hybrid electric solution, *Electrotechnology*, April/May 1994
5. Pullen and Etemad, Further developments of a gas turbine series hybrid for automotive use, EAEC paper SIA9506A22, 1995
6. Davis *et al.*, The gas turbine series hybrid vehicle – low emissions mobility for the future, Autotech paper C498/29/110, 1995
7. Strategies in electric and hybrid vehicle design, SAE Publication SP-1156, 1996
8. Thoolen, F., Dutch Centre for Construction and Mechatronics, Emafer Drive Line, 1996 FISITA conference
9. Nagasaka *et al.,* Development of the hybrid/battery ECU for the Toyota Hybrid System, SAE Special Publication SP1331, 1998

7
Lightweight construction materials and techniques

7.1 Introduction

The opening chapters of this book, on propulsion system design, demonstrate many of the possibilities for finding solutions for efficient electric traction. In this and the following chapter we consider some of the body structural, material specification and running-gear requirements which will prompt an interest in efficient platforms to receive such advanced traction systems. A fuller account, dealing with most vehicle types, can be obtained in a parallel work by the author*.

The key to lightweight construction lies in a combination of structurally efficient design, covered in Chapter 8, plus the exploitation of advanced materials technology and construction techniques. Steel, the traditional building material for the structures of volume cars, has excellent property combinations if it can be exploited in structurally efficient designs, alloyed to produce high strength sheet, then formed and fabricated by such techniques as hydroforming and laser-welded tailored blanking. Light (high specific strength and rigidity) alloys of aluminium, magnesium and titanium also have a place and advanced polymer composite systems have already been proven in race car and similar applications to provide ultra-light solutions. However, the future may lay in the wider meaning of composite construction, in a combination of metals and polymer compounds fulfilling complementary roles.

Monocoque shell structures in high strength sandwich-construction polymer composites are considered against space-frame structures in aluminium alloy.

7.2 The 'composite' approach

The use of metal platform base structures and reinforced plastic body shells is established technology for low to medium production cars, in several categories, and sometimes the weight reductions over all-steel integral sedans are not appreciated by the wider public. Indeed the wider public seems to have forgotten that 'light-cars' of early history were considerably lighter than similar sized cars of the present day, albeit now much better equipped ones. For example, the Austin Seven car when first introduced in the 1920s was a 'composite' of steel chassis and timber-framed aluminium body which scaled well below the half-tonne benchmark. Today's conventional small cars do well to scale under 1 tonne; the exceptions are the products of small companies such as Reliant whose steel chassised and GRP-bodied cars (even the four-wheelers) are still below half-tonne tare weight.

Another concept of composite construction is to use reinforced plastics in more intimate combinations with metals in such a way that permits the most efficient usage of thin-walled sheet metal structures together with plastic systems that stabilize them against buckling. Plastic-foam cored steel-skinned sandwich panels have already been exploited in this context and is discussed in Chapter 8. Now interest is being shown in truss-like plastic reinforcements for stabilizing box beams and a variety of hollow, open and closed, structural members.

Hybrid metal/plastic (stabilized core) systems, Fig. 7.1, offer considerable potential to the constructor who is prepared to depart from conventional manufacturing technology and embrace production systems geared to the technique. Promoted by Bayer[1], the principle involves bonding metal and plastic in an injection moulding process, to produce a complex, load-bearing component. The company argues that plastic plus injection moulding means economical manufacture and good integration whereas steel plus deep drawing means mass production and stiffness. The combination of the two materials and processes results in the high volume production of a complex component: place a deep-drawn and perforated piece of steel in an injection-moulding die and inject the plastic around this part.

Thermal contraction stresses also occur in a plastic/metal composite when the plastic is interlocked with the metal by both force and shape, as the melt cools in the die. For this reason, it is preferable to use semi-crystalline plastics, such as PA or PBT, which can reduce the stresses by relaxation. Glass-fibre reinforced plastics are best, since the reduced contraction (moulding shrinkage) also leads to lower thermal stresses. The relaxation of these stresses is related to time and temperature. At low temperatures (–30°C) the stresses will increase in accordance with the thermal linear expansion or contraction. With a coefficient of linear expansion of 40×10^{6} and temperature difference of 50°C, the increase in expansion will be around 0.2%.

Different torsional stiffnesses of the three test pieces at (a) are seen at (b). By ribbing the plastic/metal composite profile, the torsional stiffness of an open metal profile can be increased by a factor of 12 (geometrically). Further improvement can be obtained by using different ribbing designs and plastics. However, the torsional stiffness of the closed steel profile cannot be achieved economically by the composite. The load-bearing capacity of such composite profiles, with regard to compression in a longitudinal direction, is shown at (c). By ribbing the metal profile, premature buckling could be prevented and the load-bearing capacity increased by about 80%. The bending strength of such composite profiles could also be significantly increased. Even thin-wall closed metal profiles fail, through buckling, before bending failure of the composite profile.

In the latest version of the technique, the processes of inserting, where metal parts such as bushes are included in a polymer moulding, and/or outserting, where various functions in plastic are moulded onto a metal baseplate, are taken a stage further. Cross-sectional distortion of thin-walled metal beams can be prevented by relatively small forces applied in the new process by the presence of moulded plastic supports in the form of x-pattern ribbing. Interconnecting points between plastic and metal are preformed in the metal part before it enters the plastics mould. Either corrosion-protected steel or aluminium alloy is the normal choice of metal with glass-fibre reinforced, impact modified, polyamide-6 (Bayer's Durethan BKV) being the plastic choice. The company say it is not always desirable to have the metal 'preform' in one piece, separate sections being joined by moulding resin around the prefabricated interlocking points or by means of clinching integrated into the mould. It is said that for recycling purposes, it takes only a few seconds to break a metal/plastic composite, using a hammer mill, and that the resin element has properties akin to the virgin material, on reuse.

A research project has also been carried out on car side doors having sufficient structural integrity to transmit impact forces from A- to B-post in the closed position. At (d) is a sample door, requiring no further framework to support wing mirror, lock or other door 'furniture'. Seat frames with

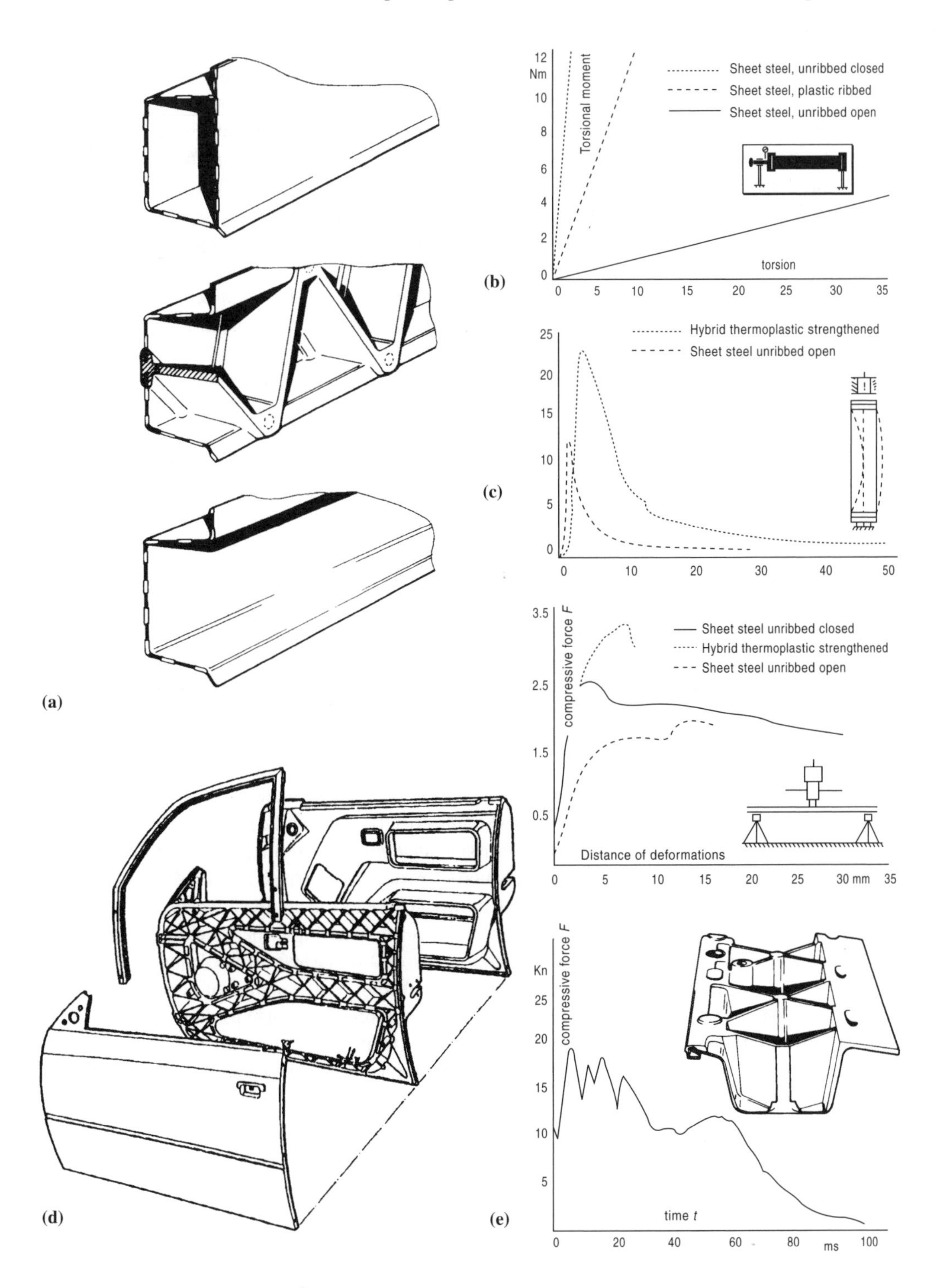

Fig. 7.1 Hybrid metal–plastic beam: (a) test sections; (b) torsional stiffness comparison for three sections; (c) compression test results; (d) metal/plastic hybrid door structure; (e) interlocking connection panels for a door with impact pendulum test results on the assembly.

integral belt anchorages have also been made for the Mercedes-Benz Viano minibus in the process. The first volume car to incorporate the technology is the Audi A6. The front-end design was developed in association with the French ECIA company. The part is injection moulded in one piece and incorporates engine mountings, together with support for radiator and headlights. In one of the door projects, bonding areas can be seen at (e, top) which shows the die head connection on the closed side of the metal profile. The plastic melt enters the countersunk openings in the metal, forming a die head between the wall of the recess in the injection-moulding die and the metal part. Effectively the die head connection takes place directly in the injection-moulding die; no additional operation is required. The view at (b, bottom) shows the force/time curve from the pendulum test on the part (pendulum weight: 780 kg, pendulum speed: 8 km/hr). The door exhibits a high degree of resilience, since a relatively constant force is exerted over a long period, followed by a sharp drop in force in a very short time.

In the Stabilized Core Composite (SCC) system devised by Gordon Wardill[2], one of the key features is that the encapsulated pieces are joined mainly by the bonding action of polyurethane, which is injected under pressure during the RIM process. Apart from making stiffer joints, this means that it is possible to manufacture relatively large body subassemblies in a 'one shot' injection process without the use of metal welding techniques and fixtures but with the use of very low cost tooling. For box-section thin-walled beams, one way of delaying the onset of buckling is to stiffen the material by increasing the thickness. This would, of course, increase the weight of the beam. However, if one halves the thickness of the metal and supports it on either side by means of rigid cellular self-skinning urethane, giving a total wall thickness of 10 x sheet thickness, then the weight of the section remains unchanged. In this case, the modulus of elasticity of the urethane is likely to be 1/250th of that of steel. Since the buckling load of the free edge of the flange varies approximately as the thickness cubed, then, ignoring secondary effects the SCC/steel buckling load ratio = 4. The net effect, in the case of a beam, is to trade off a fraction of the tensile strength for a fourfold increase in bending strength. It should be noted that for maximum efficiency, the core must be supported on both sides.

Advantages can be realized in closed section beams and vehicle joints in general, with derived effects which result in improvements to the torsional stiffness of beams: important for those which have a large influence on body torsional stiffness – such as the sills of a punt structure. Bending stiffness of longitudinal beam flanges affects body torsional stiffness. SCC flanges are stiffer in this respect than steel and therefore an improvement in beam torsional stiffness is possible. Also many conventional joints display a reduction in stiffness, due to quite large changes in overall shape. These changes can be prevented by means of internal diaphragms placed across sections of the joint. It has always been difficult if not impossible to achieve this economically in the conventional steel structures, due to welding and alignment problems during assembly. With SCC, however, it is no more difficult to include internal ribs than it is not to.

Additionally, some of the flexibility of steel joints is due to the effect of inter-spot-weld buckling. This is totally prevented with SCC because, during the RIM injection process, perfect bonded joints are produced between the simple armatures that are used – and virtually no welding need be employed in the construction, say, of a car body.

7.2.1 FOAM-CORED STEEL COMPOSITE BOX BEAMS

The concept of foam-filled box beams, as an extension of sandwich construction, can be seen as an alternative to plastic–steel stabilized core structures with latticed cores. According to Foamseal Urethane Technology, ITW and EASI Engineering[3], the results of four-point bending tests replicated by FE analysis, on foam-filled box beams, are showing that optimization based on filled areas and foam densities can achieve given weight, cost, strength and stiffness improvement targets, Fig. 7.2. In designing for improved roof crush resistance, it is shown that foam-filling can reduce pillar sections, reduce metal thickness and sometimes remove reinforcements that can compensate for the increased weight of the foam. The research also demonstrates that the FMVSS 201 Head Impact Protection upgrade can be met with a proposed design concept involving foam-filling. The view at (a) shows stress/strain curves for various densities of PUR foam between 2 and 30 lb/ft^3. In tests on B-pillar to sill joints, carried out by Ford researchers in conjunction with steel-industry engineers, 5 lb/ft^3 foam increased torsional stiffness by 250% and a 30 lb/ft^3 foam increased it by 500% over the unfilled figure. Furthermore, cyclic testing has shown the use of 25 lb/ft^3 foam in this joint was able to delay the onset of fatigue cracks from 10 to 110% of the life cycle in a typical design. The filling technique has been in use, particularly for NVH applications on volume production vehicles, since 1982 and excellent adhesion properties are claimed for both electro-coated and painted surfaces, with no foam degradation over the life of the vehicles. Now the use of foams for structural purposes is under active consideration; recent production vehicles have used foam filling of the A- and B-pillars to preserve roof strength but the relatively high density foams involved have required optimization to ensure weight limits are met.

In a four-point bending test set-up with central loads 84 mm apart reacted over a 254 mm span, 75 mm wide times 50 mm deep section steel tubes were tested with three foam densities against the base line of an unfilled tube, results confirming the effectiveness of the foam in stabilizing the

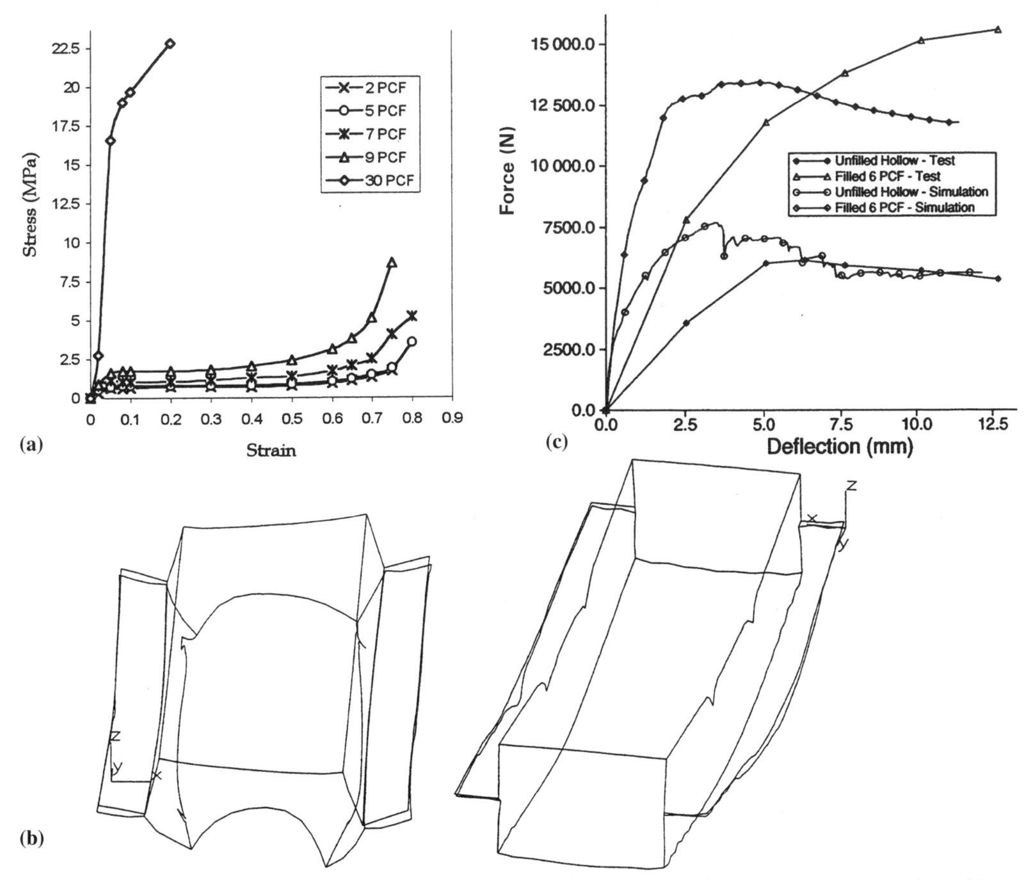

Fig. 7.2 Foam-filled tube behaviour: (a) Stress/strain curves for different density foams; (b) distorted shape for unfilled, left, and filled, right, structure; (c) deflection curves for unfilled and filled tubes.

thin-walled tube sections against buckling of the skins. Entrapment of the foam by the closed section is seen as an important criterion and doubt is expressed as to whether the foam would be effective in the case of open section beams. End section deformed shapes of the filled and unfilled tubes, obtained in the simulation, are shown at (b) while the force/deflection curves for the test and simulation results are shown at (c). Increase in bending strength and bending stiffness was found to increase almost linearly with increase in foam density.

7.3 Plastic mouldings for open canopy shells

The RIM process used in Bayer's metal/plastic composite construction, described above, also has important potential for production of structural panels for car bodies on a relatively high volume basis. In a punt-type vehicle structure, made from metal box sections stabilized by plastic cores and incorporating rollover hoopframes at the A- and C-posts, open shell sections in RIM polyurethane could be used to form the roof panel, and front/rear ends of the body superstructure enclosing windshield and backlight screen respectively. The punt-type structure also allows the possibility for a 'pillarless' sedan configuration with side doors hung from A- and C-posts, without the need for a B-post. Use of metal/plastic composite doors in conjunction with a structurally efficient sliding bolt system that would preserve the integrity of the door side-impact beams would allow unrivalled occupant access to the sedan interior. Open shells in RIM polyurethane could also be used for vehicle front and rear-end structures which would be 'canopies' suspended over purpose-designed shock-absorber systems cantilevered from the main punt structure to absorb front, rear and 'three-quarter' impacts.

7.3.1 REACTION INJECTION MOULDING (RIM) DEVELOPMENTS

Involving polymerization in the mould, the RIM technique is quite different from other plastic moulding methods and can be used for producing quite complex parts and panels without undue high tooling investment – since mould pressure is low. Two or more components flow into the mixing chamber, at relatively high pressure (100–200 bar) and are then expanded into the mould at much lower pressure. The streams impinge at high speed to obtain thorough mixing and initiate polymerization as they flow into the mould cavity at a pressure of about 100 bar. Low viscosity during mould filling is one of the key attractions of the process as a relatively small metering machine can make large parts. The low viscosity also simplifies reinforcement with, for example, the possibility of using continuous-fibre mat placed in the mould.

Some 90% of RIM production is in polyurethanes and urea-urethanes, the latter being uniquely suited to the process as they do not melt flow like normal thermoplastics and therefore conventional injection moulding is not possible.

ICI have developed a family of polyureas for body panel applications with unusually good processability and physical properties, Fig. 7.3. Gel times of 2 seconds are possible and mould temperatures are less than 93.5°C. Overall cycle time is about 1.5 minutes and further development promises 'less than 1 minute', with equipment shown at (a). Filler packages are also becoming available which allow part surface finish comparable with steel; moisture stability is high compared with competing thermoplastics and the materials can tolerate temperatures of 190.5°C. The table at (b) shows typical properties for a formulation that would suit body panels but others are available which raise the elastic modulus as high as 200 000 psi.

As well as increasing strength and rigidity of panels, the addition of glass fibre or other reinforcements considerably improves the compatibility of thermal expansion coefficient with such materials as steel and aluminium. Since polymers typically have coefficients some 10 times greater than steel, a metre-long part hung onto a steel body could change in length by 1 cm between summer and winter temperatures. S-RIM is the process, shown at (c), in which long-fibre mat is placed in the mould and reactive monomers injected onto it, the process being akin to resin-transfer moulding but with high pressure impingement mixing to accelerate the reaction. A comparison is made at (d) while typical properties of S-RIM composites are shown at (e).

7.3.2 RESIN TRANSFER MOULDING (RTM) TO INCORPORATE FOAM CORES

For covers such as bonnet and boot lid, self-supporting horizontal panels can be made with either glass-fibre laminate or foam cores, effectively automating the sandwich panel making process but allowing complex shapes and variable thickness cores in one panel. Essentially low viscosity resin is injected into a mould containing the required preformed insert. For relatively short model runs (10–20 000), RTM, Fig. 7.4, is a lower capital cost process than SMC compression moulding. The stages of the process are shown schematically at (a). First the glass reinforcement or preform is placed in the mould. Once the mould is closed the resin is injected with no or little movement of the glass. After mould filling the part is left curing in the mould until it is dimensionally stable so that it can be demoulded without losing its shape. The fact that the reinforcement is pre-placed in

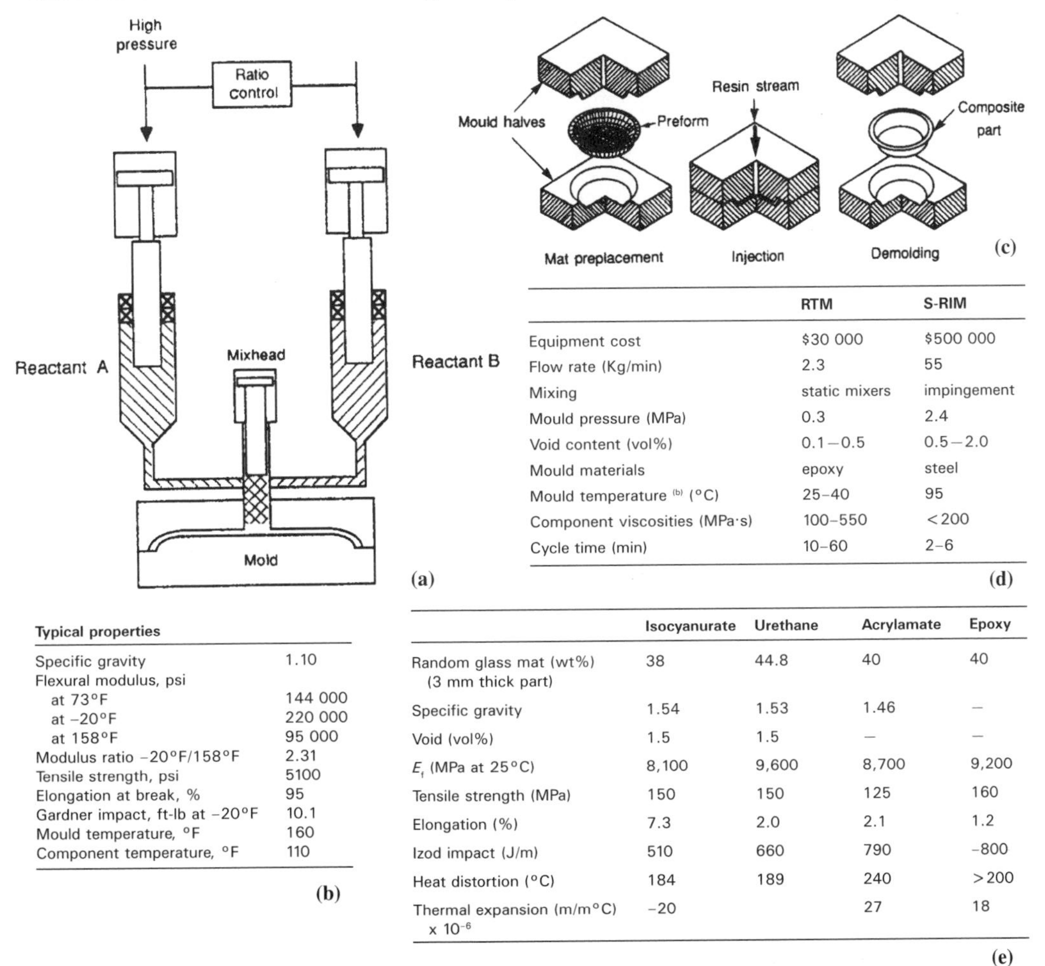

	RTM	S-RIM
Equipment cost	$30 000	$500 000
Flow rate (Kg/min)	2.3	55
Mixing	static mixers	impingement
Mould pressure (MPa)	0.3	2.4
Void content (vol%)	0.1–0.5	0.5–2.0
Mould materials	epoxy	steel
Mould temperature [(b)] (°C)	25–40	95
Component viscosities (MPa·s)	100–550	<200
Cycle time (min)	10–60	2–6

Typical properties	
Specific gravity	1.10
Flexural modulus, psi	
at 73°F	144 000
at −20°F	220 000
at 158°F	95 000
Modulus ratio −20°F/158°F	2.31
Tensile strength, psi	5100
Elongation at break, %	95
Gardner impact, ft-lb at −20°F	10.1
Mould temperature, °F	160
Component temperature, °F	110

	Isocyanurate	Urethane	Acrylamate	Epoxy
Random glass mat (wt%) (3 mm thick part)	38	44.8	40	40
Specific gravity	1.54	1.53	1.46	–
Void (vol%)	1.5	1.5	–	–
E_f (MPa at 25°C)	8,100	9,600	8,700	9,200
Tensile strength (MPa)	150	150	125	160
Elongation (%)	7.3	2.0	2.1	1.2
Izod impact (J/m)	510	660	790	-800
Heat distortion (°C)	184	189	240	>200
Thermal expansion (m/m°C) x 10^{-6}	-20		27	18

Fig. 7.3 RIM process and properties: (a) RIM machine; (b) body-panel RIM formulation; (c) steps in the S-RIM process; (d) S-RIM and RTM compared; (e) S-RIM-composite properties.

the mould gives the process the potential for making parts with better surface and mechanical properties than SMC where the fibre orientation is usually less controlled and favourable because of its flow.

During the filling stage, the resin being injected usually contains filler and can exhibit a viscosity with some shear thinning behaviour. However, for typical RTM compositions this effect is small and as a first approximation can be neglected. According to the Seger & Hoffmann subsidiary of Dow Chemical, to increase the efficiency of the moulding stage it is best to perform as many operations as convenient outside the RTM mould; the optimal placement of fibre into the RTM mould can be time consuming particularly if the shape is relatively complex. Methods available for preforming complex shapes include spraying fibres and binders onto a perforated mould or the use of mats and/or fabrics pretreated with thermoplastic binder which can be shaped by pressure when heated.

A shell of a sports seat is one example successfully produced by the company. The part is complex with a deep draw, multi-curvature shape and multi-plane shut-off. The method of preforming chosen used continuous filament random glass-fibre mats. The preforming of fibre

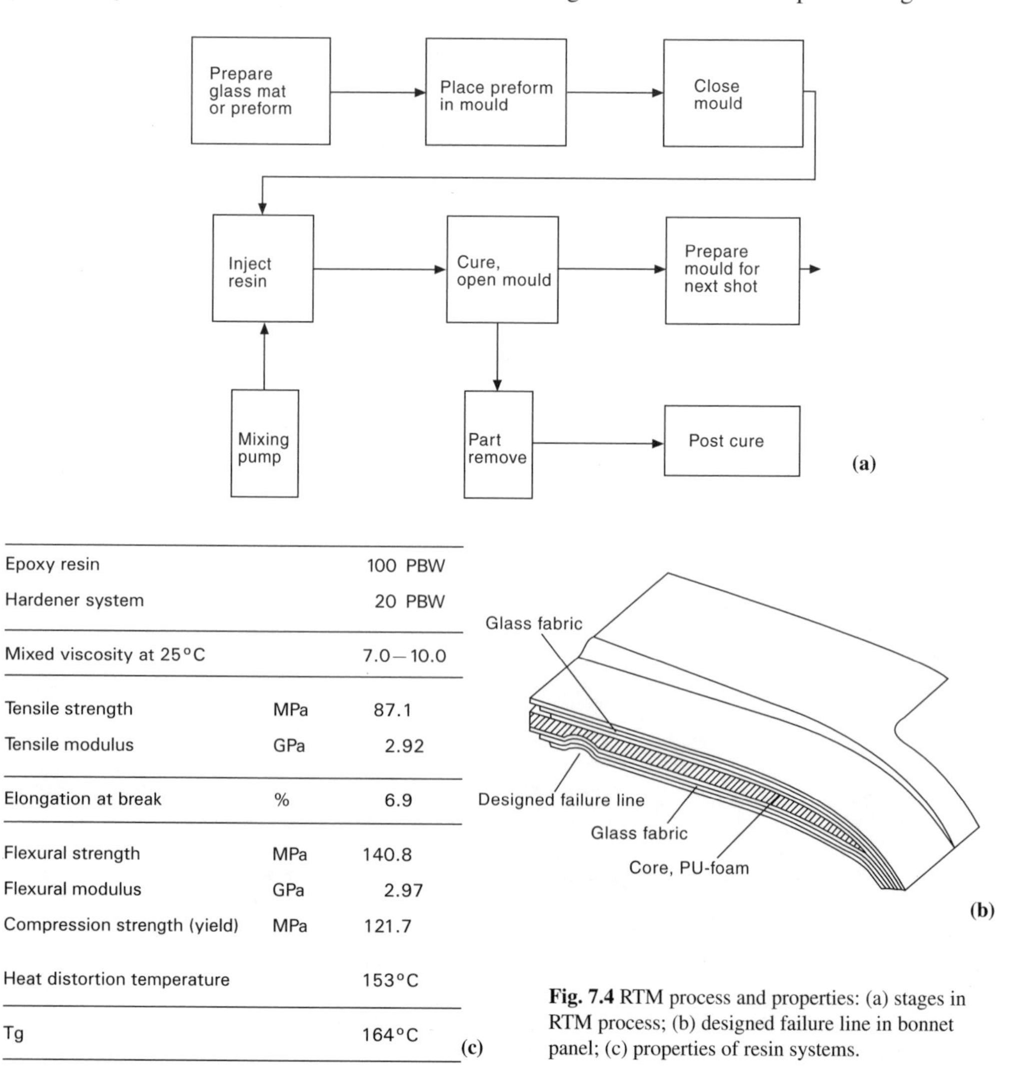

Epoxy resin		100 PBW
Hardener system		20 PBW
Mixed viscosity at 25°C		7.0–10.0
Tensile strength	MPa	87.1
Tensile modulus	GPa	2.92
Elongation at break	%	6.9
Flexural strength	MPa	140.8
Flexural modulus	GPa	2.97
Compression strength (yield)	MPa	121.7
Heat distortion temperature		153°C
Tg		164°C

Fig. 7.4 RTM process and properties: (a) stages in RTM process; (b) designed failure line in bonnet panel; (c) properties of resin systems.

reinforcement for flat mouldings such as a bonnet may not be absolutely necessary. The handling of the fibre mats, however, can be difficult as they are rarely self-supporting. A bonnet can be designed as a slim sandwich structure instead of inner and outer mouldings. The rigid foam core thus provides a convenient method of supporting the fibre reinforcement during transportation of the fibre to the RTM mould, (b).

Epoxy resin systems, with their low volume shrinkage of between 1–3% together with their good mechanical properties, are ideally suited to class 'A' surface body panel applications. Painting at temperatures up to 150°C can also be met with specifically formulated epoxy resin systems. The table at (c) gives the most important mechanical properties of a typical resin system developed for RTM. Faster curing systems based on vinyl ester resins can be used for RTM structural moulding where surface finish is less critical. Vinyl ester resin-based systems are capable of giving mould closed times of under 3.5 min. High speed mixing and dispensing technologies developed for the polyurethane industry are also being adapted for the RTM process.

Metal inserts to spread loads at high stress areas, such as hinge attachment points, can be incorporated into the RTM moulding. The shape of the polyurethane core moulding can provide a method of locating and transporting these inserts during the preform process. A particular design feature of the bonnet is the designed failure line. During high speed impacts the bonnet fails along this line.

RTM was successfully exploited by PSA in their Tulip concept battery-electric car, Fig. 7.5. The car maker worked with Sotira Composites Group to develop the body structure which comprises

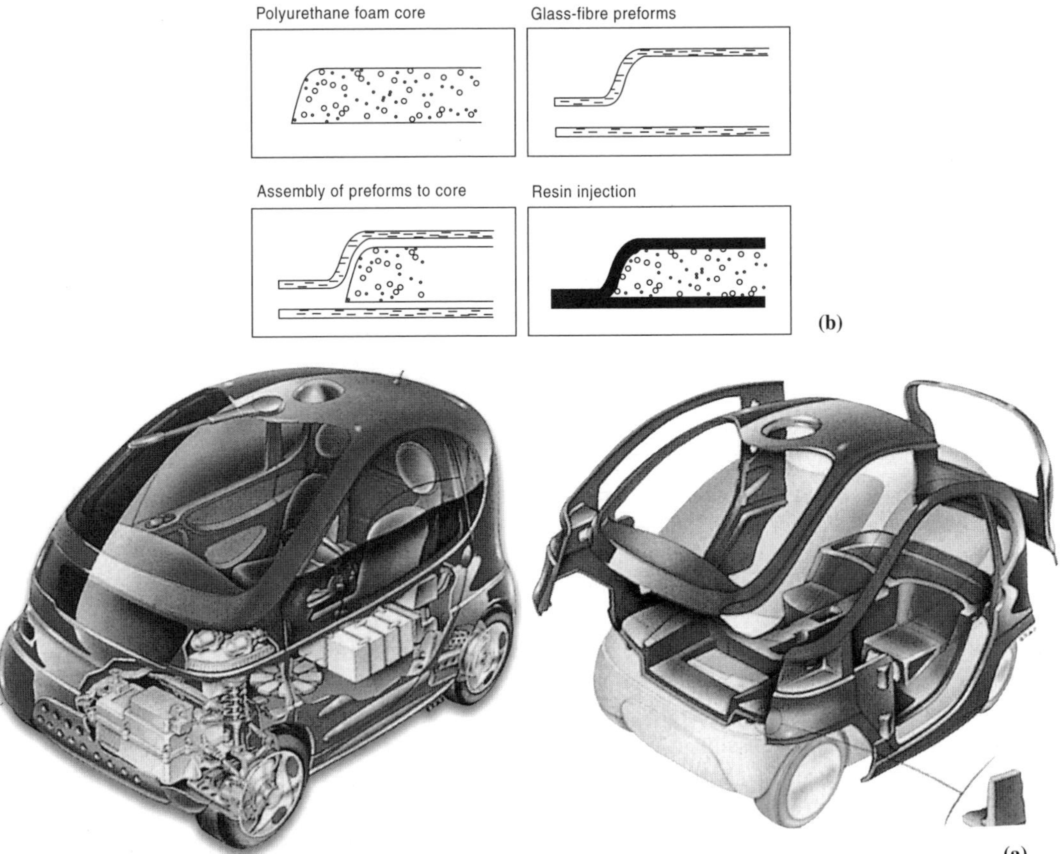

Fig. 7.5 Tulip concept car: (a) structure; (b) sandwich configuration.

just five basic elements bonded together. These five parts constitute both the exterior and interior of the vehicle bodywork, (a). In effect, the seats, dashboard, centre console and so on form an integral part of the structure which has significant benefits in terms of the rigidity of the vehicle.

Each of the five parts comprises a rigid polyurethane foam core of 110 kg/m^3 density. Glass-fibre mat is preformed and wrapped around the foam core before being placed into the low pressure injection tool. To ensure accurate location, the glass mat is retained in the part line of the tool. Polyester resin is then injected into the tool which impregnates the glass fibre and completes the sandwich construction, (b). With a 40% by weight ratio of glass reinforcement to resin, the resultant assembly weighs approximately 30% less than an equivalent steel structure. The material used is also claimed to contribute to the safety of the vehicle, both for its occupants and to pedestrians, with energy absorbing characteristics of the panels shown to be 87% higher than for standard steel parts.

To complement the precise resin injection system, the tools have a compression chamber in place of the traditional vents. The tools are also designed to maintain close temperature control across the entire surface. Typically +/-2° C is achievable to ensure consistent polymerization of the resin and the use of chromed steel or highly polished nickel shell tools allows for parts with class 'A' surface finish to be moulded. Resin supplier DSM had a key role in the project to optimize the resin system to suit the RTM process. Mould flow analysis tests have therefore been performed with the aim of reaching body panel production rates of 200 per day.

7.4 Materials for specialist EV structures

Polyester and epoxy resins have a proven record for the lower volume specialist vehicle categories. The future electric vehicle market might tend on the one hand towards localized body manufacture with considerable manual labour glass-reinforced polyester (GRP) content in developing countries and in the richer countries to the construction of ultra-lightweight bodies using techniques thus far only affordable to race-car construction, Fig. 7.6. Hand lay-up is an important factor for both of these sectors. There is also a medium volume sector in specialist vehicles which has warmed towards resin pre-impregnated sheet moulding compounds which might well be adopted for electrical commercial and passenger service vehicles as the market progresses.

7.4.1 GRP AND SMC

For particularly lightweight GRP construction, reservoir moulding is a method of producing GRP sandwich panels. Here a reservoir of flexible open-cell foam is impregnated with a resin and sandwiched between two layers of dry fibre reinforcement. The three layer sandwich is placed between dies under a pressure of just 12 kg/cm^2. The foam acts as a sponge which, during moulding, squeezes the resin into the fibre. Sandwich stiffness can be altered by varying compression pressure and the tooling can be simplified by using one rigid and one flexible die face such as a liquid-filled bag. The VARI (vacuum assisted resin injection) process was pioneered by Lotus for car body shells but is now available for a variety of licensed manufacture applications. Here the part is made between matched mould surfaces after laying up of the reinforcement by hand. Vacuum is then applied to the space between the mould faces and resin automatically drawn into the cavity. Preformed foam cores can also be placed in the mould to achieve localized box sections within the main part. It is also now possible to preform the glass reinforcement to speed the lay-up process. Another development is a technique for making metal-faced moulds which can be heated to further shorten the cure cycle. The view at (a) shows the bottom half of a Lotus car structure, made by VARI, with numbered panels indicating the weight of glass (in lb/ft^2) used in each area of the moulding.

The ability to produce sheets of glass-fibre reinforcement impregnated with a catalysed layer of polyester, within protective layers of polyethylene film, has, of course, led to sheet moulding compounds (SMC). These simplify the process of matched die moulding to give relatively high production rates. Low profile resin systems can be employed for exterior body skin panels requiring high surface smoothness. Four types of SMC reinforcement are common: short chopped glass rovings in a random pattern for equal strength in all directions; endless rovings arranged parallel to chopped fibres for unidirectional strength; medium length rovings arranged parallel but staggered to obtain better mould flow than in the last case; and a wound formation of crossed roving tapes laid at 20° again to effect directional strength.

Structural performance comparable with metals can be had with the addition of materials such as para-aramids and carbon fibres. These materials are already commonplace in racing car structures and offer possibilities for such parts as slim-line windscreen pillars. The Kevlar brand of para-aramid, by Du Pont, exists in two states: first as a 0.12 mm thick endless fibre with cut length varying from 6 to 100 mm, and second in the form of a 'pulp'. It is said to be 30% stronger than glass. The company also make a so-called meta-aramid, Nomex, which is a 'paper' that can be formed into lightweight honeycomb cores of sandwich panels. Courtaulds' Carbon Fibre Division have tabulated the properties of high strength composites, compared with metals, as at (b), which includes S- and E-glass composites as a datum.

The common thermosetting-resin systems are polyester, vinylester, epoxy, phenolic and bismalyamide. Because thermosets will not revert to their liquid state when heated, their temperature tolerance is generally higher than for thermoplastics. Compressing resin and fibres together sufficiently to get a reliable bond between them has led to systems of curing mouldings under pressure such as vacuum bag and autoclave. Designing in press-moulded GRP permits maximum parts integration into a single moulding; it is possible to increase stiffness by using double curvature and well-radiused curves (this also assists fibre distribution and air removal). As much taper as possible should be used to facilitate mould removal; use of ribs, hollow sections and lightweight cores for localized stiffness is recommended and use of large area washers and metal inserts to spread shear and bearing loads at attachment points (as stress concentrations are not relieved by yielding as with ductile metals) is recommended.

High strength composites have been proposed for car passenger compartment frameworks by VW[3]. Prior to considering the complete framework, initial studies were carried out on the front

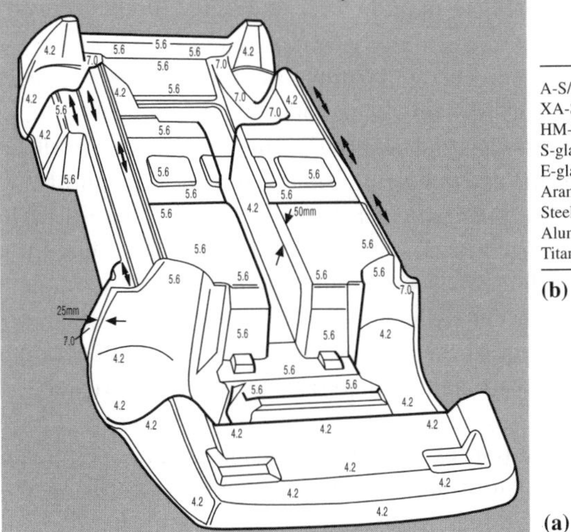

(a)

	Tensile strength (GPa)	Tensile modulus (GPa)	Specific gravity	Specific strength	Specific modulus
A-S/epoxy	1.59	113	1.5	1.06	75
XA-S/epoxy	1.90	128	1.5	1.27	85
HM-S/epoxy	1.65	190	1.6	1.03	119
S-glass/epoxy	1.79	55	7.0	0.90	27
E-glass/epoxy	1.0	82	2.0	0.50	21
Aramid/epoxy	1.29	83	1.39	1.00	60
Steel	1.0	210	7.8	0.13	27
Aluminium L65	0.47	76	2.8	0.17	26
Titanium DTD 5173	0.96	110	4.5	0.21	25

(b)

Fig. 7.6 Reinforced plastics: (a) weight of glass in moulding; (b) high strength composite properties.

door surround-frames; these, which showed 20% weight reduction against steel, could be obtained without loss of rigidity, Fig. 7.7. Normal test forces subjected to a steel frame are shown at (a) and used to test the composite frame, made from 60% (by weight) glass-reinforced SMC with two-thirds continuous filament and one-third random cut fibres. With carefully designed fibre orientation 30 GN/m^2 elastic modules can be obtained (against 210 for steel).

VW[4] has also addressed the vital topic of designing against the fundamental weakness of plastics: duration of loading causing creep and relaxation. The research analysed GRP–PUR–GRP sandwich panels in particular and the graph at (b) shows creep strain of GRP while (c) shows creep of rigid PUR under constant shear. The graph at (d) shows time and temperature dependency of GRP elastic modulus, deformation being shown against a base of loading period. Maximum normal stress in the skins can, of course, only be reached if shear stress in the core, and its adhesive bond to the skin, is not exceeded. In the case of profiled parts, the researcher points out that the skin's own bending stiffness is no longer negligible so shear stress in the core is reduced and allows higher short-time structural loading. Under all load conditions the profiled structure shows better time-dependent mechanical behaviour than the flat one. VW research has demonstrated that despite the familiar creep of plastics under sustained load, the materials possess good recovery capacities due, the author says, to their exponential and linear division of creep functions. FEM analysis of a composite car body floor (e) with 1.5 mm GRP skin and 12 mm thick rigid PUR foam core was carried out to examine these load cases: temperature-dependent creep under bending by four 100 kg passengers; static simulation of front impact at 14 *g* with and without 20 *g* seat belt forces and dynamic loading induced by the sub-idle engine shake mode. The analysis showed that compared with a steel floor the composite panel showed higher stiffness even under the constant load case of 1.5 days at 60°C.

7.4.2 HIGH STRENGTH LAMINATES: LESSONS FROM THE TRACK

The introduction into reinforced thermoset-resin laminates of reinforcements such as carbon fibre and para-aramids, together with epoxy resins, has largely been the preserve of the track race-car builder. Sandwich panels of Nomex honeycomb core materials are also a key ingredient, typically bonded to skins of high strength laminates by an adhesive film such as the Ciba-Geigy Fibredux. For curved surfaces, the core is heat formed at 170–200°C. A key property of this high performance sandwich panel, in this application, is very high resistance to buckling for very low weight.

Builders favour carbon-fibre-reinforced laminate skins for ultimate strength and para-aramids such as DuPont's Kevlar for toughness. The latter is available in a variety of cloth weaves in weights ranging from 60 to 460 grams per square metre. Typically an aramid/epoxy-resin laminate of 1.33 specific gravity has an elastic modulus of 42 N/$mm^2 \times 10^3$ and tensile strength of 760 N/mm^2. A carbon fibre/epoxy-resin laminate using a material such as Courtaulds' Grafil has a modulus up to 190 N/$mm^2 \times 10^3$ and tensile strength of up to 1900 N/mm^2 with individual fibre strengths up to 1.9 N/$mm^2 \times 10^3$. The weave of the cloth is important to laminated strength: plain weave, alternate warp and weft, is stable and easily handled while satin weave (weft passing over three yarns and under one) is suitable for draping over deep-form surfaces. For high strength applications, twill weave, in which groups of warp and weft yarns are woven alternately, is particularly suitable.

Sandwich laminates are formed either in a vacuum bag or an autoclave, an intermediate perforated membrane and blotting paper being used to absorb surplus resin squeezed out during the compression process. In the Tyrrell 011 Formula One car, for example, the main body has two full

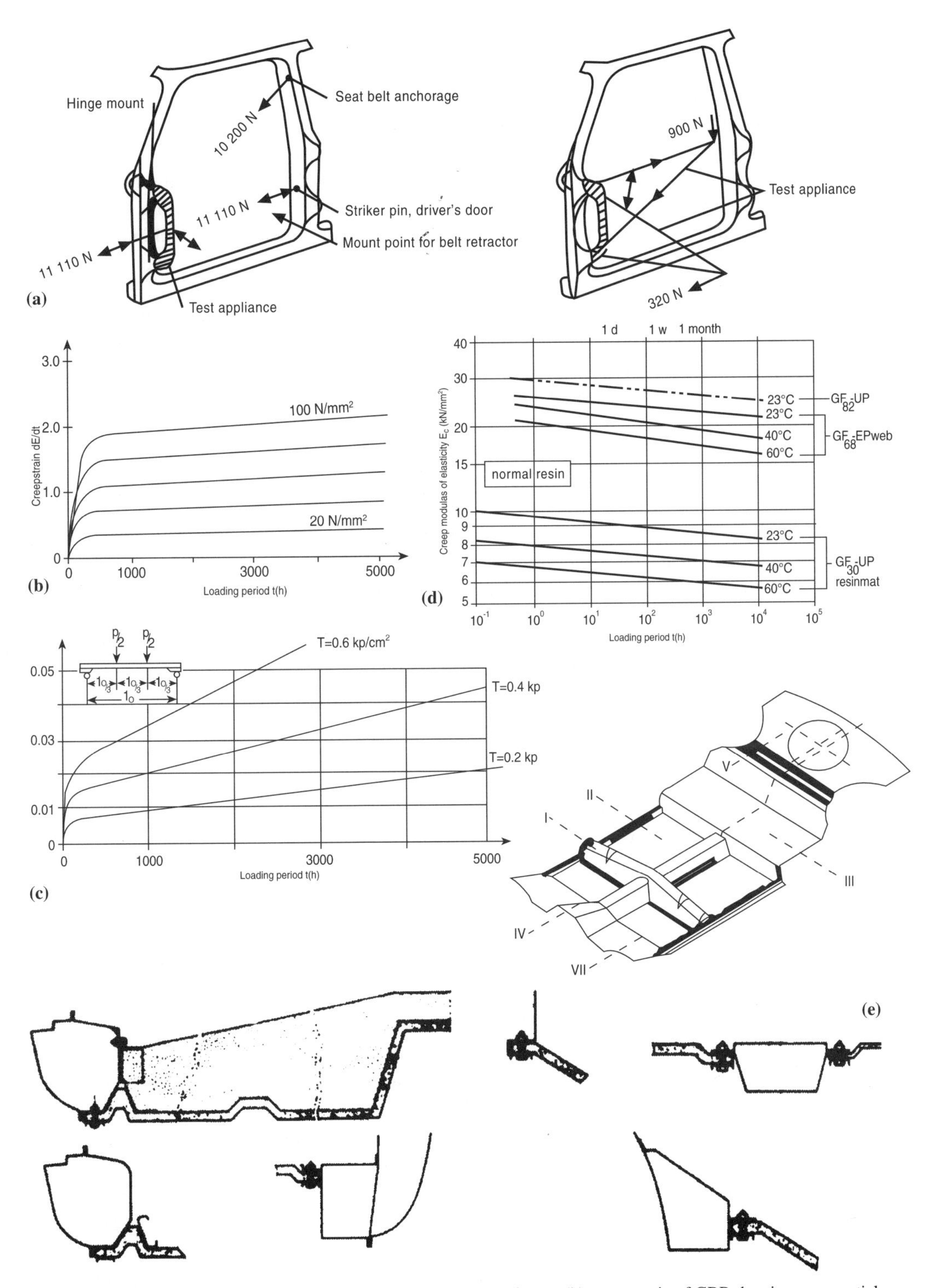

Fig. 7.7 Composites in passenger cars: (a) test forces on front door frame; (b) creep strain of GRP showing exponential and linear regions; (c) creep of rigid PUR foam under constant shear; (d) time/temperature dependency of GRP; (e) composite GRP floor design.

length side panels forming the outer skin and joined by a Nomex honeycomb core to a Kevlar reinforced inner. Bulkheads down the body are machined from aluminium alloy. The weight of the car is 585 kg, with oil and water, for an overall length of 6.65 metres. Another example is the Williams Grand Prix car, Fig. 7.8, whose main structure is shown at (a). This follows the now classic construction of a central semi-monocoque accommodating driver and front suspension; the rear-attached engine supports the gearbox casing on which, in turn, is mounted the rear suspension. The purpose of the bulkheads, (b), is to feed attachment loads (and those due to driver and fuel inertia) into the main structure. Rollover protection structures in front of and above the driver must withstand 7.5 *g* downward, 5.5 *g* rearward and 1.5 *g* sideways without collapsing. By regulation, for this formula, the car must not exceed 500 kg.

An idea of the strength of race-car structures is indicative in the test load criteria which they have to sustain under FISA regulations. As well as the nose cone being required to sustain 47 kJ under frontal impact, an additional 2000 kg lateral test load is required which checks the integrity of fastening between nose cone and the main monocoque shell. There is also a proof load test for the main rollover hoop structure involving 7.5 *g* downwards, 5.5 *g* rearwards and 1.5 *g* sideways loads, based on a mass of 780 kg. A further load of 9.42 *g* (721 kg) must not deflect the rearward part of the structure more than 50 mm. Each bulkhead has to sustain a side loading of 2000 kg in impact. Also, a Grand Prix car carries some 2000 gallons of fuel which occasionally sustains lateral loading up to 4.5 *g* even in the non-impact situation. Design load cases of 10 *g* asymmetric bump loads on individual wheel stations are now quoted, and braking loads up to 3 *g*. Aerodynamic downthrusts on the front wings are put at 5 kN maximum.

For impact integrity the fuel tank must sustain vertical and side loads of 1000 kg. A further side load in the cockpit area is also required, in a vertical plane passing through the centre of the seat belt lap strap fixing, also on a vertical plane passing half way between the front wheel axis and the centre of the dashboard hoop. In the car's safety structure two rollover protection systems must be provided, the first in front of the steering wheel not more than 25 cm ahead of the rim, the second not less than 50 cm behind the first and high enough for a line joining at the top to be 5 cm above the driver's head/helmet. Facilities at MIRA for impact testing of racing cars, (c), include a frontal impact rig which in this case does not involve an inclined plane. The graph at (d) shows examples of accelerations and durations applying to frontal impact. The 25 *g* limit for Formula One technical regulations refers to the average acceleration level throughout the test. The 20 *g*/30 ms pulse is an idealized one for car seat testing while the 60 *g*/3 ms one represents the limit adopted for chest acceleration. This is because whole vehicle decelerations of 20 *g* are typical but those of components and occupants can be much higher; high peaks can also occur due to the slack in seat belts.

7.4.3 POLYMER-COMPOSITE STRUCTURAL ANALYSIS

Design techniques for load-bearing plastics inevitably differ from those applying to elastic materials and continuous attention must be paid to creep properties when predicting stresses and deflections. Plastic material specifications must state service temperatures and duration of loads when using the so-called pseudo-elastic design method. The value of 'elastic' modulus used must be relevant to these service conditions – and because, for most plastics, modulus value is unlikely to be above a few GN/m^2 at the very best, high inertia cross-sections are the rule for minimizing deflections. Creep is defined as the strain, which is time dependent, resulting from applied load – and associated stress relaxation is the stress, which is time dependent, resulting from the applied deformation. Given a defined duration of loading, a (secant) creep modulus may be calculated as the ratio of the applied stress to the applied strain at the loading time and temperature of interest. Modulus is generally strain dependent and in carrying out material tests it is useful to represent creep data by cross-plotting to obtain isochronous stress/strain curves for different times under load. The design approach is first to identify maximum service temperature and duration of load; next to calculate maximum stress applying to the particular component; then to read the strain from the appropriate creep data and calculate the value of creep modulus; finally to use this value in elasticity calculations to predict deformations as stress relaxation in the component – arriving at the final dimensions by an iterative process, Fig. 7.9.

The first graph, at (a), presents the challenge of metals and structural timbers to polymer-composites while that at (b) shows creep resistance of unreinforced polypropylene. In designing with this material, ICI consider deformation, fracture after dynamic fatigue and resistance to impact are the important mechanical properties and the company supply data based on both creep, (c), and stress-relaxation tests. The first involves observation of strain as a function of time when the specimen is held at a constant load while the second involves measuring stress as a function of time while maintaining constant strain.

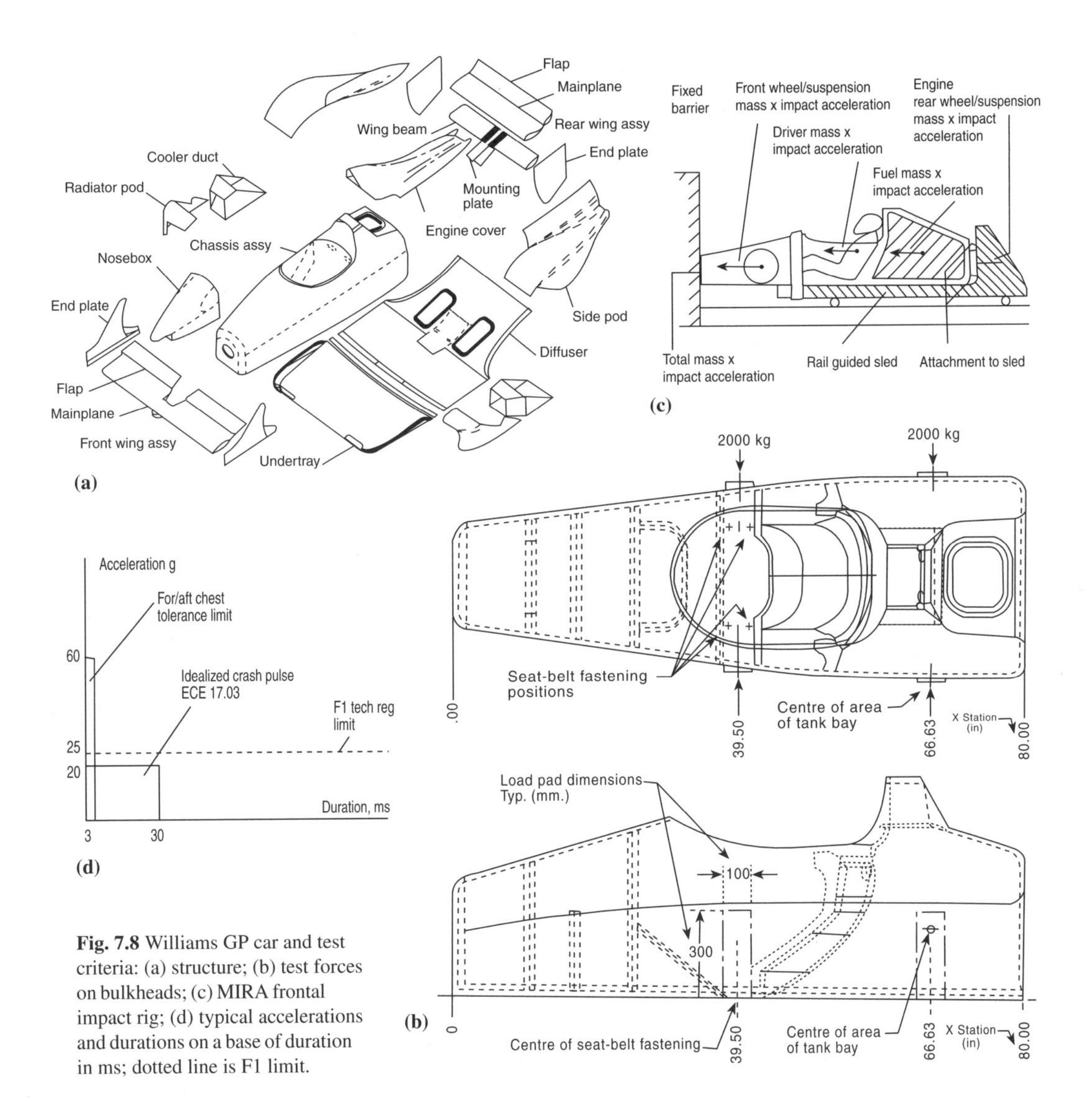

Fig. 7.8 Williams GP car and test criteria: (a) structure; (b) test forces on bulkheads; (c) MIRA frontal impact rig; (d) typical accelerations and durations on a base of duration in ms; dotted line is F1 limit.

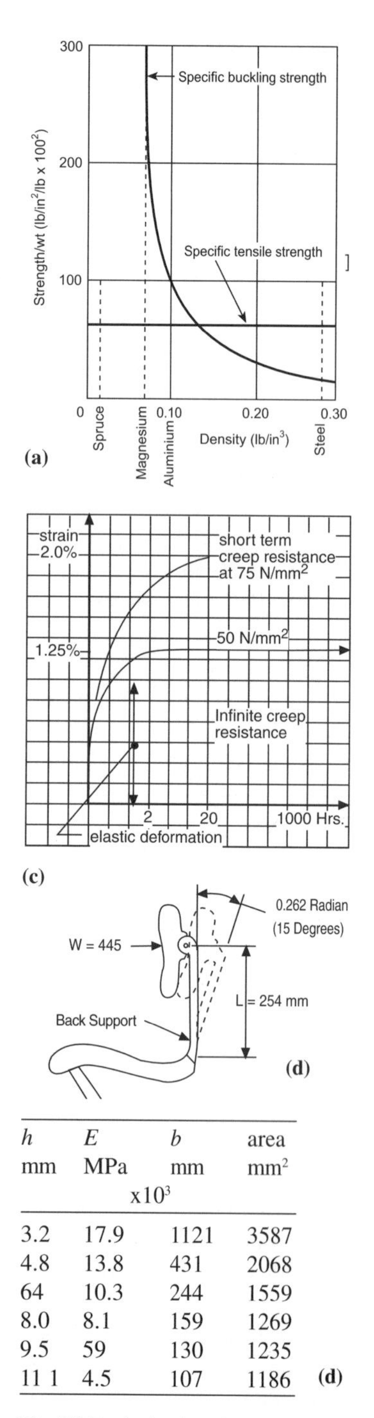

h mm	E MPa x10³	b mm	area mm²
3.2	17.9	1121	3587
4.8	13.8	431	2068
64	10.3	244	1559
8.0	8.1	159	1269
9.5	59	130	1235
11 1	4.5	107	1186

(d)

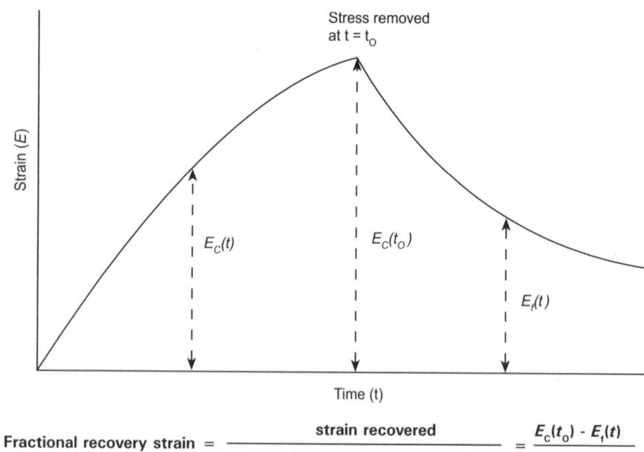

$$\text{Fractional recovery strain} = \frac{\text{strain recovered}}{\text{creep strain when load is removed}} = \frac{E_c(t_o) - E_f(t)}{E_c(t_o)}$$

$$\text{Reduced time} = \frac{\text{recovery time}}{\text{duration of creep period}} = \frac{t - t_o}{t_o}$$

(b)

Property			Compression flange	Tension flange	Web and diaphragm
Module					
— Longitudinal	E_{C1}	GNm^{-2}	36.5	36.6	19.75
— Transverse	E_{C2}	"	5.75	5.75	19.75
— Shear	G	"	2.14		7.53
Poissons ratio					
— Longitudinal	ν_{C1}		0.275	0.275	0.312
— Transverse	ν_{C2}		0.043	0.043	0.312
Strength					
— Longitudinal	GNm^{-2}		$0.42(X_{c1})$	$1.26(X_{r1})$	0.395
— Transverse	"		$0.050(X_{c2})$	$0.025(X_{r2})$	0.395

(e)

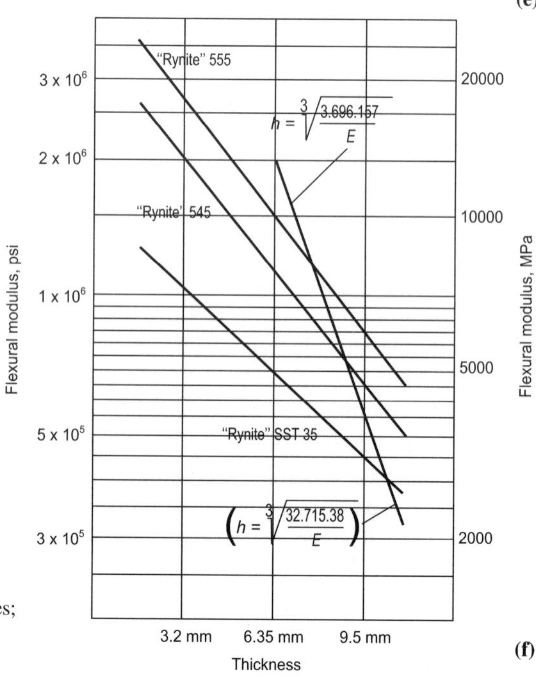

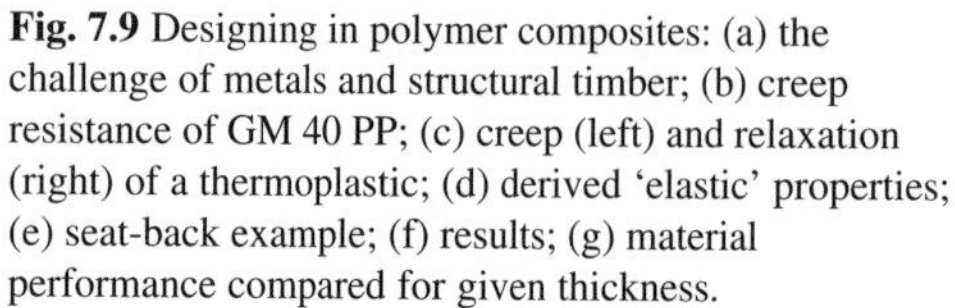

Fig. 7.9 Designing in polymer composites: (a) the challenge of metals and structural timber; (b) creep resistance of GM 40 PP; (c) creep (left) and relaxation (right) of a thermoplastic; (d) derived 'elastic' properties; (e) seat-back example; (f) results; (g) material performance compared for given thickness.

Du Pont have applied pseudo-elastic design methods for such structures as rectangular section seat backs, (d), using the expression for breadth: $b = 6WL^2/Eqh^3$. Putting various thicknesses and the relevant moduli from the table in (d) for different materials into the equation gives a series of widths; from these the amounts of material used and an idea of relative part costs is obtained. Thinner sections will have shorter moulding cycles, important in determining part cost. For the stiffest resin, Rynite 555, the flexural modulus at a thickness of 3.2 mm is = 17.9×10^3 MPa, giving breadth of:

$$[16 \times 445 \times (254)^2]/[17.9 \times 0.262 \times (3.2)^3] = 1121 \text{ mm}$$

and the con-sectional area – proportional to the mass of material used – is $3.2 \times 1121 = 3587$ mm^2

This width is obviously too great, so Du Pont repeat the exercise with some thicker sections/ smaller widths which look more acceptable: for $h = 4.8$ mm, $E = 13.8 \times 10^3$ MPa then $b = 431$ and the con-sectional area is 2069 mm^2, and so on for $h = 6.4$, 8.0, 9.5 mm, and by extrapolating the experimental data slightly, for $h = 11.1$ mm. All the results are in the table at (e). To find the thickness to satisfy the 15° (0.262 radian) deflection criterion, using the same equation, but this time solving for h, to determine the two unknowns h and E, we insert the known values for load, deflection, length and width then plot this in (f) – for example, by using two values of E and the corresponding calculated values of h ($E = 13\ 800$ MPa, $h = 6.4$ mm; $E = 4140$ MPa, $H = 9.7$ mm) and connecting the points by a straight line. This intersects the modulus/thickness line for Rynite 555 to give the solution for h as 7.8 mm. The lines for Rynite 545 and SST 35 intersect the equation plotted at greater thicknesses, and since these resins can take more impact than Rynite 555, one has the choice of playing safe – at the cost of more resin in the back support. But the stress in the thicker section would be lower, and this could be especially important. Keeping stresses low in the back support will help to extend fatigue life, reduce creep, and ensure that the part can safely take overload several times the design load.

Allowing for reinforcements in the analysis is complex, Fig. 7.10, when the reinforcement results in directional properties. Structural composites can be based on a variety of resin and reinforcement systems, the properties of some of the main ones being given by Phillips[4] at (a). The effectiveness of fibre reinforcement depends on the fibre aspect ratio of length/diameter which should be a minimum of 100. Most commercial fibres are less than 20 microns in diameter but the area of each available for adhesion to the resin is relatively high compared with the compressed bulk of the individual fibres. Resins, in general, are an order of magnitude lower in strength/ stiffness than fibre reinforcements.

The graph at (b) provides a useful relationship between fibre and resin matrix to show percentage strength against ratio of moduli. The view at (c), however, provides a more detailed examination from which basic calculation formulae can be derived. Between the two extremes of submerged depth shown, for the fibre inside the resin, there is a critical length L_c where fibre failure may be either pull-out or tensile fracture and the failing load is balanced between shear and tension. These can be equated to give $\tau = \sigma_f r/2l_c$. At the end of the fibre tensile stress is zero and rises to a constant maximum once l_c is exceeded.

In the case of carbon-fibre reinforcement, low molecular weight epoxy resins have become the favoured matrix. Properties of commercial grades are shown in column 3 of the table at (d), the first two columns showing potential properties available when more exotic types 1 and 2 carbon-fibres are used. The table at (e) shows the properties available with the available forms of aramid fibre.

Tensile modulus of a composite laminate is:

$$E_k = E_f V_f B + E_m V_m$$

Fibre/material	Specific gravity density (g/cm³)	Tensile strength (MPa)	Tensile strength (x10³ lbf/in²)	Young's modulus (GPa)	Young's modulus (x10⁶ lbf/in²)
E-glass	2.54	2410	349	69	10
S-glass	2.49	2620	380	87	12.6
Carbon type 2	1.75	2410	349	241	35
Aramid (high modulus)	1.44	3450	500	124	18
Boron	2.63	2760	400	379	55
Alumina	3.30	3000	435	297	43
Asbestos (chrysotile)	2.40	1490	216	183	26.5
Wood	0.50	69	10	7	0.95
Light alloy	2.69	476	69	72	10.50
Steel (structural)	7.85	413	60	207	30
Titanium alloy	4.52	711	103	117	17

(a)

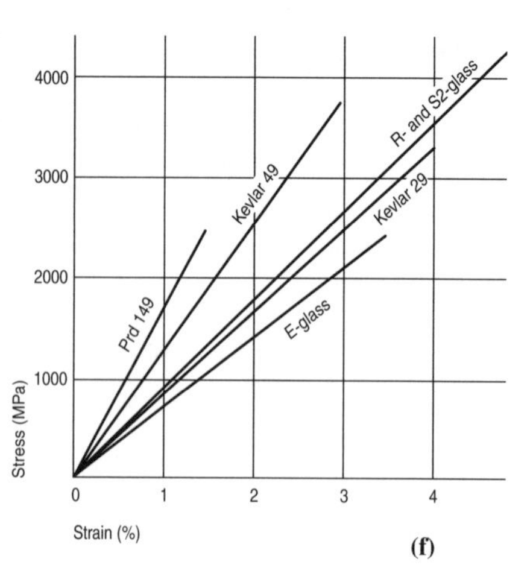

Fibre value \ Fibre type	1		2		3	
Young's modulus, lbf/in² x 10⁶ (GPa)	50	(345)	35	(241)	29	(200)
Ultimate tensile strength, lbf/in² (GPa)	290 000	(2.00)	350 000	(2.41)	320 000	(2.21)
Specific gravity	1.92	–	1.75	–	1.70	–
Unidirectional composite (V_f = 60%)						
Young's modulus, lbf/in² x 10⁶ (GPa)	26	(179)	17.5	(121)	14.5	(100)
Ultimate tensile strength, lbf/in² (GPa)	145 000	(1.00)	200 000	(1.38)	175 000	(1.21)
Specific gravity	1.72	–	1.55	–	1.52	–

(d)

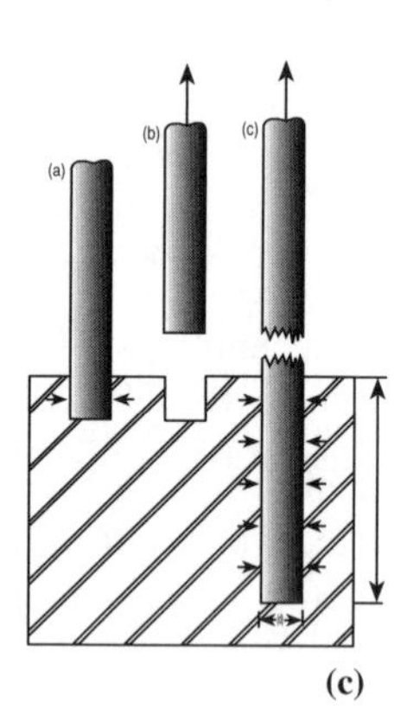

	Tensile strength (MPa)	Tensile strength (x10³ lbf/in²)	Young's modulus (GPa)	Young's modulus (x10⁶ lbf/in²)	Specific gravity	Elongation of break (%)
Lower-modulus fibre	2650	384	59	8.6	1.44	4
Higher-modulus fibre	2650	384	127	18.4	1.45	24

(e)

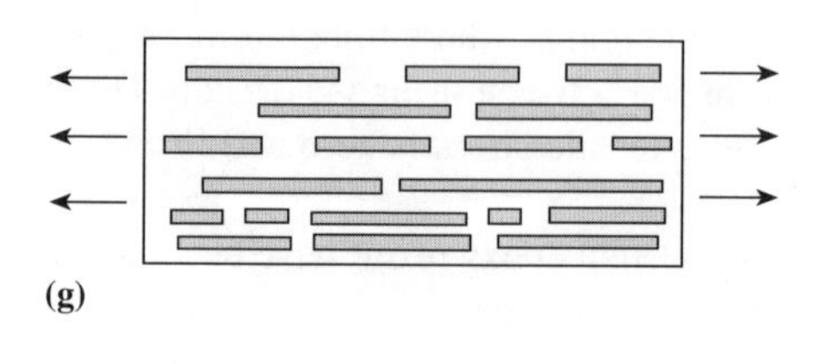

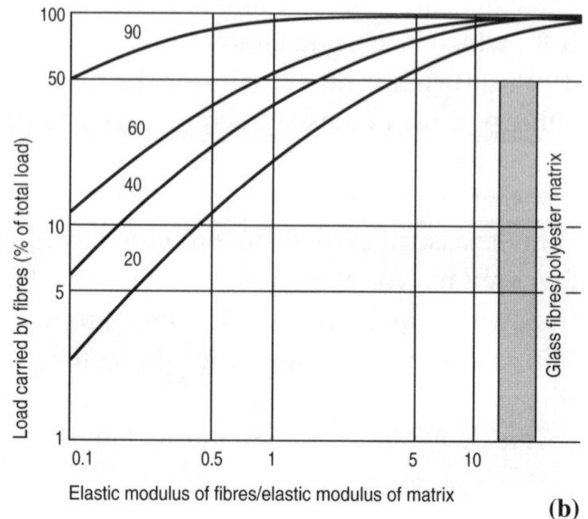

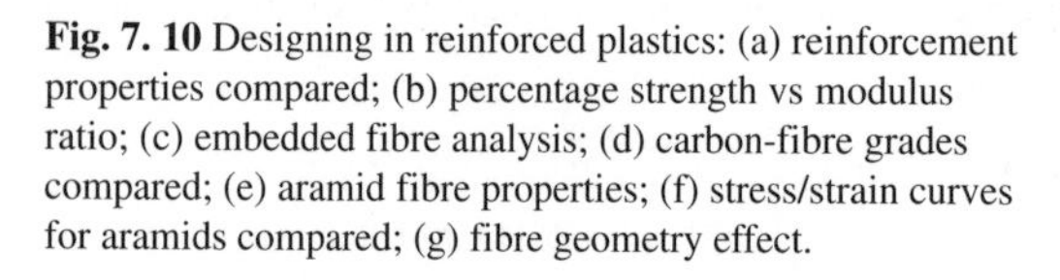

Fig. 7. 10 Designing in reinforced plastics: (a) reinforcement properties compared; (b) percentage strength vs modulus ratio; (c) embedded fibre analysis; (d) carbon-fibre grades compared; (e) aramid fibre properties; (f) stress/strain curves for aramids compared; (g) fibre geometry effect.

where E is modulus, k composite, f fibre and m the matrix, in an expression due to Hollaway[5]. V_m is the volume fraction of the matrix and the lay-up factor $B = 1$, 1/2, 3/8 for unidirectional, bidirectional or random (in-plane) orientation respectively. Stress/strain curves for glass and aramid fibres are shown at (f), with effect of fibre geometry at (g).

7.5 Ultra-lightweight construction case study

The General Motors Ultralite concept car is an encapsulation of advanced carbon-fibre reinforced epoxy resin structure, combined with aerodynamic optimization, that would suit a high-performance electric car. An EC energy saving study[6] put forward the forecast of local manufacturing units that would make the hand-building of such vehicles a reality. The study suggests that despite some notable lightweighting initiatives, there is little sign yet of affordable super-light production cars. While vehicle makers continue to spend fortunes on refinement of vehicles that will need revolutionary redesign if they are to meet supercar standards, it is suggested that the move is akin to refining the typewriter after the desktop computer had been invented. The real fuel economy gains are to come from extensive weight and aero-drag reductions rather than further engine development on systems that might soon succumb to hybrid drives. There is also substantial potential in reducing tyre rolling resistance; when best experimental-car fitment figures are compared with those for production cars, it is clear that substantial savings are available.

The study suggests that the problem for steel is not necessarily its high density, because specific strength can be high with appropriate alloying. The problem is that its low raw-material cost blinds people to the massive tooling and processing costs required for its successful volume exploitation. In pressworking alone, thousands of engineers are employed over a 2 year period in producing expensive dies for new model introductions.

By contrast it is interesting to note that the GM Ultralite concept car, Fig. 7.11, involved around 50 technologists for a 100 day introduction lead-time for this polymer-monocoque structured vehicle. By achieving a drag coefficient of 0.192, forward-projected cross-section area of 1.71 m^2, mass of 635 kg for this 165.6 inch long car, and road-speed rolling resistance of 0.007 at 4.4 bar tyre pressure, it is still a standard against which other new cars can be measured.

The six-piece 191 kg monocoque body involved foam sandwiched between laminates of carbon-fibre reinforced epoxy resin. The CF cloth was interwoven with single-strand roving at high stress areas and press critics foolishly dismissed the car for its carbon-fibre composite material cost, per

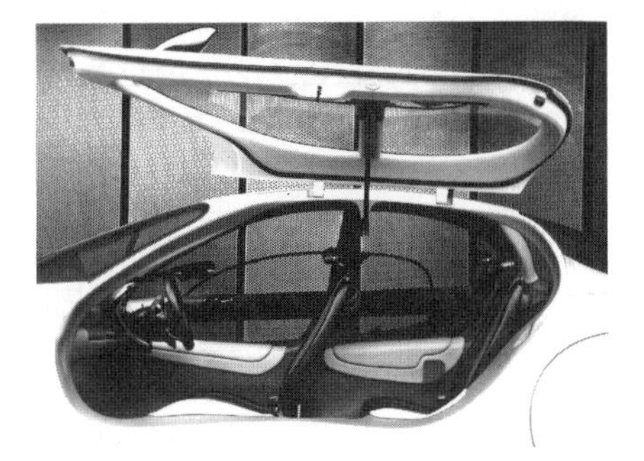

Fig. 7.11 GM Ultralite concept.

lb, being greatly higher than steel without consideration of the enormous tooling/process cost of steel fabrication or the fact that much less weight of material was used per car.

The use of advanced materials for interior equipment – thin Duofix seats – allowed designers to create a spacious passenger compartment; the front-seat passengers have 37.2 inches of head room, 57.1 inches of shoulder room, and 42.8 inches of leg room. Rear-seat passengers have 36.3 inches of head room and 33 inches of leg room. Besides the C_dA of just 0.329, another aerodynamic challenge was to provide proper air flow through the vehicle as the structural requirements prevented any openings in the front or rear of the car which have an interchangeable, modular 'power pod' located in the rear of the structure. This allowed designers to minimize the car's frontal area, but posed a challenge to get sufficient air flow to the engine, radiator and heat exchangers. Air flow channels, designed as NACA ducts, were incorporated into the vehicle's underbody structure to deliver air to these components. Further weight savings are engineered into the brakes with the use of aluminium silicon carbide rotors. The aluminium composite contains 20% silicon carbide particulates which provide an excellent friction surface. The material offers a 50% weight reduction compared to an iron rotor. Lightweight aluminium callipers decrease brake drag, while the entire system is designed so that power assist is not needed. Self-sealing tyres contribute weight savings by eliminating the need for a spare tyre. Goodyear decreased the rolling resistance of the tyres to about 2.5 to 3 lb per tyre at highway speeds; production cars have about 9 lb of rolling resistance per tyre. The 18-inch tyres are large for the car's weight class but are optimized for low rolling resistance.

7.6 Weight reduction in metal structures

7.6.1 STEEL USED INTELLIGENTLY LEADS TO SUBSTANTIAL WEIGHT SAVINGS

A car body weight reduction of 25%, without cost penalty, plus a 20% reduction in part count, has been the result of the final phase of the USLAB project for lightweighting steel automotive structures carried out by an international consortium of steel companies supervised by Porsche Engineering Services, Fig. 7.12. The objective of a 'feasible design, using commercially available materials and manufacturing processes', has also been met. The elegant shell at (a) does indeed show some important advances in structural efficiency within the confines of this brief but of course radical redesign away from the conventional sedan shape and use in the main of existing pressworking machinery were not allowed.

Nevertheless, an 80% gain in torsional rigidity and 52% in bending have also been recorded and estimated body shell cost of $947 compares with $980 for a year 2000 comparison figure with a conventionally constructed body to a similar specification. The 203 kg body shell, of 2.7 metres wheelbase, has a torsional rigidity of 20 800 Nm/° and a first torsional vibration mode of 60 Hz, using the well-documented construction techniques.

A supercomputer analysis of crashworthiness has shown that frontal NCAP and rear moving-barrier levels have been achieved at 17% higher than the required speed. The body has also satisfied 55 kph/50% offset AMS, European side impact and roof crush requirements. The 35 mph frontal NCAP showed a peak deceleration of 31 *g*, considered satisfactory in that stiffer body sides are required to meet 50% AMS offset requirements. The offset AMS frontal impact deceleration showed a peak at 35 *g*, considered a good result in relation to the severity of the test. The simulation was carried out at 1350 kg kerb weight, plus 113 kg luggage and 149 kg for two occupants. At the final count, high strength steel was used for 90% of the structure, ranging from 210 to 550 MPa yield, in gauges from 0.65 to 2 mm. Around 45% of the structure involved laser-welded tailored blanks, including the monoside panels, (b), which ranged in gauge from 0.7 to 1.7 mm and were made up of steel elements having yield strengths from 210 to 350 MPa.

While it was decided to stay with more traditional concepts, known manufacturing and design technology was taken to its limits by combining the unibody structure with hydroforming and maximum use of structural adhesives, as well as laser welding, tailor-welded steel blanks, and roll forming. The process was computer interactive all along, optimizing sections, improving joints and using different steel qualities and material thicknesses at appropriate points in the design to create the optimum structure.

A number of parts that would traditionally be created by sheet metal stampings were replaced by one-piece hydroformed tubes. This required tight control during design of section perimeters and transitions and some unique joint designs. To attach other body-in-white components to the tube requires a one-sided attachment method and in this case laser welding was chosen. The view at (c) shows the hydroformed side roof rail. Particular attention was also paid to the connection from the A-, B- and C- pillars and to the integration of the rear shock towers into the rear rail in order to provide optimal load transfer into the structure, (d).

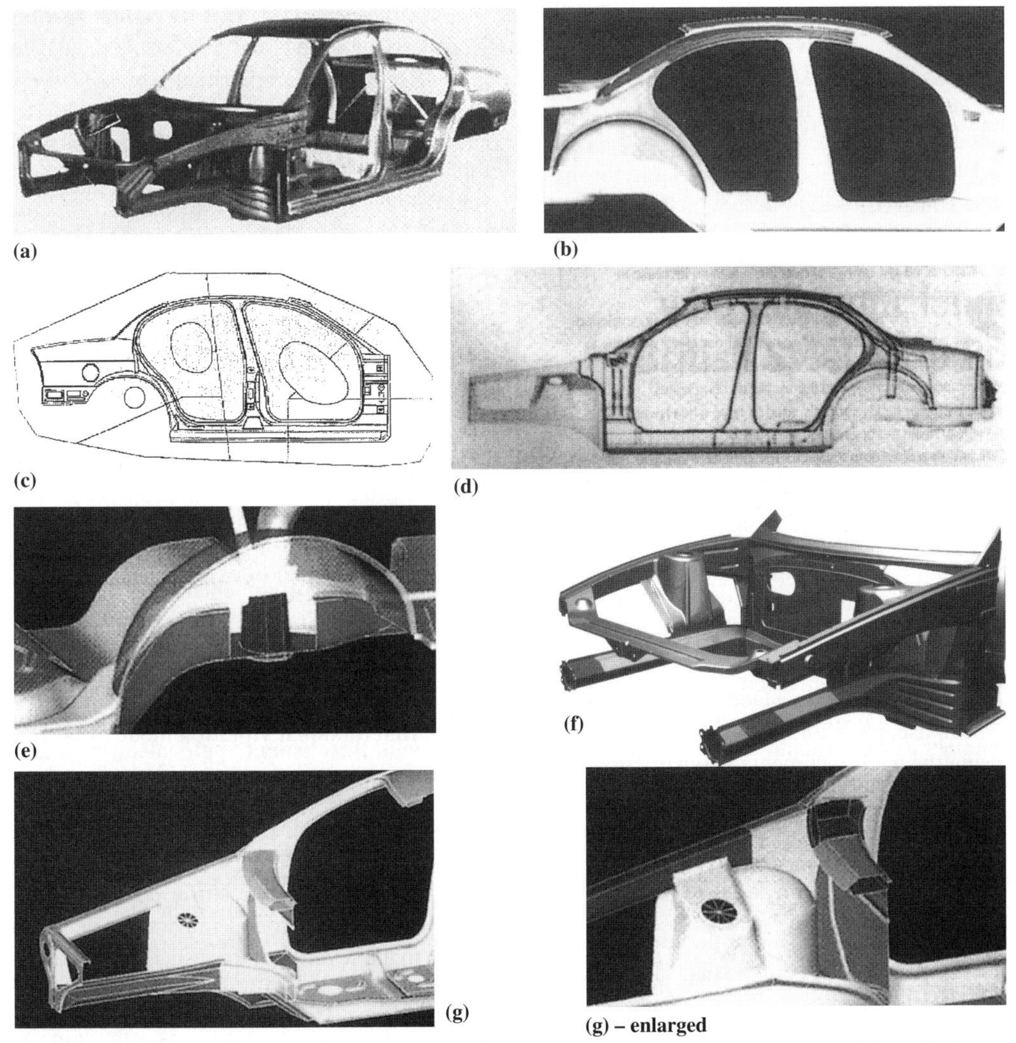

Fig 7.12 ULSAB steel weight-reduction project: (a) body shell; (b) monoside panel; (c) hydroformed side roof rail; (d) principal application areas for bonding; (e) integration of rear shock absorber tower; (f) running the lower cowl through the A-pillar; (g) integration of front shock absorber tower.

To improve the rigidity of the ULSAB, the body side inner assembly is weld bonded to the body side outer assembly. Combined use of welds and high temperature adhesive provides continuous bonding. The view at (e) shows the two principal application areas for bonding. For structural efficiency reasons, a non-traditional approach to the cowl to the A-pillar joint was taken by running the lower cowl section through the A-pillar inner and outer creating a hole, (f). This produced a large rigidity gain for the structure. It worked because the centres of some traditional joints become a structurally dead region that can be removed. The front-area/shock-tower region is integrated into the skirt, which is itself laser welded to the fender support rail, (g). The wheel house is spot welded to the front rail welding flange and, in the shock tower area, to the front rail's lower flange, hence forming a well-integrated structure.

In order to reduce the part count by eliminating reinforcement, and also to reduce subassembly welding and to increase the structural efficiency, a number of tailor-welded blank parts are used in the design. These include the front and rear rails, the rocker, the A- and B-pillars, and the wheel house fronts and inner wing panels. About 67% of the structure of the ULSAB is made from special steels, either dual phase or hard baked dependent upon the difficulty of forming and the strength requirements.

7.6.2 LIGHT ALLOYS FOR IMPROVING SPECIFIC RIGIDITY

The non-ferrous light alloys such as those of aluminium and magnesium can, of course, produce panels which are inherently less prone to buckling without the same need for stabilizing reinforcement required of steel panels, Fig. 7.13. Comparatively thicker skinned structures are possible and high specific rigidities are obtainable in boxed punt structures such as the hydro aluminium one, (a), used on the BMW E1 experimental electric car. This is likely to be a more rewarding approach than the direct substitution of aluminium alloy for steel illustrated in Fig. 0.1 in the Introduction. The need is again for designing to obtain full advantage from the material.

According to researchers at Raufoss Automotive Structures, there are a number of different structural requirements for vehicle body shell materials. In normal road use, bending and structural stiffness are the key parameters. However, in low speed impact (4–9 kph) elastic deformation energy is most important; in a mild crash, elastic deformation energy is the key whereas in a severe crash, structure integrity is all important, with no breakage or fragmentation occurring. In comparing structural performance with steel it is argued that cost per unit volume rather than per unit weight should be considered. Examination of individual parts of the structure can still show some surprises when beam elements, for example, are found to be subjected to bending and torsional moments, as well as shear and axial loads. The response/displacement characteristic of the beam element to these moments is a measure of the contribution of the element to global stiffness and strength, while contribution from shear and axial loads are minor in comparison.

With a straight substitution for steel in a sill member, (b, i), and assuming equivalence of yield strength, there is a one-third weight saving, offset by a two-thirds reduction in stiffness; however, the beam profile will absorb three times more elastic energy before permanent set. When optimized for bending, (b, ii), there is now more than one-third the stiffness but three times the displacement and more than three times the elastic energy absorbed. There is also more elastic-plus-plastic energy absorbed. When optimized for torsion, (b, iii), again there is more than one-third the stiffness and three times the displacement and more plastic-plus-elastic energy absorbed. However, when optimized for stiffness, generally, (b, iv), there is a 40–50% weight saving, stiffness is the same as steel and approximately 2.5 times the displacement applies. There is also much more elastic-plus-plastic energy absorbed, (c).

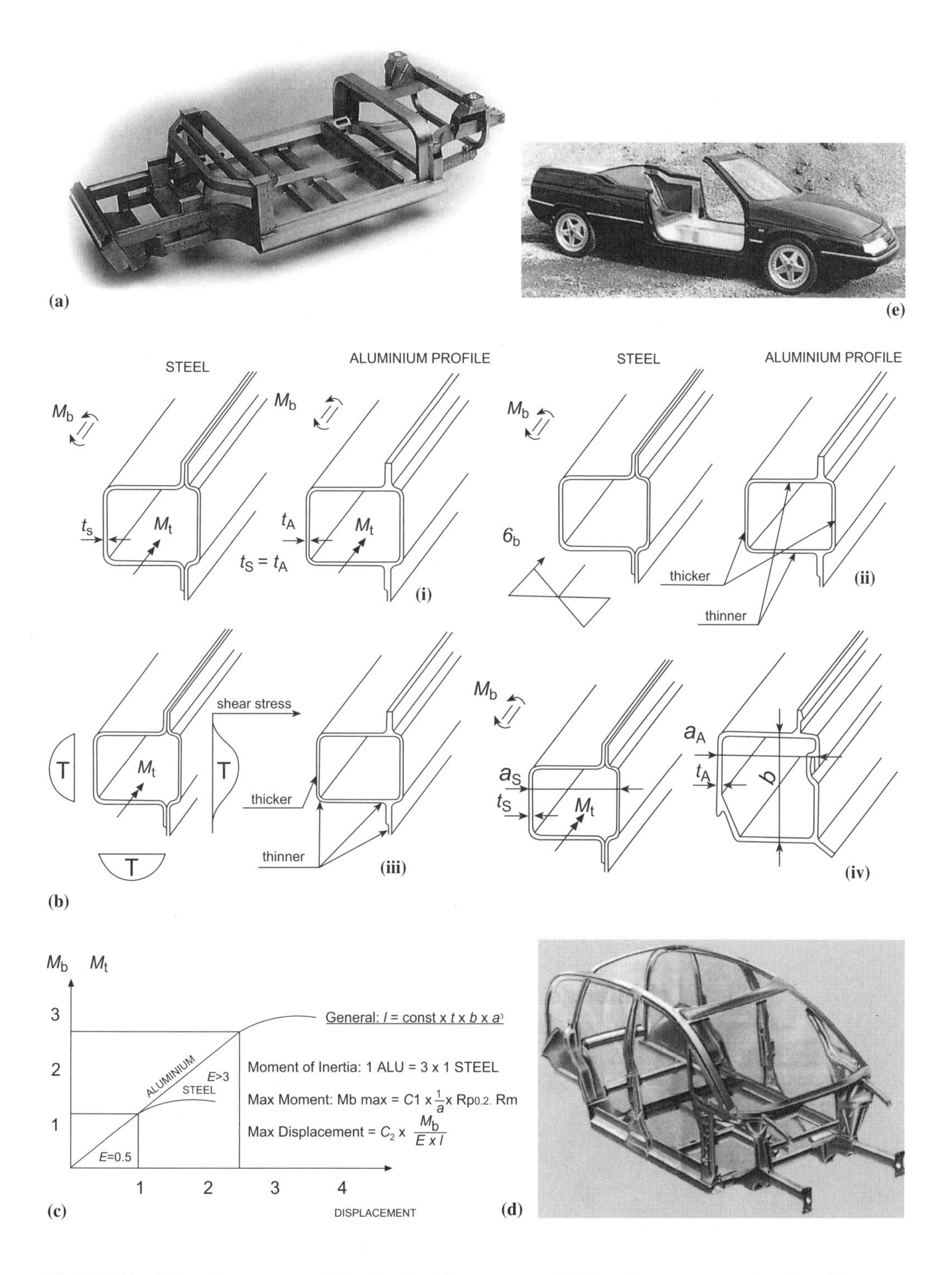

Fig. 7.13 Aluminium alloy structures: (a) hydro aluminium structure; (b) sill section parameters when (i) straight substitution of aluminium for steel; (ii) optimized for bending; (iii) optimized for torsion; (iv) optimized for stiffness; (c) Applied moments vs displacement for sill member optimized for stiffness; (d) Al_2 concept structure; (e) XMple concept car using aluminium skinned sandwich panels.

Significant gains can only be made when profile dimensions are increased until the section moment of inertia is three times that of steel. But as the moment of inertia increases in relation to the cube of the depth of section, an increase in beam thickness of 30–40%, together with somewhat thicker walls and increased height, will achieve the optimum section properties. As section moment of inertia increases, the section modulus increases by the square of the amount and there is much more moment carrying before permanent set.

Audi's aluminium alloy body construction programme, which involved cooperation with ALCOA in making structures from extruded section members joined by die cast nodes, has matured into the Al_2 concept car, (d). The 3.76 metre long times 1.56 metre high car weighs just 750 kg in 1.2 litre engined form, some 250 kg less than a conventional steel body vehicle. The number of cast nodes has been reduced compared with the phase one aluminium alloy structure of the Audi A8. Most of the nodes are now produced by butt welding the extruded sections. High level seating is provided over a sandwich-construction floor. While a structural punt is not employed an approach has been made to turn the A-post into a structural member by making it an extension of the cant-rail so that the roof-level structure can play some part in obtaining overall rigidity.

Aluminium alloys like steel can be used for the face skins of sandwich panels. An interesting 'thin' sandwich material Hylite (aluminium/plastic/aluminium) developed by Hoogovens Groep is claimed to be the lightest bodywork material outside exotic polymer composites. It consists of two layers of 0.2 mm aluminium and a core of 0.8 mm polymer material. For equal flexural rigidity it is said to be 65% lighter than steel sheet, 50% lighter than plastics and 30% lighter than aluminium sheet. It can be deep drawn on existing presses and is form stable up to 150°C, which is of importance for painting. When mass produced, Hylite is more expensive than steel but is in the same range as aluminium and cheaper than plastic.

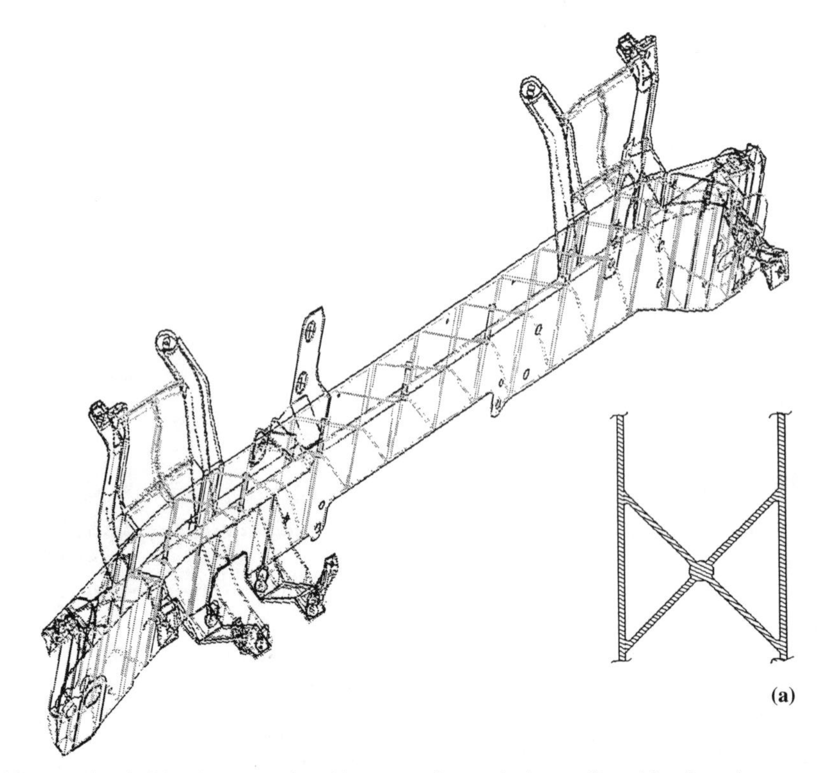

Fig. 7.14 Magnesium bulkhead crossmember: (a) open-section carrier beam; (b) multi-web section.

An early application was the concept convertible based on the Citroen XM shown at (e). It also uses a combination of steel and aluminium extrusions and aims to resolve the main problem in engineering a convertible, ensuring the rigidity of the body by using a structured floor section made from hollow, thin-walled aluminium extrusion sections. The central part of the steel floor is replaced by aluminium, while the front and rear ends and the sills remain in steel. The hollow floor section on which the platform is based is 300 mm wide by 50 mm deep, with cross-ribs on the inside. An enclosed tunnel section, two side sections and two cross-sections complete the welded structure. The complete floor weighs only 50 kg, and the steel is joined to the aluminium floor with adhesive and monobolts, overall rigidity being similar to the original XM sedan.

The plastic core material in Hylite is specified in a way that allows the shear modulus to retain an acceptable value over the operating range from –30 to +85°C; and compared with sandwich cores made from visco-elastic materials, sound deadening properties are preserved throughout the operation range. Hoogovens have ascertained that a skin yield stress of 130 kN/m^2 minimum is required and have chosen an appropriate aluminium alloy grade, 5182, containing 4.5% magnesium. The polypropylene core chosen is tolerant of paint stoving temperatures. This combination is also said to have satisfactory drawability with standard press tools modified to account for the lower tear strength compared with a solid aluminium alloy. Forming radii also must not be less than 5 mm and slightly curved drawbeads are needed to increase stretch level. Thus far the product is tolerant of pressing rates up to 50 mm/sec. A warm bending technique has also now been developed to enable radii down to 2 mm to be achieved on the outer skin for flanging the edges of parts such as bonnet panels. For recycling the product a technique of cooling parts to –100°C, using liquid nitrogen, is proposed at which temperature plastic and aluminium can be separated using a hammer mill, the materials having sheared apart by the differential thermal expansion effect.

Typically, a magnesium alloy body panel would be twice as thick as a steel one but less than half its weight. The thickness will give benefits both in vibration reduction and resistance to denting by minor impacts. Fiat engineers have described an interesting structural application of magnesium alloy, Fig. 7.14, in which a single-piece carrier beam under a dashboard which replaced an 18-part spot-welded assembly. A robust design in terms of section thicknesses and blend radii minimized the influence of the relatively low modulus on structural performance of the part which supports the dashboard, passenger-side air bag, electronic system controllers, steering column and heater–radiator matrix. The pressure die-casting production process necessitated the use of an open-section member, (a), in place of the fabricated box section of the original spot-welded steel assembly. A multi-web reticular structure, (b), was the result; it achieved 50% weight reduction; 80% increase in XY bending stiffness; 30% increase in YZ bending stiffness and 50% increase in torsional stiffness.

References

* Fenton, J., *Handbook of automotive body construction and design analysis,* Professional Engineering Publishing, 1998

1. Rink and Pugh, The perfect couple – metal/plastic hybrids making effective use of composites, IBCAM Conference, 1997
2. Wardill, G., The stabilised core composite, IMechE Autotech Congress, 1989
3. Lilley and Mani, Roof-crush strength improvement using rigid polyurethane foam, SAE paper 960435
4. Phillips, L., Improving racing car bodies, *Composites,* 1(1)50
5. Hollaway, L., *Glass reinforced plastics in construction,* Surrey University Press

6. Lovins and Barnett, Supercars: the coming light-vehicle revolution, Rocky Mountain Institute, Summer Study, European Council for an energy-efficient economy, Denmark, 1993

Further reading

Houldcroft, P., *Which Process?,* Abington, 1990
Gibson/Smith, *Basic welding,* Macmillan, 1993
Pitchford, N., Adhesive bonding for aluminium structured vehicles, IMechE seminar, Materials – fabricating a novel approach, 1993
McGregor *et al.,* The development of a joint design approach for aluminium automotive structures, SAE paper 922112, 1992
Pearson, I., Adding welded/mechanical fastening to adhesive-bonded joints, *Automotive Engineer,* Aug./Sept. 1995
Timings, R., *Manufacturing technology,* Longmans, 1993
Structural Adhesives in engineering, IMechE conference report 6, 1986
James, P., *Isostatic pressing technology,* Applied Science Publishers, 1983
Schonberger, R., *World class manufacturing casebook,* Collier Macmillan
Institute of Materials conference, Moving forward with steel automobiles, 1993
Mann, R., *Automotive plastics and composites,* Elsevier Advanced Technology
West, G.H., *Manufacturing in plastics,* PRI
Goldbach, H., IBCAM Boditek conference, 1991
Hartley, J., *The materials revolution in the motor industry,* Economist Intelligence Unit, 1993
Young and Shane (eds), *Materials and processes,* Marcel Dekker, 1985
Gutman, H., New concept bumper in plastic, SITEV conference, 1990
Thermoplastic matrix composites, Profile series, Materials Information Service, DTI
Wood, R., *Automotive engineering plastics,* Pentech Press, 1991
Design in composite materials, IMechE conference report 2, 1989
Maxwell, J., *Plastics in the automobile,* Woodhead Publishing, 1994
Data for design in Propathene polyurethane, Tech Service Note PP110, ICI
Ashley, C., Weight saving in steel body structures, *Automotive Engineer,* December 1995
British Standard 8118. 1991: *Structural Use of Aluminium, Parts 1 and 2*
Vaschetto *et al.,* A significant weight saving application of magnesium to car body design, paper SIA9506B02, EAEC congress, 1995
Engineering steels, Profile series, Materials Information Service, DTI
Nardini and Seeds, Structural design considerations for bonded aluminium structured vehicles, SAE paper 890716/7
Ruden *et al.,* Design and development of a magnesium/aluminium door frame, SAE paper 930413
User manual for 3CR12 steel, Cromweld Steels
Cowie, G., The AISI automotive steel design manual, SAE paper 870462

8

Design for optimum body-structural and running-gear performance efficiency

8.1 Introduction

Both structural and performance efficiencies are considered, in turn, within this chapter which examines first the body-structural shell and second the running gear of the vehicle. The approach to designing for lightweight, recommended in the introductory chapter, is implicit in the design calculation formulae provided for the several aspects of structural design. Second, the optimization of running gear is based on the most efficient exploitation of the special features of electric and hybrid-drive vehicles in again applying calculation techniques for accelerative performance and weight distribution; ride and handling evaluation; electric steering and braking; also CVT and drivetrain for parallel hybrid-drive vehicles. Again a fuller account of vehicle structural design can be obtained from the author's work*, referenced at this point in Chapter 7, together with a parallel work on running gear design, applying to a wider range of vehicles.

Structural design for optimum efficiency involves the best utilization of the body shell in reacting to passenger, cargo and road load inputs with minimum weight penalty. The evolving design packages for electric vehicles, described in Chapters 5 and 6, suggest that (with the possible exception of the parallel hybrid drive) layout of the electromechanical systems is not constrained by the mechanical drive from power unit to drive wheels and, particularly in the case of battery-electric vehicles, the principal mass can be spread uniformly over a wide platform area between the steered and driven wheels. This chapter examines the monocoque tubular shell and open-integral punt structure as possible structural solutions to two different EV requirements.

In approaching EV structural design, an interesting departure would be for automotive body engineers to put themselves in the shoes of aerospace fuselage designers in trying to elevate the status of structural efficiency above those of passenger convenience and use of existing production equipment, both of which are important in conventional road-vehicle design. A brainstorming approach which does not rule out any possible solution would be in order for the conceptual designer of an electric vehicle. How can efficient thin-walled tubular structures be exploited? How can occupant access solutions be devised that will minimize reductions to the structural integrity of the vehicle? How can ultra-lightweight combinations of metal and polymer-composite materials be exploited in the construction? Could sandwich construction be used to obtain a load-bearing skin without need for supporting pillars and rails? These are the sorts of questions that might be asked at the concept stage of the body structure.

For the concept running-gear and chassis-systems designers many other radical solutions might be possible and could prompt the following questions. Can ultra-lightweight materials be used to decrease the rotating mass factors of the vehicles transmission and axle-hub/wheel-assembly components?; can wiring harness weight be substantially pared by adopting a multiplex system? How can components be integrated into one-shot consolidated assemblies which achieve weight as well as build-cost savings?

New interior packaging initiatives for both the occupants and mechanical/electrical systems might be exploited to advantage. Handling investigations which will find the best possible compromise between not just ride and handling, but also minimal rolling resistance, is already a fruitful field for tyre designers. The interior systems for climate control, window regulation, noise reduction, occupant protection and seating comfort might well be integrated in ways that achieve lighter-weight and more efficient, if unorthodox, solutions.

8.2 Structural package and elements

The trend over the past century has been a move from separate chassis frames and 'panelled' bodies, first to integral and then monocoque construction of road-vehicle body structures – in the interest of light weight and high rigidity. The process has always been frustrated by the demands of conventional occupant-access arrangements and the resulting 'shell' structures are a mass of cutouts whose shapes would horrify an aerospace designer in terms of structural efficiency. It is quite hard to visualize a car body behaving as an efficient tubular member reacting and transmitting the forces applied by its control surfaces as does the fuselage of an aircraft in flight. When brainstorming new concepts, however, it might well be worthwhile considering the car body as such a box tube, without side openings, window areas being flush bonded structural glass and front/rear access being obtained via bubble-car style full-width hinged doors with integral windshields. These might perhaps be built on strong ring frames, incorporating buffer systems to protect against impact, that would engage with ring frames in the ends of the main body tube over a series of conical plugs which would ensure structural integrity of the closure. By considering the classic, aircraft-structural, analysis for such structures a feel for their behaviour becomes possible and most importantly some quantitative indication of the considerable increases in bending and torsional stiffness over conventional car structures becomes possible.

8.2.1 BOX TUBES IN BENDING AND TORSION

While the semi-trailing road tanker is a good example of such a structure, the effectiveness in bending efficiency would relate to the aspect ratio of the box tube and so might not fit the desired relative dimensions of a passenger car. In bending and torsion the tube would be considerably stiffer than most monocoque sedan body designs but careful design of the end ring frames would be necessary, in relation to the cross-section dimensions of the main tube, so as to minimize the tendency to axial warping and ensure as far as possible that input loads are applied across the whole section, Fig. 8.1. Axial warping can be visualized by imagining the deflection of a large section box beam, such as the cantilevered example at (a). The cause is shear lag, in the case of the horizontal panels, and its effect is to increase overall bending of the beam as well as an increase in longitudinal stress at the web/flange intersection, together with a decrease at midsection. In design it is usual to replace actual breadth by an equivalent breadth as set out in British Standard BS 5400 Part 3. Shear lag is particularly dependent on plan dimensions of the flange B/L as at (b). Effective flange breadth in a continuous box beam can be estimated by treating each portion, between points of contra-flexure, as an equivalent simply supported span. Thus the beam becomes an assemblage of L-section beams having flange and web plates, for analysis. Effective width of

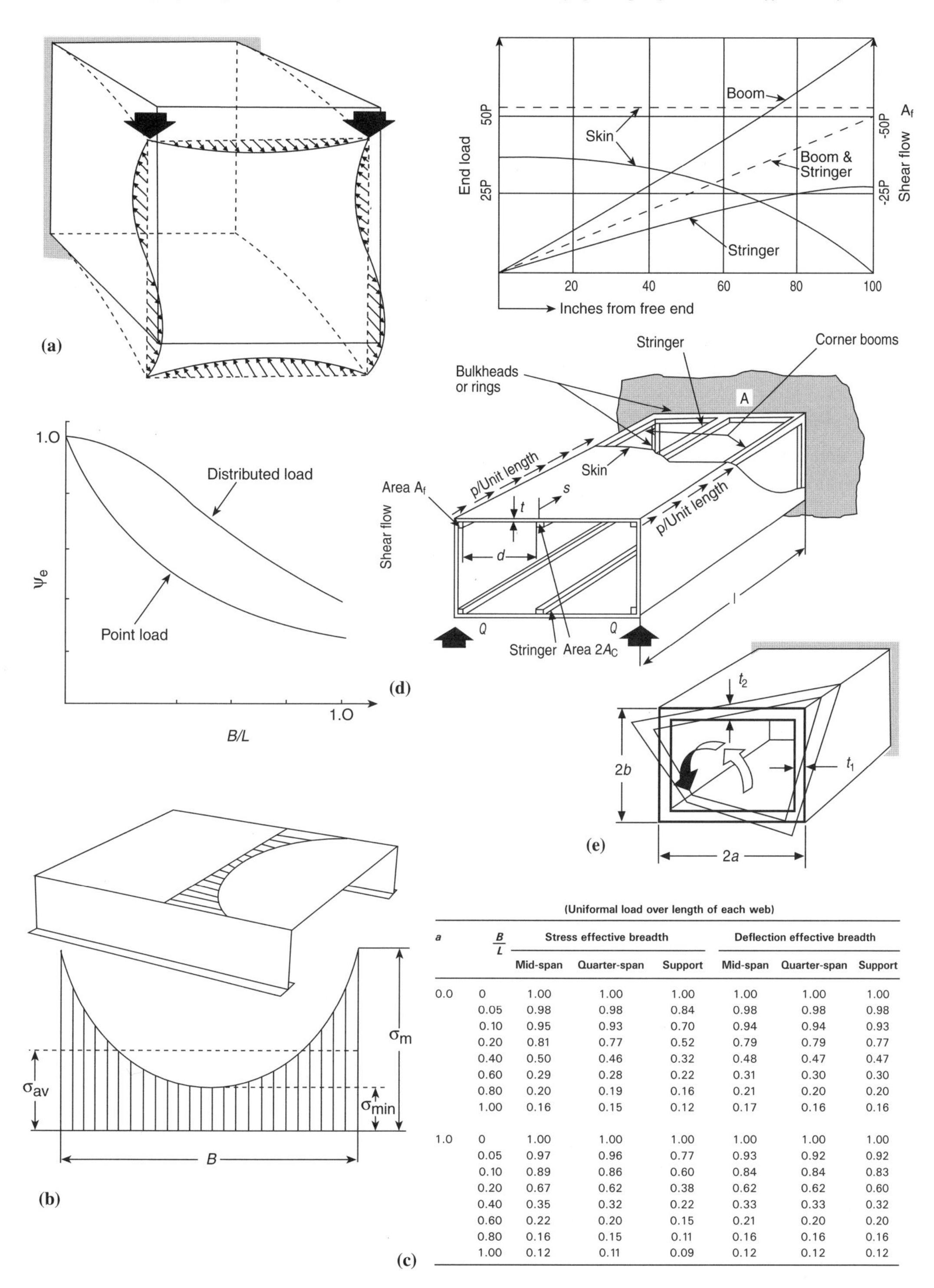

(Uniformal load over length of each web)

a	B/L	Stress effective breadth: Mid-span	Quarter-span	Support	Deflection effective breadth: Mid-span	Quarter-span	Support
0.0	0	1.00	1.00	1.00	1.00	1.00	1.00
	0.05	0.98	0.98	0.84	0.98	0.98	0.98
	0.10	0.95	0.93	0.70	0.94	0.94	0.93
	0.20	0.81	0.77	0.52	0.79	0.79	0.77
	0.40	0.50	0.46	0.32	0.48	0.47	0.47
	0.60	0.29	0.28	0.22	0.31	0.30	0.30
	0.80	0.20	0.19	0.16	0.21	0.20	0.20
	1.00	0.16	0.15	0.12	0.17	0.16	0.16
1.0	0	1.00	1.00	1.00	1.00	1.00	1.00
	0.05	0.97	0.96	0.77	0.93	0.92	0.92
	0.10	0.89	0.86	0.60	0.84	0.84	0.83
	0.20	0.67	0.62	0.38	0.62	0.62	0.60
	0.40	0.35	0.32	0.22	0.33	0.33	0.32
	0.60	0.22	0.20	0.15	0.21	0.20	0.20
	0.80	0.16	0.15	0.11	0.16	0.16	0.16
	1.00	0.12	0.11	0.09	0.12	0.12	0.12

Fig. 8.1 Box beams in bending and torsion: (a) axial warping of box beam in bending; (b) shear lag effect on breadth; (c) effective width ratios for simply supported box beams; (d) reinforced box beam with skin, stringer and boom load distribution above right supported box beams; (e) axial warping in torsion.

these are found from the table at (c). The effect of reinforcing the box tube, using central stringers and corner booms, is seen at (d).

Generally, the analysis of box tubes by classic beam theory is possible if plane cross-sections remain plane after bending deflection, and there should be no swelling out or bowing in of their planar outline. The 'essing' deflection of the side walls arises from the parabolic distribution of shear stress across the vertical panels. There is also axial warping in torsion if cross-sectional dimensions are not appropriately chosen. At (e) this effect is again shown, much exaggerated, for a cantilevered box tube. For the dimensions shown in this view of the figure, with shear modulus G and applied torque T, axial warping of the corner point of the box section is given by

$$(T/16abG)[(b/t_1) - (a/t_2)] \qquad \text{and if} \qquad b/t_1 = a/t_2$$

no axial warping will occur. Notation for symbols appears at the end of the chapter.

Torsion of box tubes, Fig. 8.2, involves shearing deformation as the dominant mode and it is useful to reconsider the basics of the process before embarking on analysis. Any element under shear stress, (a, top), τ is subject to a complementary shear stress τ' arising from the equilibrium of forces on the element (important in timber which is weak in shear along the grain), as

$$\tau_{xz} y = \tau'_{yz} x \qquad \text{while} \qquad \tau/r = Ga$$

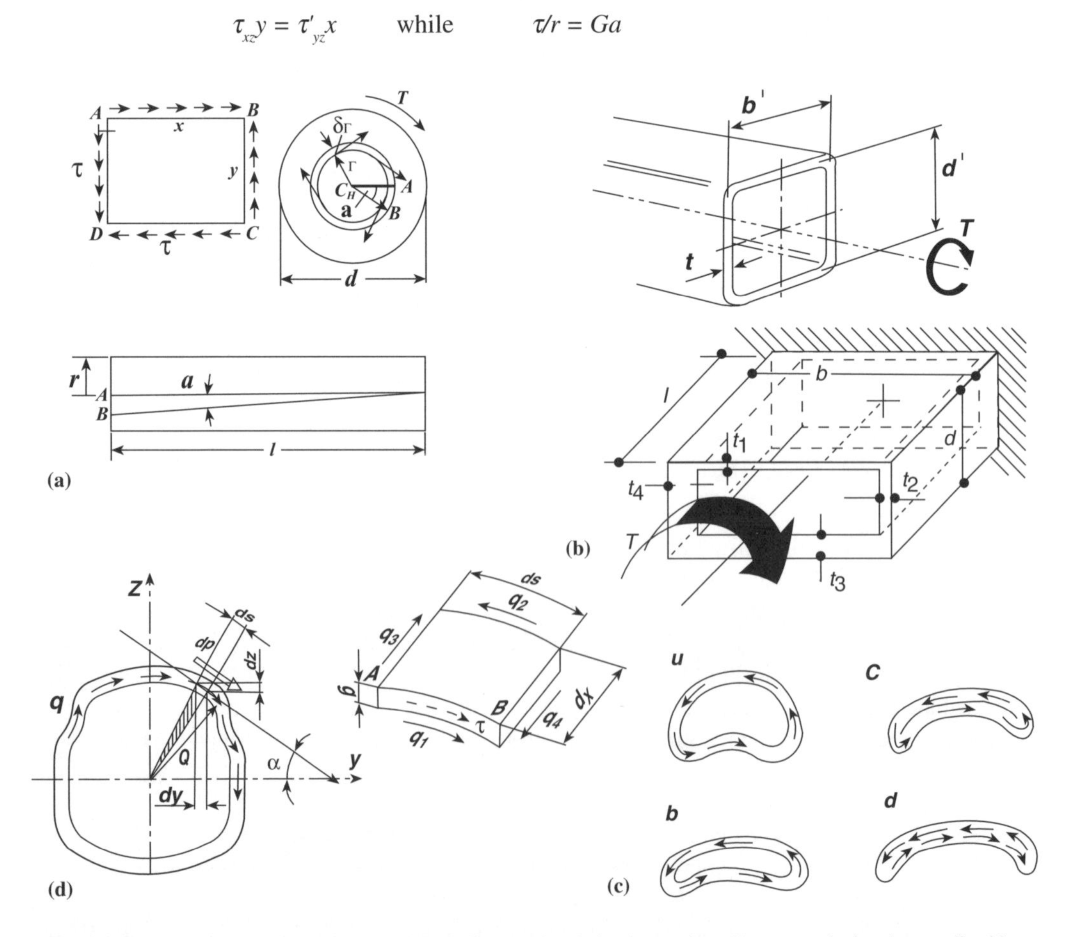

Fig. 8.2 Shear development in torsion: (a) shear in flat panel and circular section; (b) symmetrical and generalized box sections; (c) transition from closed to open section; (d) shear flows in beam element.

expresses the effect of complementary shear in the longitudinal plane while in torsion, (a, bottom), as primary shear is applied to the transverse planes. Torque T can be equated to the sum of all these tangential stresses and when the formula for polar moment of inertia is substituted

$$T/J = \tau/r \qquad \text{and} \qquad T/a'' = GJ/l$$

expresses the torsional stiffness. A conventional car structure has torsional stiffness around 6500 Nm/°. The importance of closed as opposed to open cross-sections can be seen experimentally in a thin-walled tube slit along its length. Under torsion the effect of complementary shear can readily be seen as relative displacement either side of the split, and gross weakening of the section as a consequence. For the rectangular hollow section (b, top), angle of twist per unit length is given by

$$T(b' + d')/2Gb'^2d'^2t \qquad \text{while} \qquad T/2tb'd'$$

expresses the average shear stress. A generalized section box beam (b, bottom) has a twist/unit length of

$$\{\tau/2[A]g\}(\Sigma ds/t) = [T/4(bd_2)G](b/t_1 + d/t_2 + d/t_3 + d/t_4)$$

For the transition from a closed to an open section, shown at (c), the open section can be seen as a limiting case of the closed tube as it collapses. Transverse shear stress distribution has an important effect on the torsional rigidity of the cross-section, the closed tubes resisting torsion by the area of the cross-section enclosed by the mean thickness of the wall. Shear flows around an elemental section of the wall, (d), are equal and constant and by integrating along the section circumference, the sum of horizontal and vertical components can be summed to zero and their tangential components equated to T in

$$q = T/2[A] \qquad \text{while} \qquad T/4GA^2(\Sigma ds/t)$$

expresses twist per unit length, found by equating the shear energy per unit volume with the work done in the twisted cross-section, for $[A]$ bounded area by the mean thickness line, shear modulus G and wall thickness t.

8.2.2 'THIN-WALLED' STRUCTURAL ANALYSIS

Because the wall thickness of the car body panel is very small compared with the cross-section of the body shell, thin-walled structural analysis developed by aeronautical engineers can provide a surprisingly uncomplicated analysis which can give the concept designer a very good 'feel' for the structural behaviour. The structure is idealized in a way which assumes panels can only carry shear and tensile stresses in their own plane and it is taken that all loads can be ignored to which the thin wall is flexible or would collapse by buckling, Fig. 8.3. Flanged edges of panels and intermediate stiffening swages can be represented by end-load carrying area added to the cross-section of panels used in the idealization. In bending of thin-walled beams principal panel loads are shear ones and small elements, (a), can be considered for design purposes. The magnitude of panel shear stress $\tau = S/ht$ for a height h and thickness t of the section. As shear flow is force per unit length $q = Sh = \tau t$. The bending moment M causes longitudinal stress in the 'edge bar' of $\sigma = P/A = MhA$. The section second moment of area is $I = \Sigma Ay^2$ where y is the distance from the neutral bending axis to the edge boom. At any distance L along the beam from the point of application of S, since $M = SL$, boom end load is qL and for incremental length $dP/dz = S/D = q$ or the shear flow in the web is equal to the rate of change in end load in the boom.

Peery[2] uses the assumption of constant shear flow in the web, for the generalized curved section of a beam, (b), to show that twice the area enclosed by the curved web [A] divided by the 'chord' of the curve gives the moment arm of the resultant of shear flow in the web. This is helpful, for example, in finding the 'shear' centre of an asymmetric section beam, or that distance offset from the section that a shear force can be applied without causing any twist of the section. For a beam section such as that in Fig. 8.1(d), but with corner booms, area A_f, only and no central stringers, for section depth d and breadth b, with corresponding symmetrical wall thicknesses t_d and t_b, this would be idealized by a four boom box beam having boom section areas $A_f + (bt_f/2) + (dt_b/6)$. The analysis technique is to make an imaginary cut which makes the closed section momentarily open, and finding 'open-section' shear flows. The next stage is to apply an opposite sense shear flow q_o on all elements (between booms) which equals the 'resultant shear flow' q^* on the cut element. The final stage is to balance moments of the combined shear flows against the moment arm of the externally applied load, the q_o flows being assumed to set up torque $T = q_o 2[A]$. Because, in this case, the beam is subjected to bending loads only, and not external torque, the twist per unit length can be equated to zero in $\sigma q^*(s/t) = 0$.

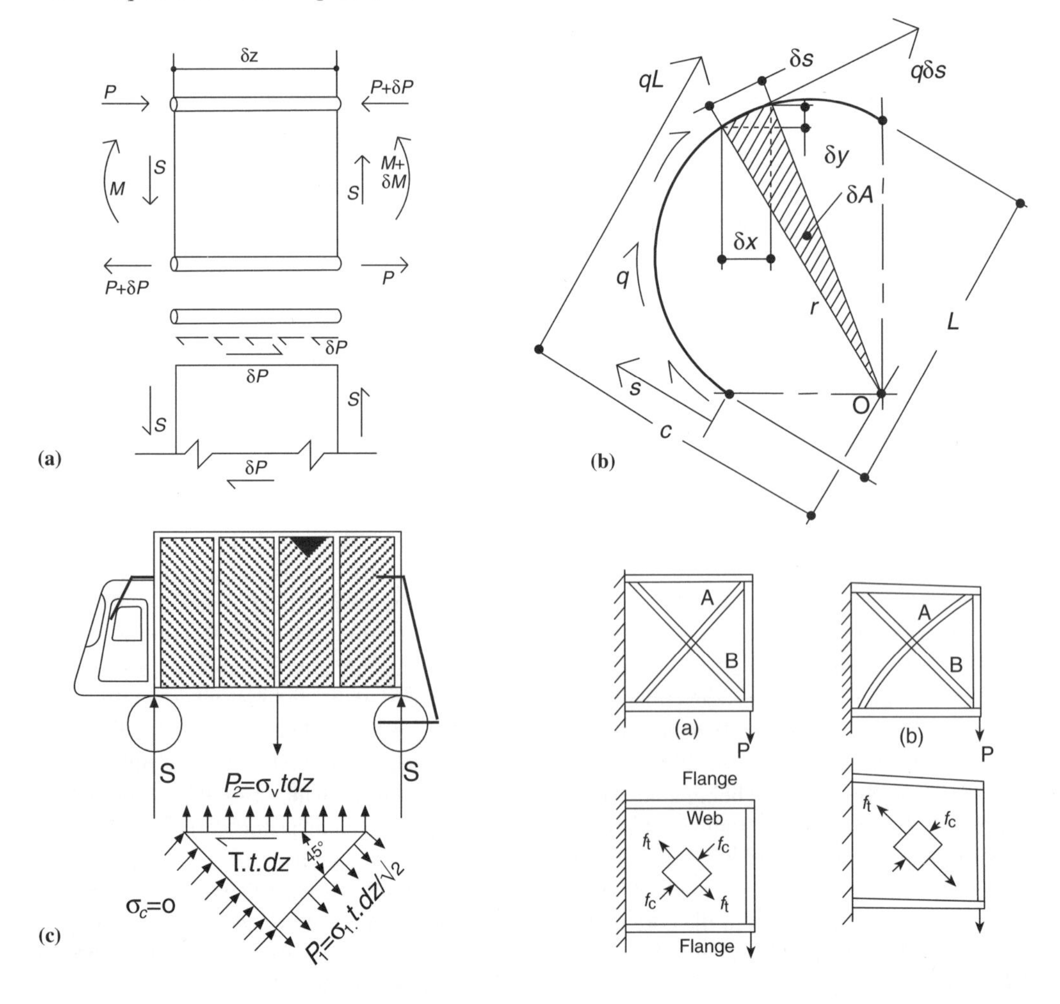

Fig. 8.3 Thin-walled structural idealizations: (a) flat shear panels and end-load bars; (b) shear flow in curved section beam; (c) tension field in sidewall to examine shear buckling.

8.2.3 AVOIDING SHEAR BUCKLING

The assumption of panels being infinitely stiff in shear would require that the panels were stabilized in some way to avoid buckling under shear force, recognizable by diagonal creasing, easily demonstrated on a stiff sheet of paper. In conventional practice this is achieved by pillars, rails and/or swages which divide up a vehicle sidewall, for example as in the view at (c, left), into panels which are small enough to avoid the possibility of buckling. The pioneering aircraft structural designers used tension-field beam analysis to model the tension field which develops diagonally in the sheared panel. It is best described by the framework analogy of (c, right) where the buckling of the diagonal compression strut results in the loading of the tension brace, analogous to the diagonal tensile and compressive stresses set up in the web of a short flanged beam. In the idealized vehicle sidewall (c, left) the central load *2S* creates tension bands represented by broken lines in the figure. Assuming the dimensions of the reinforcing frame are such as to load the panels in pure shear then forces acting on a triangular element of the panel are

$$\sigma_v tdz - \sigma t(tdz/2^{1/2})1/2^{1/2} = 0 \qquad \text{so that} \qquad \sigma t = 2\tau \qquad \text{and} \qquad \sigma_v = \tau$$

is obtained by equating direct stress σ with shear stress τ. If h is the vertical distance between upper and lower rows of spot-welds securing the panel to its top and bottom rails then $\tau = S/ht$ and spot-welds are subject to loading τtdz horizontally and vertically, with resultant force $2^{1/2}\tau tdz$ or spot-weld load per unit length of $1.41S/h$. With rails held by pillars spaced d apart then the compressive load in them is $P_1 \sin 45$ or Sd/h.

For a much idealized box-tubular structure, matrix force analysis carried out by Tidbury[3] showed that for a simulated framed structure incorporating window cutouts on either side, due to the shear deformation in the panels the relatively short beam departed from the theoretical ideal of equal stress developed in the top and bottom rails by showing an actual maximum stress distribution top to bottom of more like 1:3. In a similar analysis, (c), carried out by the author[4] it was shown that by avoiding the cutouts, and using sandwich panels to fully stabilize the panels against buckling, this distribution was improved to a ratio of 3.9:4.7.

8.2.4. SANDWICH CONSTRUCTION FOR PANEL STABILITY

In determining preferred dimensions and properties of the sandwich panels in terms of the face and core materials, advantage again can be taken of pioneer work used in aircraft structural analysis. Choice of skin and core modulus, density and sustainable working stress is made so that the lightest possible panel is stable with respect to the external loading, Fig. 8.4. Aircraft designers found that the most elaborate 'sheet/stringer' construction was only able to stabilize a panel up to 50% of the maximum stress sustainable by a panel whereas with sandwich construction it is theoretically possible to stabilize the skins so they can support their full yield stress. However, the sandwich panel has to be stable against overall 'column' buckling as well as local 'wrinkling' of the skins. The core material being able to transmit shear force between the skins effectively acts as the 'web' between the two 'flanges'. The efficiency of a sandwich panel is thus defined as the ratio maximum load per given width of panel by the weight of the given area of panel – divided by the same ratio for an equi-sized panel which has faces fully stabilized by a weightless core. This points to the key properties of the core material as specific stiffness and specific strength and the failure modes under end load being local instability of the faces or overall column buckling of the panel (or, if overstabilized, direct overstress of facings or core). In the post World War 2 period, researchers at the Royal Aircraft Establishment (Farnborough) used a strain-energy method to derive the critical end load for wrinkling failure as:

$$0.63E_f^{1/3}E_c^{2/3}$$

where E is the elastic modulus for face and core.

For column buckling of the panel, the parameters of panel length and edge support must also be considered – together with thicknesses of face and core plus the bending stiffness of the combination.

To calculate sandwich panel bending stiffness it is generally assumed that shear stress is uniformly distributed across the core. Central deflection for a uniformly loaded (w per unit length) sandwich panel of moderate width, simply supported at its ends, is, for dimensions seen at (a), given by:

$$(5wl^4/384E_fI) + (wl^2/8G_cA)$$

where l is the length between supports, A is the core cross-section area, I the section second moment of area and G_c the shear modulus of the core material. To allow simply supported panels to be designed without resort to complex calculation, correction factors shown at (b) have been

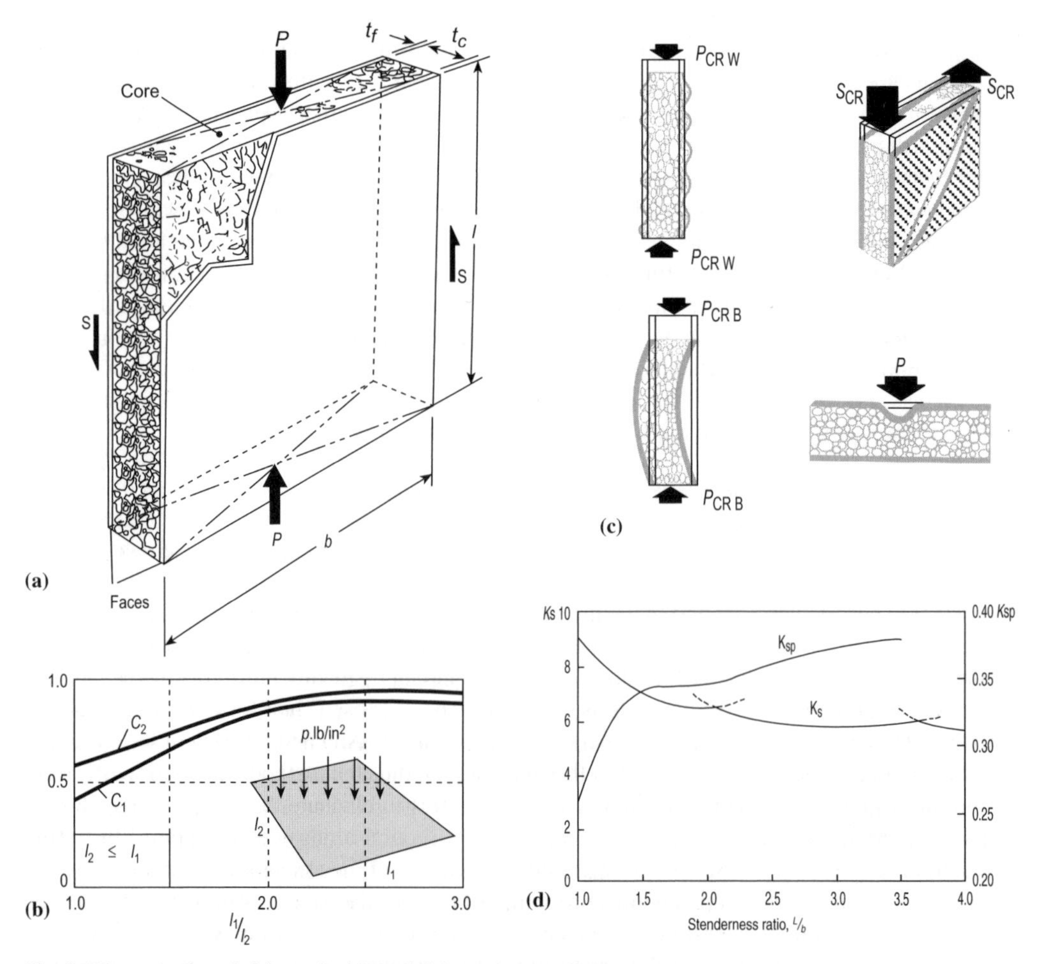

Fig. 8.4 Foam-cored sandwich panels: (a) sandwich panel parameters; (b) correction factors; (c) failure modes; (d) K factors.

derived for use with simple beam theory to give good approximations. The deflection, or bending moment, is first calculated by simple beam theory, above, and then multiplied by the appropriate factor, C_1 in the case of central deflection and C_2 in the case of bending moment. The critical end load P_{cr} for unit width of panel, in overall column buckling, can then be found from the formula:

$$(1/P_{cr}) = (2a^2/\pi^2 E_f t_f t_c^2) + 3/2Gt_c \qquad \text{factored by} \qquad (1 + a^2/b^2)^2$$

for panel length *a* and width *b,* where dimensions shown at (a) and (c) apply.

For shear instability of the panel depicted in Fig. 8.4, critical shear stress in the panel is:

$$(2.75K_s E_f t_c^2/b^2)/(1 + 5.5E_f t_f t_c/[b^2 k_{sp} G_c])$$

where K_s and K_{sp} are factors relating to single faces and sandwich panels respectively as obtained from (d). The above calculation formulae apply to isotropic core materials such as rigid foams for which it can be assumed that physical and mechanical properties are uniform in all directions.

8.2.5 *REINFORCED SHELLS*

While sandwich shells have been realized in semi-trailing insulated van structures, where the core material has served as a heat insulant as well as a skin stabilizing material, passenger vehicles have conventionally involve cutouts for which reinforcing frameworks have been found necessary. Although the monocoque box tube with bonded-in structural glass windows, discussed earlier, could be a way of reaching substantial weight reduction for electric car structures the configuration would be confined to single-box styles of 'multipurpose vehicle' or 'minibus' category, of quite high projected frontal area. The aerodynamic demands of higher speed vehicles might dictate profiles involving greater curvature. These, too, can of course be structurally optimized for lightweight construction by analysing the load paths through them and gauging skin and reinforcement thickness to even out stress distribution over the structure, Fig. 8.5.

Pawlowski's SSS method, described in Section 8.3, can be extended to encompass curved panels, modelled as shown at (a). Shear flow in the panel is found from the sum of the projections on the related reference axis. Thus on the *x* and *y* axis respectively it is

$$\int_0^h q\,ds \,.\, \cos a \qquad \text{and} \qquad \int_0^h q\,ds \,.\, \sin a$$

Shear flow does not depend on the shape of the wall but only on height of the element *h* which is effectively the depth of the beam which the element represents. Location of the resultant shear, through the shear centre, is found by taking moments, which transform to give $a = 2A/h$ where *A* is the area bound by the mean width of the panel.

Cumulative combinations of this basic beam element can be made to form open 'vee' section beams, closed tubes and open channels as depicted at (b). In closed shell form the multi-stringer arrangement at (c) is rigid to a force *Q* which causes shear, bending and twisting of the structure. The bending is statically determinate but the shear not so; thus the methods are used which were introduced in the earlier section on thin-walled structures to evaluate the flows – a figure then shown depicting the shear flow diagram for the 'cut' section alongside a diagram illustrating calculation of the unknown shear flow q_0. Full exploitation of the strength and rigidity of shells built up in this way depends upon maintenance of the cross-section shape by the use of bulkheads.

Bending of a multi-stringer tubular shell can be analysed relatively easily with the aid of some simplifying assumptions. For the arbitrary section at (d), stringer cross-section areas *A* are enlarged by allowing for the contribution made by adjacent panelling. For width of skin panel *b*, thickness

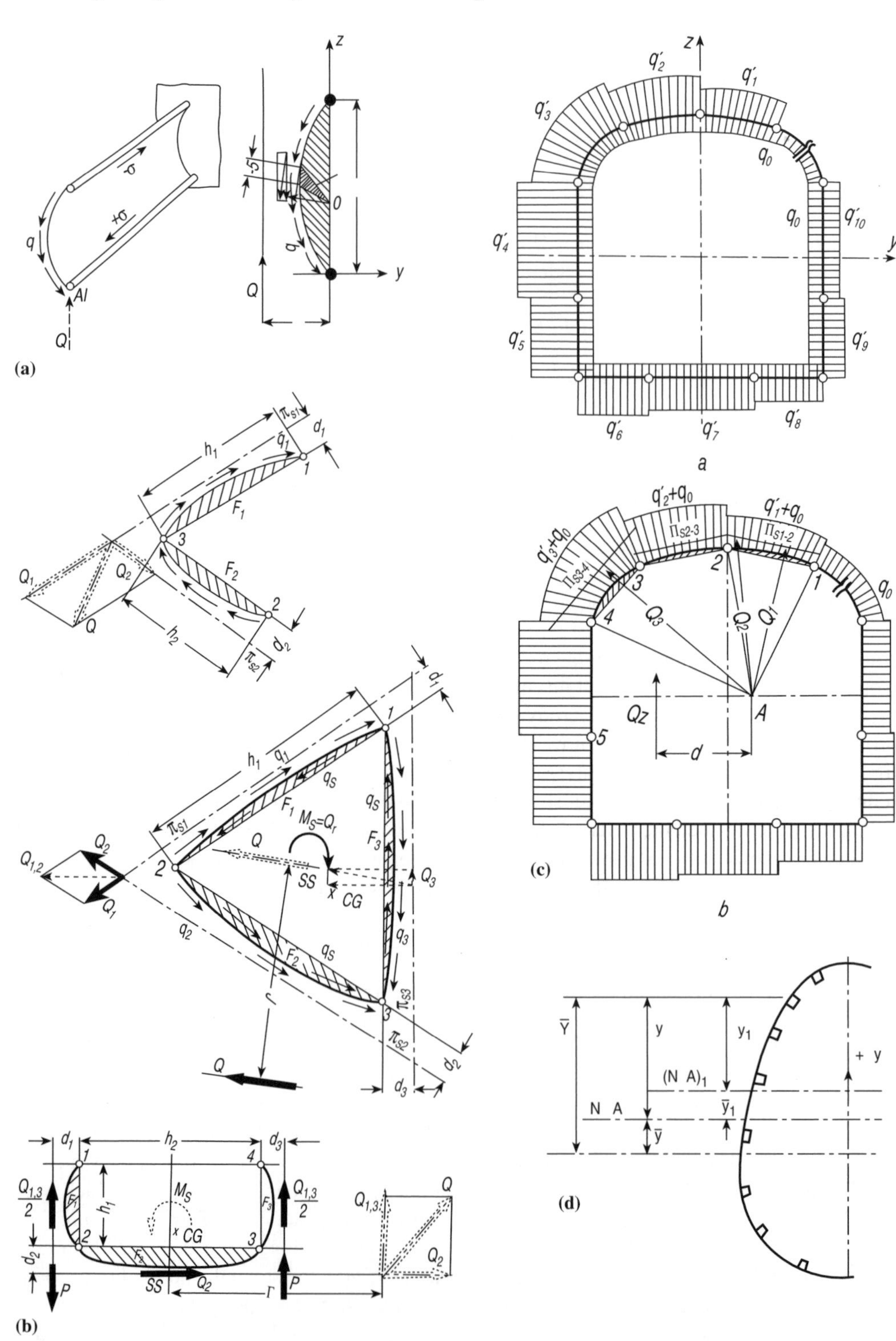

Fig. 8.5 Curved shell structures: (a) idealization of curved panel; (b) shell made from curved panels; (c) multi-stringer shell beam; (d) reinforced generalized curved shell.

t and distance of effective stringer area from neutral axis y, direct stress is notated as σ and buckling stress as σ_b. When there is no skin buckling, effective stringer area on the tension side of the shell:

$$A = A_s + (1/2)b_u t_u + (1/2)b_l t_l = A_s + b_m t$$

when $t_u = t_l = t_s$ meaning stringer; u is upper and l lower. After the onset of buckling, the effective area of the panel is assumed to be:

$$(1/3)bt + (2/3)bt(\sigma_b/\sigma)$$

and the effective stringer area is approximately:

$$A_1 = A_s + (1/3)b_m t = (2/3)b_m t(\sigma_{bm}/\sigma_s)$$

where m is mean value between upper and lower panels. The method involves tabulating A and Y from a chosen datum and finding its distance from the neutral axis as sAY/sA and the section second moment of area about the neutral axis as $I = \Sigma AY^2 - y\Sigma A$. If some panels are assumed to buckle then their effectiveness will have been overestimated and a reduction of area is necessary of $dA = A - A_1$. A new neutral axis can then be found from $y_1 = -\Sigma dAy/\Sigma A_1$ and a new value of area moment $I_1 = I - \Sigma dAy^2 - y_1\Sigma A1$ and the stress becomes $\sigma_1 = M(y - y_1)/I_1$. Successive further approximations can also be made following the same procedure.

8.3 'Punt'-type structures

The optimally efficient shell-beam structure, typified by an aircraft fuselage, is argued above to fit a relatively limited range of automotive body configurations. It also requires radical solutions to the mounting of wheel suspension and drive systems to the end bulkheads or ring frames, particularly so pending the development of lightweight wheel motors with low enough mass to keep unsprung weight within reasonable bounds for ultra-light vehicles; see Section 4.7.2. Where conventional closed-sedan or open-sports bodywork is a market imperative a structurally efficient compromise between a closed-shell and chassis-platform structure is the open-integral punt, a three-dimensional chassis with rigid vertical posts allowing firm anchorage for suspension springs, door hanging, engine mounting and as the basis for strong transverse bulkhead ring frames for enhancing torsional stiffness, Fig. 8.6.

At (a) is such a structure designed by a former chief chassis engineer for Bristol Cars who had been trained in the associated British Aerospace company. The design was produced for a firm of consultants serving the volume-production automotive industry for whom he worked in a similar capacity. This design would be much more successfully produced using current techniques for laser-welding plane blanks as with spot welding equipment available at the time it was difficult to fabricate stabilizing diaphragms at intervals along the box tubes and to insert kink-strut diaphragms at the corner joints, and spot-welding flanges compromised low mass. Furthermore, the industry was not prepared to compromise on styling and interior layout for improved structural design, to the extent that is now required for the realization of ultra-light structures suitable for electric cars. Though the concept did not reach series production at the time embodiments can be seen in such designs as the hydro aluminium structure of Fig. 7.13(a).

Cantilevered to the vertical posts in this case are engine bay and luggage compartments. This sort of base-structure is also particularly suited for the attachment of rollover bars at the A- and C-posts and the mounting of ultra-light plastic superstructure bodywork. Its key feature, however is

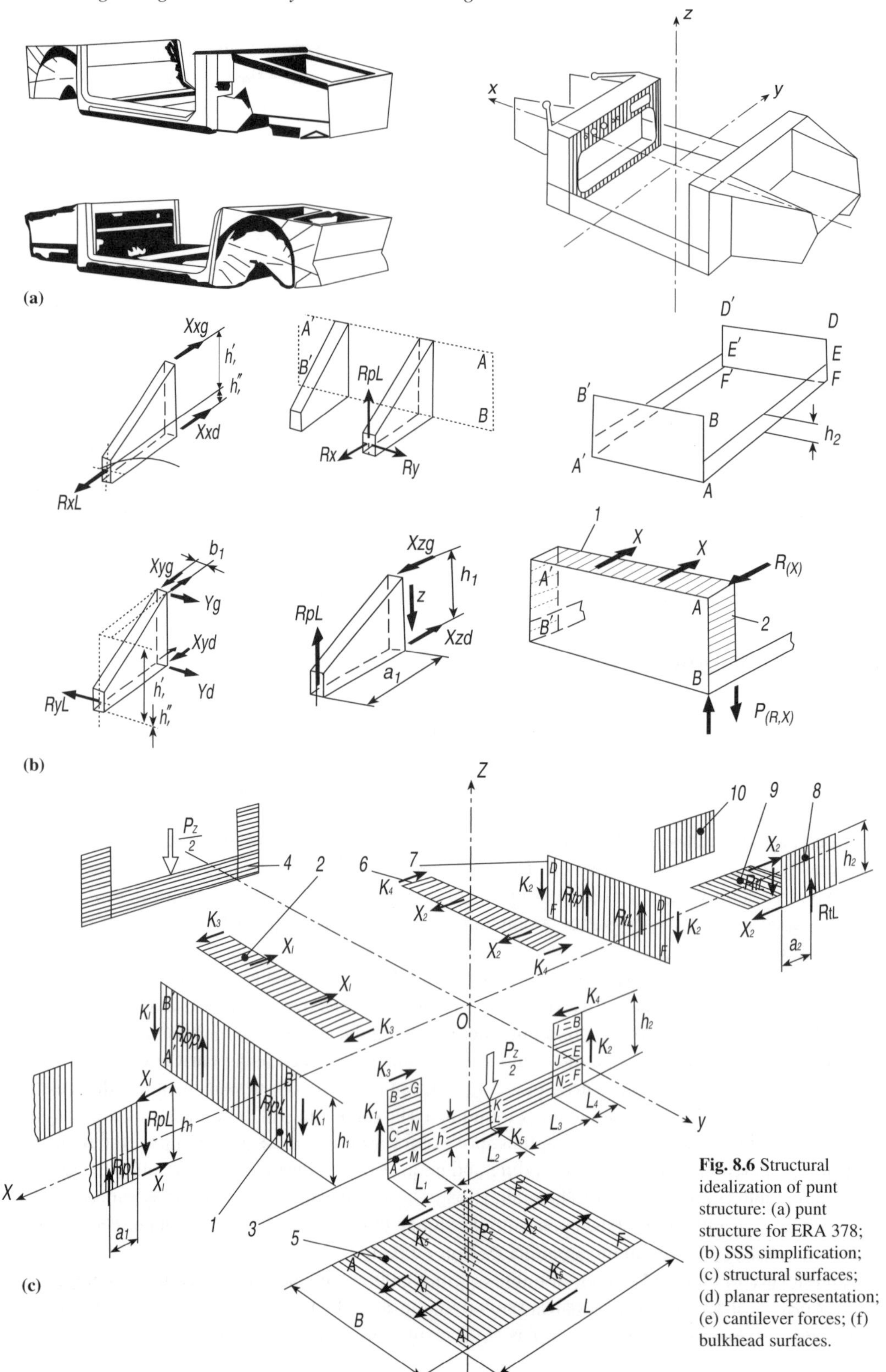

Fig. 8.6 Structural idealization of punt structure: (a) punt structure for ERA 378; (b) SSS simplification; (c) structural surfaces; (d) planar representation; (e) cantilever forces; (f) bulkhead surfaces.

the use of large uniform-section box tube for the sills, A/C-posts and their upper and lower crossmembers. The use of a sandwich floor panel, with such a punt structure, would also allow central positioning of propulsion-battery packs in a structurally efficient manner for a mid-size electric car.

8.3.1 IDEALIZATION OF THIN-WALL STRUCTURES

Initial analysis of this type of optimized lightweight structure, simplified for analysis as at (b top right), is best carried out by a method such as Pawlowski's[5] Simple Structural Surface (SSS) technique which gives a better feel for structural efficiency than the full finite-element analysis, carried out at a later stage. Pawlowski's contribution was to adopt aeronautical thin-walled structural analysis techniques for automotive structures and his SSS method (nowadays much more readily applied with the assistance of computer simulation) involves dividing the key elements of the structure into systems of plates, assumed to carry shear loads, surrounded at their edges by bars assumed only to carry end load. These systems are represented in (c) by structural surfaces, the edge forces in which are those that occur in bending of the structure during, say, symmetrical impact of the front wheels against a ramp in the road, causing an upward force on the front-end reacted by a downward inertia force applied by the combined mass of the occupants and battery pack, assumed to act through the centre-line. The following key elements are involved: (1) front bulkhead; (2) dash upper panel; (3) left-hand side wall; (4) right-hand side wall; (5) floor; (6) rear parcel shelf; (7) rear bulkhead; (8) LH boot side-panel; (9) boot floor; (10) RH boot side panel.

8.4 Optimizing substructures and individual elements

The connection points between the major structural elements (Fig. 8.7) are worthy of detailed attention so that the strength and stiffness of the connected members is preserved at the joint. At a joint between box members such as the important rear quarter to sill junction of a car, (a), consideration should be given to load transfer at the corner. In figures taken from the design of small car body structure, using aircraft thin-walled structural techniques, by T.K. Garrett[6], a bending moment of 51 000 lbf in (5.76 kNm) induced end loads of 33 000 lbf (146.7 kN) in the top and bottom flanges of the 5 in (127 mm) sill. The 18.5 MN/m^2 safe working stress steel used in this application dictated an area of 0.123 in^2 (0.795 cm^2) edge material – a 3.4 in (86.5 mm) wide flange of 20 gauge sheet. The webs of sill and quarter panels overlap as shown by the dotted lines to provide the necessary transfer of the 1900 lbf (8.45 kN) shear force applying in this example. It is also important to maintain the cross-section shape of thin-walled box beams, both along their length and, crucially, at the corner joint where a diagonal diaphragm (bulkhead) might be required.

The importance of structural continuity, in order to transfer loads from one part of a structure to another, is clear from the example shown. Here it is necessary to transfer both shear and bending moment between the sill and the rear end of the vehicle. The sill outer panel is joggled under the rear quarter panel and there is a vertical joint line CD. At the rear end the sill is closed by flanging outwards of the outer panel and its connection to the heel board at EF. End loads to be transferred from the top and bottom of the section to the rear quarter panel are found by dividing the section depth into the bending moment. These loads will govern the number of spot-welds required at the connection. The direction of the loads will determine whether a 'kink strut' is required at the joint, its dimensions being found by resolving the forces along the line of the strut.

In the case of a curved corner, or angular, joint between box beams, (b), the web is subject to direct stress M/hRt where R is the radius of curvature and t the web thickness. For the lightly angled joint shown in the same figure, a 'kink strut' may be required at the junction to maintain the cross-section against collapse. Load in the strut will be $M/d(\cos a + \cos b)$. If the beam has particularly deep webs, it may be necessary to provide vertical stiffeners which help stabilize shear panels against buckling. One technique is to swage the panel rather than fasten on additional parts. General rules for swaging are that all swages should be straight and not intersect, and that they should run along the shortest distance between supported edges of the panel. In calculation they can be represented by end-load carrying bars.

Attachment of mechanical units to an integral body needs special attention to the spreading of load, in avoiding local distortion and failure also in limiting undesirable relative movement. The rear spring front hanger bracket is an important area in leaf sprung front-engine/rear-drive layouts on MPVs and minibuses. This mounting has to transmit braking and drive loads horizontally as well as vertical loads due to weight and braking/driving torque reaction, and side loads due to cornering. The example illustrated at (c) shows that the boxed (longitudinal) body member has an internal stiffener and side plates with through bolts.

Where mechanical units are cantilevered out from the main structure by triangulated frameworks, the theory of space trusses can be used to analyse loads and deflections. For a three-dimensional frame, resolving forces is a little more difficult than with two-dimensional frames and a method

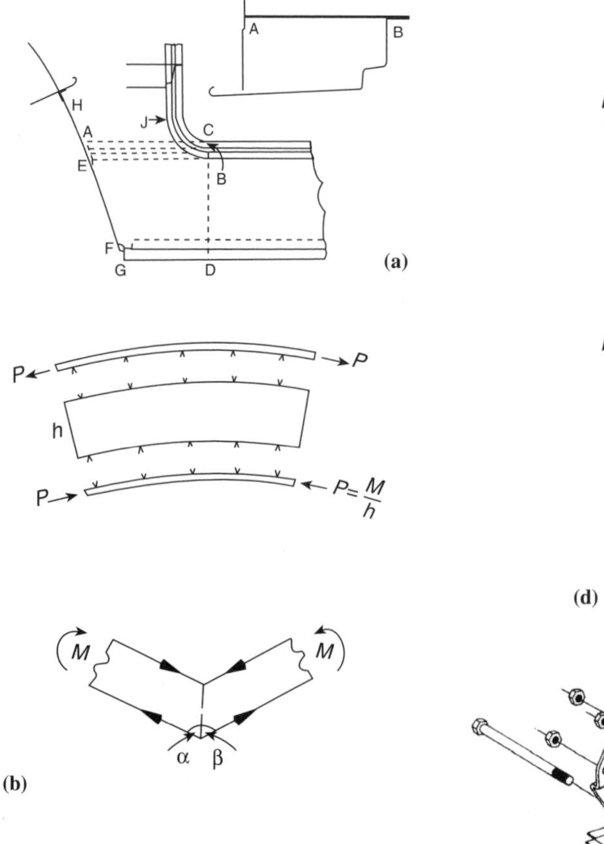

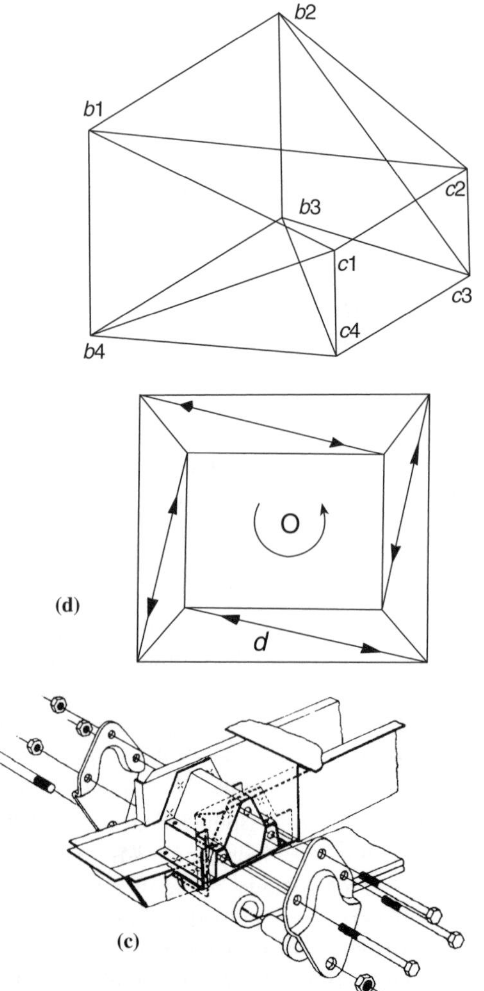

Fig. 8.7 Joints and sub-structures: (a) load-carrying joints; (b) curved or angular joint; (c) structural attachment to box member; (d) single bay of space truss.

known as tension coefficients is applied. This is based on the fact that proportionality exists between both resolved components of length and of force. A tension coefficient

$$S/L = F_x/X = F_y/Y = F_z/Z$$

for member force S, in a length L, has projecting force vectors and length components, F_x, F_y, F_z and X, Y, Z on perpendicular coordinate axes.

The view at (d) shows one bay of a space truss, that might be an extension used to support an engine/gearbox unit behind a monocoque bodyshell structure. For analysis, the frame is assumed to have rigid plates at $b_{1,2,3,4}$ and $c_{1,2,3,4}$ which offer no force reaction perpendicular to their own planes (zero axial warping constraint). When considering the torsional load case, for the bay, b_1c_1 and b_4c_4 member forces are zero by resolving at b_1 and c_4, observing zero axial constraint of the truss. Remaining members (the envelope) have equal force components F_z at the bulkheads and tension coefficients for all of them are F_z/Z. This common value can be determined by projecting envelope member forces onto the bulkhead planes, as shown. By taking moments about any axis O, for any one member (c_3b_4, say, with projected length d), contribution to torque reaction is = rtd (half the area of the triangle formed by joining O to c_3 and b_4). Thus total torque is $2tA$, by summation.

8.4.1 STIFF-JOINTED FRAMES

Where structural members are curved or cranked over wheel arches or drive shaft tunnels the unit load method of analysis applied to stiff-jointed frames is particularly useful, Fig. 8.8. It is best understood by considering a small elastic element in a curved beam (a) which is otherwise assumed to be rigid. An imaginary load w, at P, is then considered to cause vertical displacement Δ; the beam deflects under the influence of the bending moment and it can be shown that $\Delta = \int Mm/EI \,.\, ds$ where m is bending moment due to imaginary unit load at P and M due to the real external load system. An example might be a car floor transverse crossbearer having a 'tunnel' portion for exhaust-system and prop-shaft clearance, (b). To find the downward deflection at B, the imaginary unit load is applied at that point and it develops reactions of two-thirds at A and one-third at D. The bending moments for the real and imaginary loads are shown at (c) which also illustrates the subsequent calculation to obtain the deflection due to the driver and seat represented by distributed load w over length l. Another example is the part of the frame at (d) subject to twisting and bending, deflection due to twist = $\int Tt \, dx/GJ$. This deflection would then be added to that due to bending, obtained from the formula given in the earlier section. Usually the deflection due to axial straining of the elements is so small that it can be discounted.

A good example of a frame subject to twisting and bending is the portal shape frame at (e) having loading and support, in the horizontal plane, indicated. If it is required to find the vertical deflection at D for the load P applied at B, equating vertical forces and moments is first carried out to find the reactions at the supports. Next step is to break the frame up into its elements and determine the bending moments in each – ensuring compatibility of these, and that associated loads are 'transferred' across the artificially broken joints. Integrations have to be carried out for the three deflection modes as follows: $\int Sb \,.\, ds/AE + \int Mm \,.\, ds/EI + \int Ff \,.\, ds/AG$ to obtain the combined deflection. Both vertical and horizontal components can be obtained by applying vertical and horizontal imaginary unit loads, in turn, at the point where the deflection is to be determined. For the example shown in the figure, elastic and shear moduli, E and G, are 200 and 80 GN/m^2 respectively. S is axial load and F the shear load, the latter being negative in compression. The values of these, with the bending moments, are as shown at (f right). These can readily be substituted in the expression (f left) to determine deflections as a factor of P.

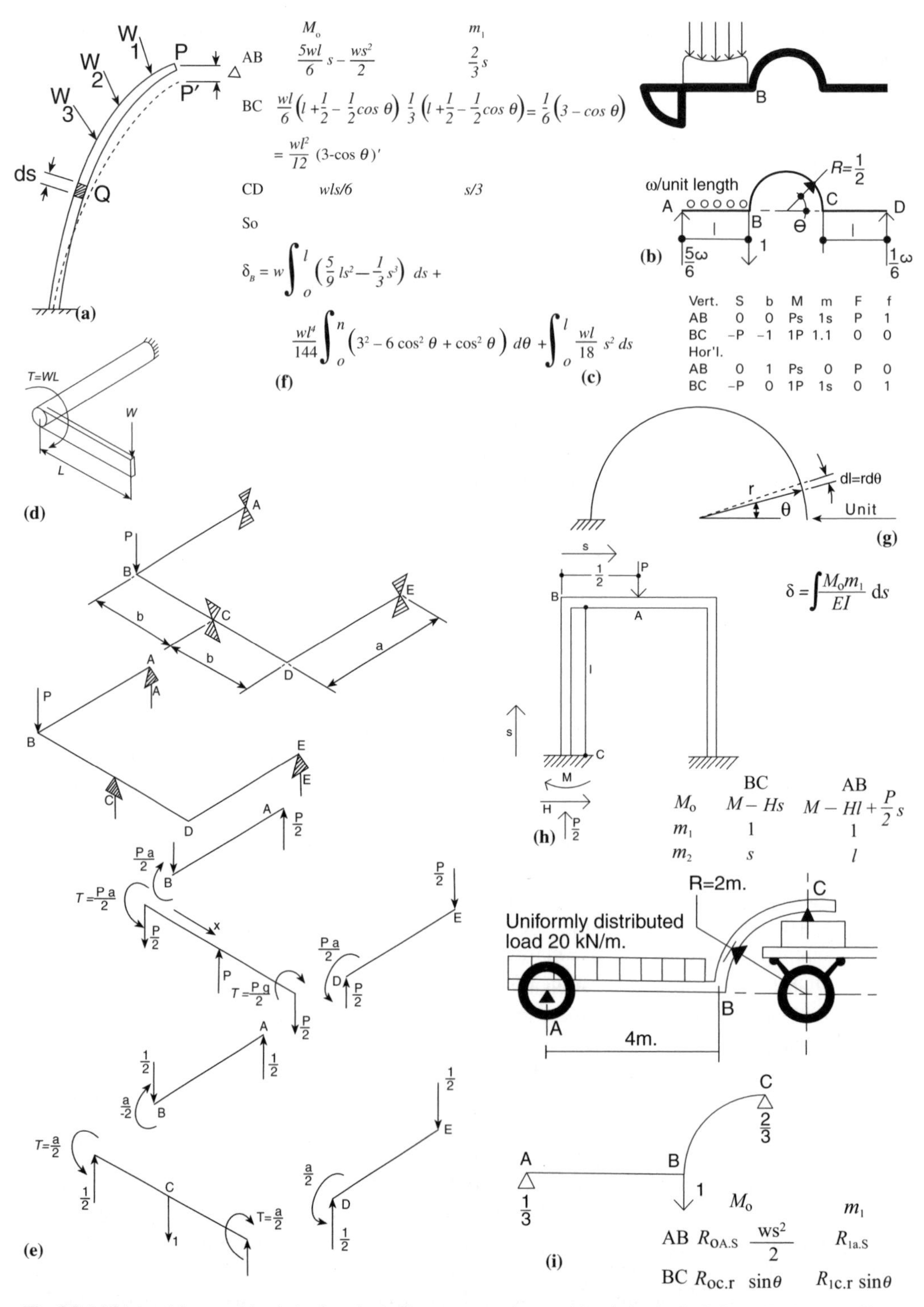

Fig. 8.8 Stiff-jointed frames: (a) unit-load method; (b) seat support bearer; (c) unit-load calculation; (d) element subject to twisting and bending; (e) portal frame under combined loading; (f) loads in members; (g) semi-circular arch member under load: (h) front-end frame as portal; (i) battery-tray adjacent to wheel arch.

The method can also be used for 'continuum' frames incorporating large curved elements provided curvature is high enough for the engineer's simple theory of bending to be applicable. In the semicircular arch at (g), at any point along it defined by variable angle θ, bending moment is given by $Wr \sin \theta$ due to the external load and $r \sin \theta$ due to the imaginary unit load at the position and in the direction of the deflection which is required. In this case it is the same as that of the external load and the deflection due to bending is equal to the integral of the product of these between $q = 0$ and p, divided by the flexural rigidity of the section EI, product of elastic modulus and section second moment of area.

A common example of a portal frame in a horizontal plane is the front crossmember and front ends of the chassis sidemembers of a vehicle imagined 'rooted' at the scuttle. The load case of a central load on the crossmember, perhaps simulating towing, can be visualized at (h). An example of a frame having both curved and straight elements is the half sill shown at (i) supporting a battery tray, with stout crossmembers at mid-span and above the wheel arch reacting the vertical load in this case. Again a table of bending moments can be written, as shown, and the vertical deflection at B calculated by using the unit load formula. Given the value of I for the sidemember of 0.25×10^4 m^4 and E of 210 GN/m^2, the deflection works out to be 53 mm. It is generally best to integrate along the perimetric coordinate for curved elements and, as above, when requiring a translational deflection, apply unit force, and for a rotational displacement, apply unit couple. In the case of redundancies, remove the redundant constraint by making a cut then equate the deflection of the determinate cut structure with that due to application of the redundant constraint.

8.4.2 *BOX BEAMS*

The panels and joints in box-membered structures can be treated differently, Fig. 8.9. In the idealized structure at (a), the effect on torsional stiffness of removing the panels from torsion boxes can be seen in the accompanying table, due to Dr J. M. Howe of Hertfordshire University. Torsional flexibility is 50% greater than that of the closed tube (or open tube with a rigid jointed frame of similar shear stiffness to the removed panel surrounding the cutout) if the contributions of flanges and ribs are neglected.

The effect of joint flexibility on vehicle body torsional resistance must also be taken into account. Experimental work carried out by P. W. Sharman at Loughborough University has shown how some joint configurations behave. The importance of adding diaphragms at intersections of box beams was demonstrated, (b). Without such stiffening, the diagram shows the vertical webs of the continuous member are not effective in transmitting forces normal to their plane so that horizontal flanges must provide all the resistance. The distortions shown inset were then found to take place if no diaphragms were provided.

8.4.3 *STABILITY CONSIDERATIONS*

Applying beam theory to large box-section beams must, however, take account of the propensity of relatively thin walls to buckle. Smaller box sections such as windshield pillars may also be prone to overall column buckling. Classic examples of strut members in vehicle bodywork also include the B-posts of sedans in the rollover accident situation. Such a B-post section, idealized for analysis, is shown at (c). To determine its critical end load for buckling in the rollover situation, its neutral axis of bending has first to be found – using a method such as the tabular one at (d). Assuming the roof end of the pillar to impose 'pinned' end fixing conditions (so that $L = 2l$) and that the pillar is 1 metre in length, then critical load is $10.210.10^9.8.4.10^{-8}/22$. Taking E for steel as 210×10^3 N/mm^2, the stress at this load is $44.10^3/280.10^6$ = 157 MN/m^2 since A = 250 mm^2 – which is above the critical buckling stress. If, however, the cant rail is assumed to provide lateral

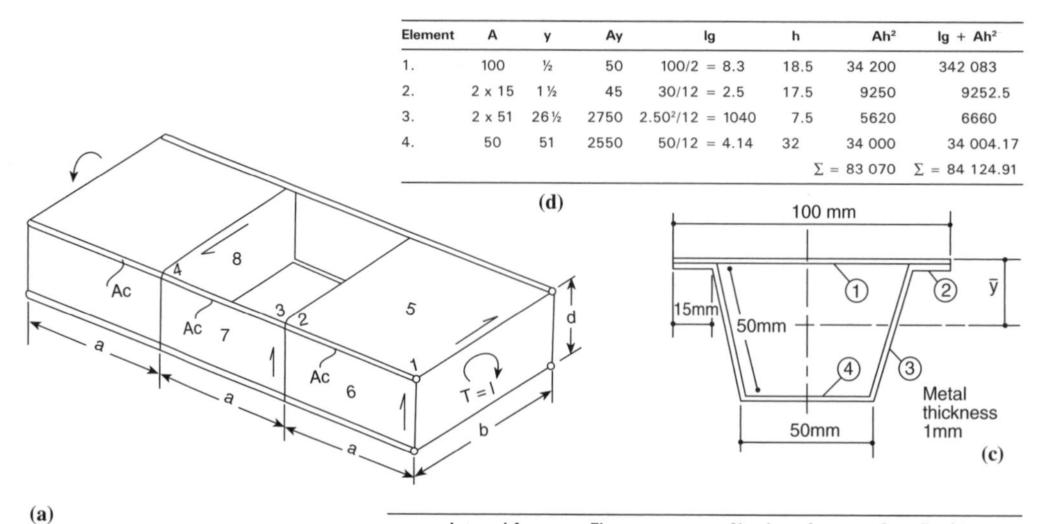

Element	A	y	Ay	Ig	h	Ah²	Ig + Ah²
1.	100	½	50	100/2 = 8.3	18.5	34 200	342 083
2.	2 x 15	1½	45	30/12 = 2.5	17.5	9250	9252.5
3.	2 x 51	26½	2750	2.50²/12 = 1040	7.5	5620	6660
4.	50	51	2550	50/12 = 4.14	32	34 000	34 004.17
						Σ = 83 070	Σ = 84 124.91

	Internal force due to T = 1 b	Element flexibility matrix	Number of identical elements	Contribution to structural flexibility $b^T fb$
1	o	$\frac{a}{6EAc}\begin{bmatrix}2 & 1\\1 & 2\end{bmatrix}$	8	$8a^3/12b^2d^2EAc$
2	$a/2bd$			
3	$a/2bd$	$\frac{a}{6EAc}\begin{bmatrix}2 & 1\\1 & 2\end{bmatrix}$	4	$4a^3/12b^2d^2EAc$
4	$-a/2bd$			
5	$3/4bd$	ab/Gt	4	$9ab/4b^2d^2Gt$
6	$1/4bd$	ad/Gt	4	$ab/4b^2d^2Gt$
7	$1/bd$	ad/Gt	2	$2ad/b^2d^2Gt$
8	$3/4bd$	db/Gt	2	$9db/8b^2d^2Gt$
				$\frac{9(b + d)a}{4b^2d^2Gt}$

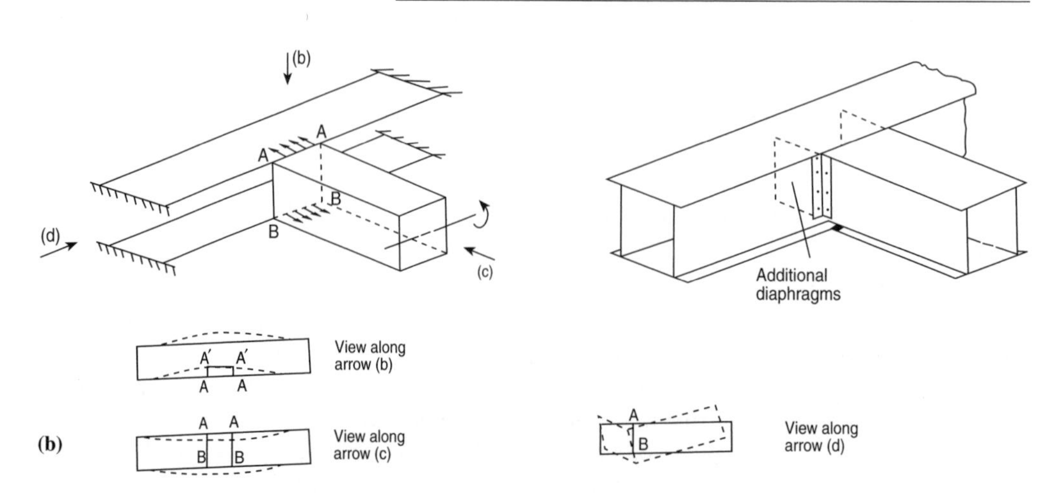

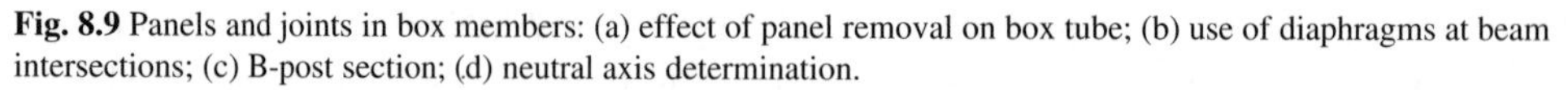

Fig. 8.9 Panels and joints in box members: (a) effect of panel removal on box tube; (b) use of diaphragms at beam intersections; (c) B-post section; (d) neutral axis determination.

support at the top end of the pillar then $L = 0.7l$ and critical buckling stress is 1280 MN/m^2 and the strut would fail at the direct yield stress of 300 MN/m^2. Other formulae, such as those due to Southwell and Perry-Robertson, will allow for estimation of buckling load in struts with initial curvature.

8.5 Designing against fatigue

Dynamic factors should also be built in for structural loading, to allow for travelling over rough roads. Combinations of inertia loads due to acceleration, braking, cornering and kerbing should also be considered. Considerable banks of road load data have been built up by testing organizations and written reports have been recorded by MIRA and others. As well as the normal loads which apply to two wheels riding a vertical obstacle, the case of the single wheel bump, which causes twist of the structure, must be considered. The torque applied to the structure is assumed to be 1.5 × the static wheel load × half the track of the axle. Depending on the height of the bump, the individual static wheel load may itself vary up to the value of the total axle load.

As well as shock or impact loading, repetitive cyclic loading has to be considered in relation to the effective life of a structure. Fatigue failures, in contrast to those due to steady load, can of course occur at stresses much lower than the elastic limit of the structural materials, Fig. 8.10. Failure normally commences at a discontinuity or surface imperfection such as a crack which propagates under cyclic loading until it spreads across the section and leads to rupture. Even with ductile materials failure occurs without generally revealing plastic deformation. The view at (a) shows the terminology for describing stress level and the loading may be either complete cyclic reversal or fluctuation around a mean constant value. Fatigue life is defined as the number of

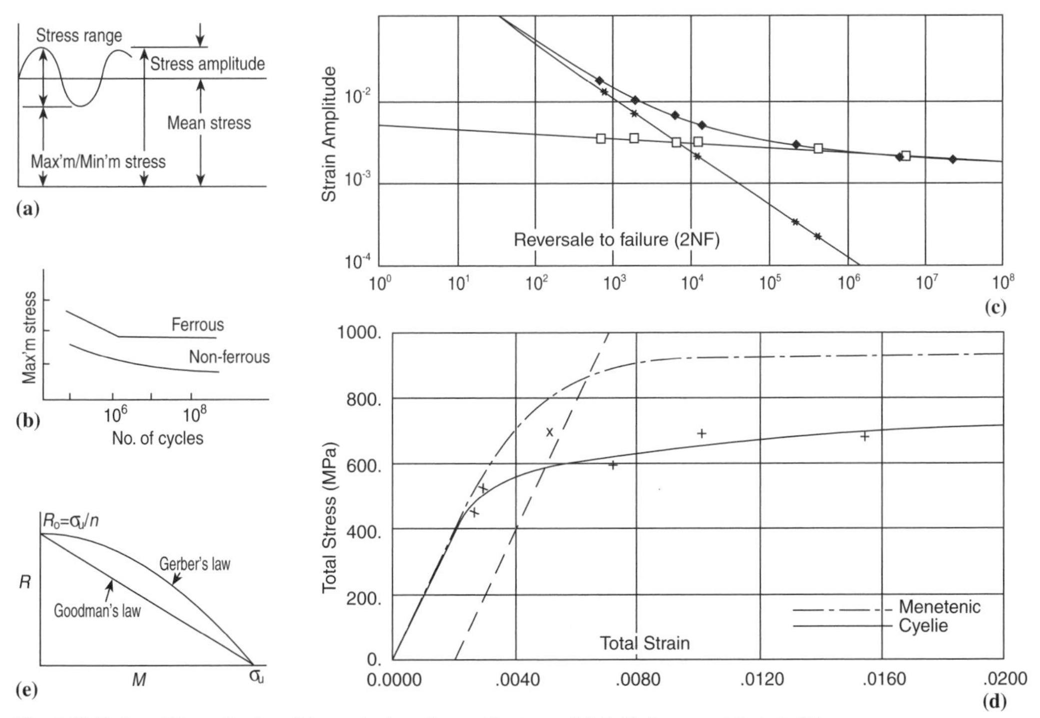

Fig. 8.10 Fatigue life evaluation: (a) terminology for cyclic stress; (b) S–N diagram; (c) strain/life curves; (d) dynamic stress/strain curves; (e) fatigue limit diagrams.

cycles of stress the structure suffers up until failure. The plot of number of cycles is referred to as an *S–N* diagram, (b), and is available for different materials based on laboratory controlled endurance testing. Often they define an endurance range of limiting stress on a 10 million life cycle basis. A log–log scale is used to show the exponential relationship $S = C \,.\, Nx$ which usually exists, for C and x as constants, depending on the material and type of test, respectively. The graph shows a change in slope to zero at a given stress for ferrous materials – describing an absolute limit for an indefinitely large number of cycles. No such limit exists for non-ferrous metals and typically, for aluminium alloy, a 'fatigue limit' of 5×10^8 is defined. It has also become practice to obtain strain/life (c) and dynamic stress/strain (d) for materials under sinusoidal stroking in test machines. Total strain is derived from a combination of plastic and elastic strains and in design it is usual to use a stress/strain product from these curves rather than a handbook modulus figure. Stress concentration factors must also be used in design.

When designing with load histories collected from instrumented past vehicle designs of comparable specification, signal analysis using rainflow counting techniques is employed to identify number of occurrences in each load range. In service testing of axle beam loads it has been shown that cyclic loading has also occasional peaks, due to combined braking and kerbing, equivalent to four times the static wheel load. Predicted life based on specimen test data could be twice that obtained from service load data. Calculation of the damage contribution of the individual events counted in the rainflow analysis can be compared with conventional cyclic fatigue data to obtain the necessary factoring. In cases where complete load reversal does not take place and the load alternates between two stress values, a different (lower) limiting stress is valid. The largest stress amplitude which alternates about a given mean stress, which can be withstood 'infinitely', is called the fatigue limit. The greatest endurable stress amplitude can be determined from a fatigue limit diagram, (e), for any minimum or mean stress. Stress range R is the algebraic difference between the maximum and minimum values of the stress. Mean stress M is defined such that limiting stresses are $M +/- R/2$.

Fatigue limit in reverse bending is generally about 25% lower than in reversed tension and compression, due, it is said, to the stress gradient – and in reverse torsion it is about 0.55 times the tensile fatigue limit. Frequency of stress reversal also influences fatigue limit – becoming higher with increased frequency. An empirical formula due to Gerber can be used in the case of steels to estimate the maximum stress during each cycle at the fatigue limit as $R/2 + (\sigma_u 2 - nR\sigma_u)^{1/2}$ where σ_u is the ultimate tensile stress and n is a material constant = 1.5 for mild and 2.0 for high tensile steel. This formula can be used to show the maximum cyclic stress σ for mild steel increasing from one-third ultimate stress under reversed loading to 0.61 for repeated loading. A rearrangement and simplification of the formula by Goodman results in the linear relation $R = (\sigma_u/n)[1 - M/\sigma_u]$ where $M = \sigma - R/2$. The view in (e) also shows the relative curves in either a Goodman or Gerber diagram frequently used in fatigue analysis. If values of R and σ_u are found by fatigue tests then the fatigue limits under other conditions can be found from these diagrams.

Where a structural element is loaded for a series of cycles $n1$, $n2$ ... at different stress levels, with corresponding fatigue life at each level $N1$, $N2$ ... cycles, failure can be expected at $\Sigma n/N = 1$ according to Miner's law. Experiments have shown this factor to vary from 0.6 to 1.5 with higher values obtained for sequences of increasing loads.

8.6 Finite-element analysis (FEA)

This computerized structural analysis technique has become the key link between structural design and computer-aided drafting. However, because the small size of the elements usually prevents an overall view, and the automation of the analysis tend to mask the significance of the major structural scantlings, there is a temptation to by-pass the initial stages in structural design and perform the structural analysis on a structure which has been conceived purely as an envelope for the electromechanical systems, storage medium, passengers and cargo, rather than an optimized load-bearing structure. However, as well as fine-mesh analysis which gives an accurate stress and deflection prediction, course-mesh analysis can give a degree of structural feel useful in the later stages of conceptual design, as well as being a vital tool at the immediate pre-production stage.

One of the longest standing and largest FEA software houses is PAFEC who have recommended a logical approach to the analysis of structures, Fig. 8.11. This is seen in the example of a constant-sectioned towing hook shown at (a). As the loading acts in the plane of the section the elements chosen can be plane. Choosing the optimum mesh density (size and distribution) of elements is a skill which is gradually learned with experience. Five meshes are chosen at (b) to show how different levels of accuracy can be obtained.

The next step is to calculate several values at various key points – using basic bending theory as a check. In this example nearly all the meshes give good displacement match with simple theory but the stress line-up is another story as shown at (c). The lesson is: where stresses vary rapidly in a region, more densely concentrated smaller elements are required; over-refinement could of course, strain computer resources.

Each element is connected to its neighbour at a number of discrete points, or nodes, rather than continuously joined along the boundaries. The method involves setting up relationships for nodal forces and displacements involving a finite number of simultaneous linear equations. Simplest plane elements are rectangles and triangles, and the relationships must ensure continuity of strain across the nodal boundaries. The view at (d) shows a force system for the nodes of a triangular element along with the dimensions for the nodes in the one plane. The figure shows how a matrix can be used to represent the coefficients of the terms of the simultaneous equations.

Another matrix can be made up to represent the stiffness of all the elements $[K]$ for use in the general equation of the so-called 'displacement method' of structural analysis:

$$[R] = [K] \,.\, [r]$$

where $[R]$ and $[r]$ are matrices of external nodal forces and nodal displacements; the solution of this equation for the deflection of the overall structure involves the inversion of the stiffness matrix to obtain $[\mathrm{K}]^{-1}$. Computer manipulation is ideal for this sort of calculation.

As well as for loads and displacements, FEA techniques, of course, cover temperature fields and many other variables and the structure, or medium, is divided up into elements connected at their nodes between which the element characteristics are described by equations. The discretization of the structure into elements is made such that the distribution of the field variable is adequately approximated by the chosen element breakdown. Equations for each element are assembled in matrix form to describe the behaviour of the whole system. Computer programs are available for both the generation of the meshes and the solution of the matrix equations, such that use of the method is now much simpler than it was during its formative years.

Economies can be made in the discretization by taking advantage of any symmetry in the structure to restrict the analysis to only one-half or even one-quarter – depending on degree. As well as planar symmetry, that due to axial, cyclic and repetitive configuration, seen at (e), should be considered. The latter can occur in a bus body, for example, where the structure is composed of identical bays corresponding to the side windows and corresponding ring frame.

Element shapes are tabulated in (f) – straight-sided plane elements being preferred for the economy of analysis in thin-wall structures. Element behaviour can be described in terms of 'membrane'

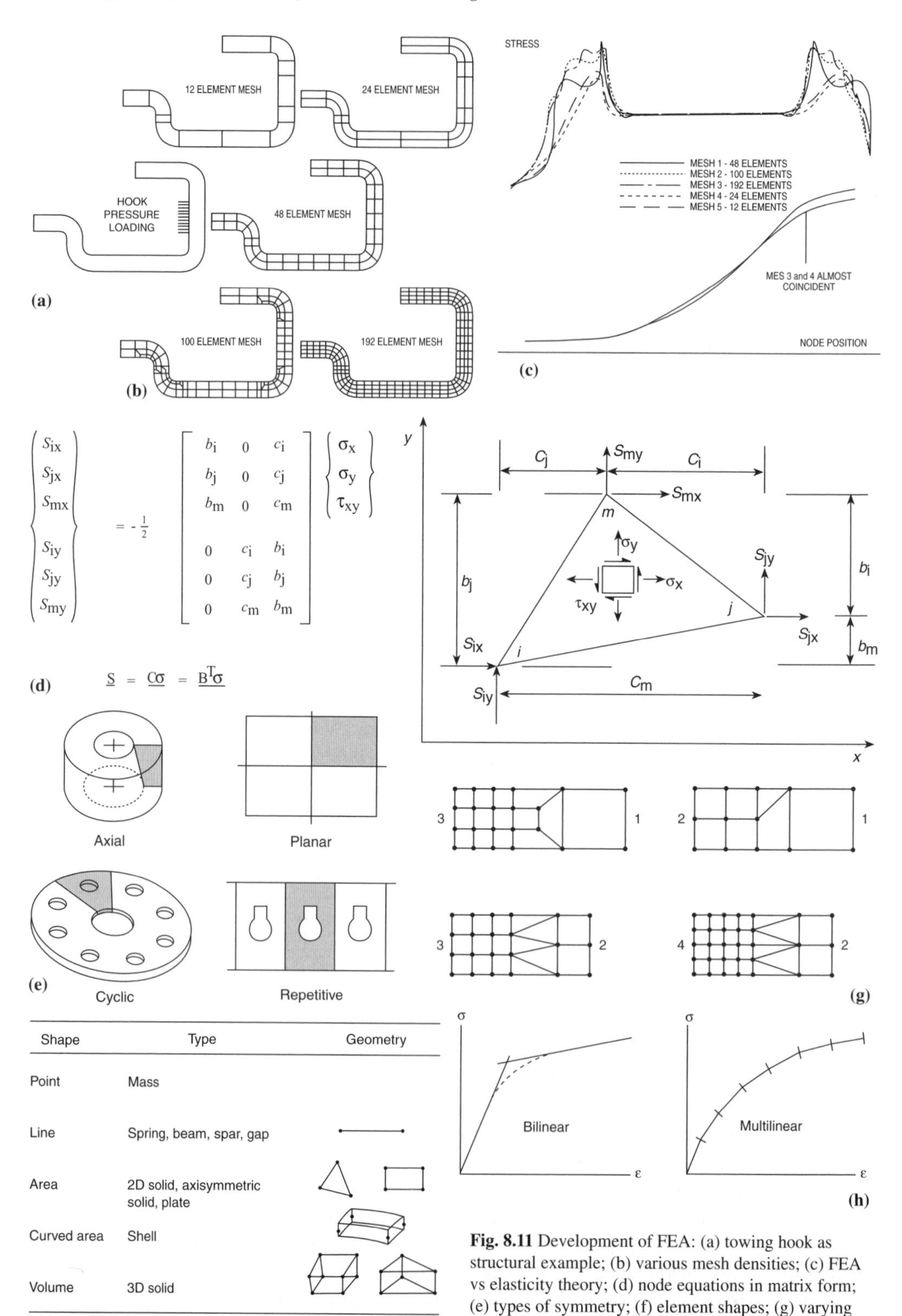

$$\begin{Bmatrix} S_{ix} \\ S_{jx} \\ S_{mx} \\ S_{iy} \\ S_{jy} \\ S_{my} \end{Bmatrix} = -\frac{1}{2} \begin{bmatrix} b_i & 0 & c_i \\ b_j & 0 & c_j \\ b_m & 0 & c_m \\ 0 & c_i & b_i \\ 0 & c_j & b_j \\ 0 & c_m & b_m \end{bmatrix} \begin{Bmatrix} \sigma_x \\ \sigma_y \\ \tau_{xy} \end{Bmatrix}$$

$$\underline{S} = \underline{C\sigma} = \underline{B^T\sigma}$$

Fig. 8.11 Development of FEA: (a) towing hook as structural example; (b) various mesh densities; (c) FEA vs elasticity theory; (d) node equations in matrix form; (e) types of symmetry; (f) element shapes; (g) varying mesh densities; (h) stress–strain curve representation.

(only in-plane loads represented), in bending only or as a combination entitled 'plate/shell'. The stage of element selection is the time for exploiting an understanding of basic structural principles; parts of the structure should be examined to see whether they would typically behave as a truss frame, beam or in plate bending, for example. Avoid the temptation to over-model a particular example, however, because number and size of elements are inversely related, as accuracy increases with increased number of elements.

Different sized elements should be used in a model – with high mesh densities in regions where a rapid change in the field variable is expected. Different ways of varying mesh density are shown at (g), in the case of square elements. All nodes must be interconnected and therefore the fifth option shown would be incorrect because of the discontinuities.

As element distortion increases under load, so the likelihood of errors increases, depending on the change in magnitude of the field variable in a particular region. Elements should thus be as regular as possible – with triangular ones tending to equilateral and rectangular ones tending to square. Some FEA packages will perform distortion checks by measuring the skewness of the elements when distorted under load. In structural loading beyond the elastic limit of the constituent material an idealized stress/strain curve must be supplied to the FEA program – usually involving a multilinear representation, (h).

When the structural displacements become so large that the stiffness matrix is no longer representational then a 'large-displacement' analysis is required. Programs can include the option of defining 'follower' nodal loads whereby these are automatically reorientated during the analysis to maintain their relative position. The program can also recalculate the stiffness matrices of the elements after adjusting the nodal coordinates with the calculated displacements. Instability and dynamic behaviour can also be simulated with the more complex programs.

The principal steps in the FEA process are: (i) idealization of the structure (discretization); (ii) evaluation of stiffness matrices for element groups; (iii) assembly of these matrices into a super-matrix; (iv) application of constraints and loads; (v) solving equations for nodal displacements; and (vi) finding member loading. For vehicle body design, programs are available which automate these steps, the input of the design engineer being, in programming, the analysis with respect to a new model introduction. The first stage is usually the obtaining of static and dynamic stiffness of the shell, followed by crash performance based on the first estimate of body member configurations. From then on it is normally a question of structural refinement and optimization based on load inputs generated in earlier model durability cycle testing. These will be conducted on relatively course mesh FEA models and allow section properties of pillars and rails to be optimized and panel thicknesses to be established.

In the next stage, projected torsional and bending stiffnesses are input as well as the dynamic frequencies in these modes. More sophisticated programs will generate new section and panel properties to meet these criteria. The inertias of mechanical running units, seating and trim can also be programmed in and the resulting model examined under special load cases such as pot-hole road obstacles. As structural data is refined and updated, a fine-mesh FEA simulation is prepared which takes in such detail as joint design and spot-weld configuration. With this model a so-called sensitivity analysis can be carried out to gauge the effect of each panel and rail on the overall behaviour of the structural shell.

Joint stiffness is a key factor in vehicle body analysis and modelling them normally involves modifying the local properties of the main beam elements of a structural shell. Because joints are line connections between panels, spot-welded together, they are difficult to represent by local FEA models. Combined FEA and EMA (experimental modal analysis) techniques have thus been proposed to 'update' shell models relating to joint configurations. Vibrating mode shapes in theory and practice can thus be compared. Measurement plots on physical models excited by vibrators are made to correspond with the node points of the FEA model and automatic techniques in the computer program can be used to update the key parameters for obtaining a convergency of mode shape and natural frequency.

An example car body FEA at Ford was described at one of the recent Boditek conferences, Fig. 8.12, outlining the steps in production of the FEA model at (a). An extension of the PDGS computer package used in body engineering by the company – called FAST (Finite-Element Analysis System) – can use the geometry of the design concept existing on the computer system for fixing of nodal points and definition of elements. It can check the occurrence of such errors as duplicated nodes or missing elements and even when element corners are numbered in the wrong order. The program also checks for misshapen elements and generally and substantially compresses the time to create the FEA model.

The researchers considered that upwards of 20 000 nodes are required to predict the overall behaviour of the body-in-white. After the first FEA was carried out, the deflections and stresses derived were fed back to PDGS-FAST for post-processing. This allowed the mode of deformation to be viewed from any angle – with adjustable magnification of the deflections – and the facility to switch rapidly between stressed and unstressed states. This was useful in studying how best to reinforce part of a structure which deforms in a complex fashion. Average stress values for each

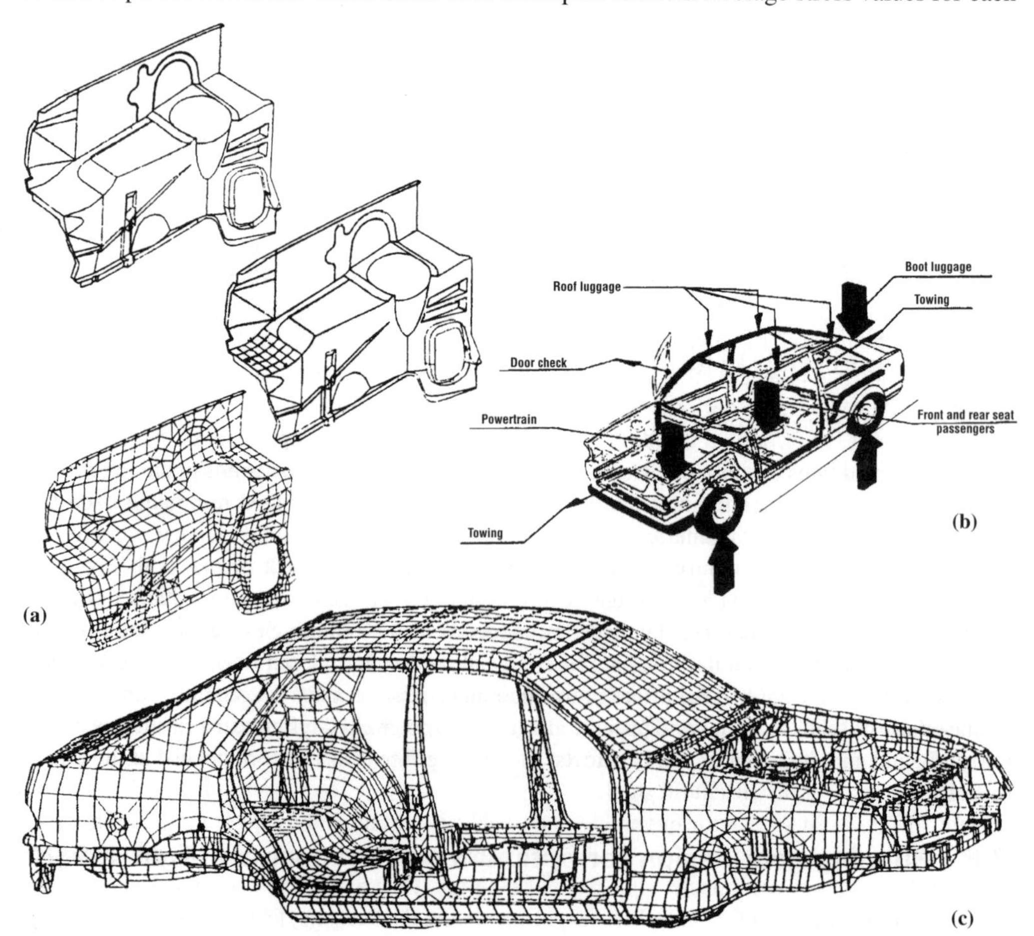

Fig. 8.12 FEA of Ford car: (a) steps in producing FEA model; (b) load inputs; (c) global model for body-in-white (BIW).

element can also be displayed numerically or by graduated shades of colour. Load inputs were as shown at (b) and the FE model for the BIW at (c).

8.7 Case study of FEA for EVs and structural assemblies

Daewoo's DEV3 electric vehicle has used structural analysis techniques applied by the Institute for Advanced Engineering in the design of its space frame[7]. The frame comprises welded extruded aluminium alloy members of identical cross-section and the design was optimized by setting up design variables for each member in differential element thicknesses. The structure incorporates a large battery tray over the floor area, the design of which was also optimized such that a lightweight EV build was possible even with fitted batteries. Its kerb weight of 1187 kg compares with 2100 for the pioneering Lucas Hybrid EV, for example; while battery weight for the DEV3's Ni–MH units was 373 kg against 550 kg for the Lucas lead–acid units.

For the floor tray, upper and lower skins over the extruded frame members formed a strong sandwich construction, with a box frame around its periphery. The material used is AL 6061-T6 of 310 MPa ultimate tensile strength, joined by TIG welding. Target weight for the space frame was 130 kg and a torsional stiffness of 7940 Nm/° was sought to give first torsional frequency of 25 Hz and 140 MPa stress under service loads. Figure 8.13 shows the finite-element model, with structural member selection, and in testing the physical structure a torsional stiffness of 8682 Nm/° was obtained which was 31% stiffer than that of the GM Impact vehicle against which it was compared, considered to be the effect of integrating the tray structure; torsional frequency was 26.5 Hz. Service stresses exceeded the target values, however, for 3 *g* bump loading of both 'together' and 'individual' front wheels, necessitating thickening of the strut member supporting the front of the battery tray.

Of the 34 extruded members comprising the space frame (a), those at (b) were chosen, in addition to the tray members, (c), for the purpose of optimizing the structure by thickness variation from a base size of 2 mm. The cross-section of the tray frame member is shown at (d). A table of the combined members chosen is at (e) and the results of the sizing optimization tabled at (f), leading to a final weight of 96.39 kg for the space frame.

8.7.1 PROGRAM FOR DYNAMIC ANALYSIS

A variety of programs exist for carrying out dynamic as well as static analysis. For the foam-filled structural construction technique discussed in Section 7.2.1, the transient non-linear code LS-DYNA3D was used in the numerical FEA with Material Number 57 solid elements adopted for the modelling. The B-pillar and rear roof header were modelled for a pick-up truck cab, Fig. 8.14, to simulate the FMVSS 2116 roof crush test, the view showing the undeformed and deformed shapes of the unfilled structure, alongside the model of the cab frame. Several simulations were carried out in order to evaluate foams of different densities and identify critical areas of foam application so as to establish the effect of using reduced foam density in certain regions[8].

The other books in this series will provide greater detail on chassis-system performance including the study and simulation of dynamic effects. Further reading is also recommended at the end of this chapter.

8.8 Running gear design for optimum performance and light weight

8.8.1 INTRODUCTION

The viability of pollution control by electric traction rests not only on energy storage, propulsion technology, body design and construction, but also on light weight and low-drag vehicle running

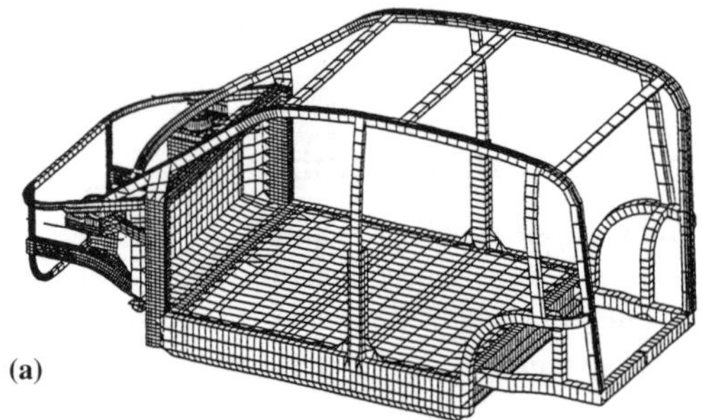

(a)

(b)

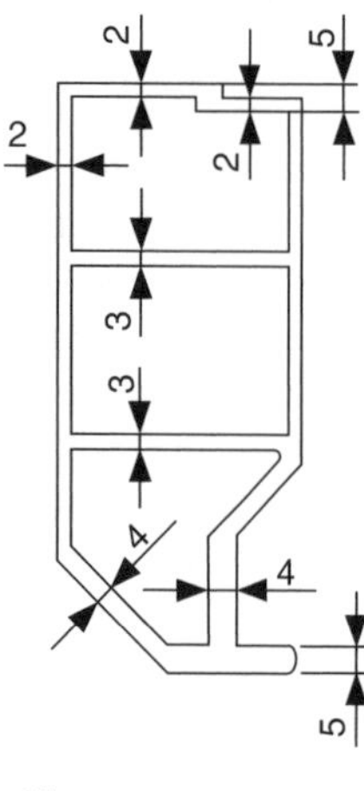

(d)

No.	Initial thickness (mm)	Lower bound (mm)	Upper bound (mm)	Final thickness (mm)
1	3	2	5	4.11
2	2	1.5	5	1.50
3	3	2	5	2
4	3	2	5	2
5	3	2	5	2
6	3	2	5	2
7	2	1.5	5	1.55
8	3	2	5	3.03
9	4	2	5	2
10	5	2	7	2
11	8	2	10	2
12	3	2	5	2

(f)

No.	Label	Part Name	Geometry
1	strut	strut	3mm panel
2	A	A-pillar	2mm panel
3	support	front longi. support column	w50 x h50 x t3 extrusion
4	upper B	upper B-pillar	w50 x h30 x t3
5	lower B	lower B-pillar	w80 x h70 x t3
6	rear	rear longl. frame	w60 x h70 x t3
7	roof	side roof rail	t2 extrusion
8	tray 1	tray frame	t3
9	tray 2	tray frame	t4
10	tray 3	battery seperator	t5
11	tray 4	floor supporter	t8 panel
12	tray 5	tray floor	t3 panel

(e)

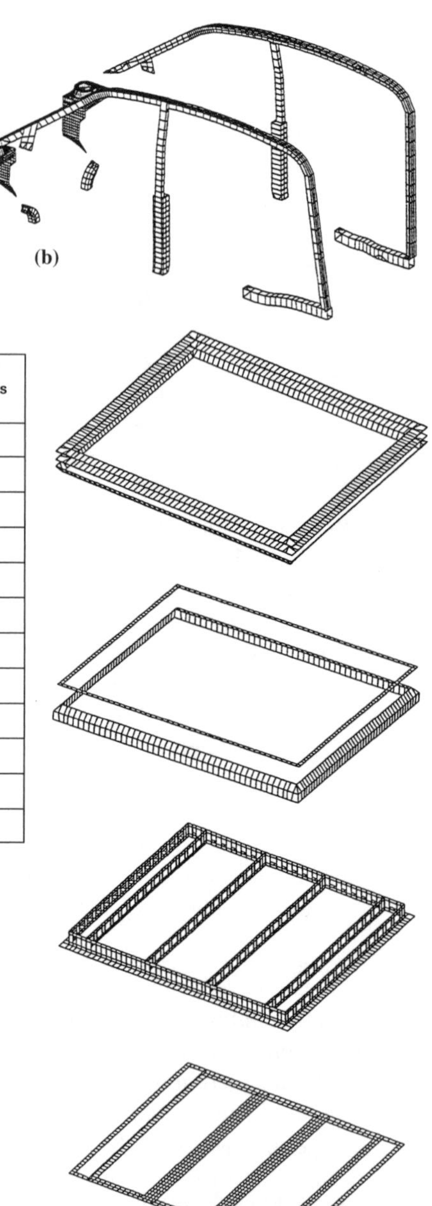

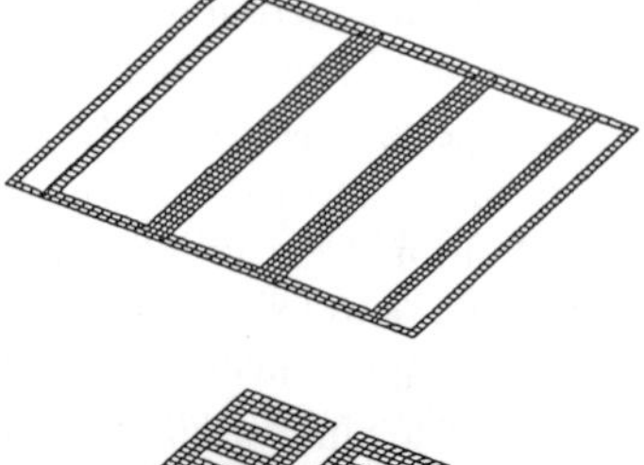

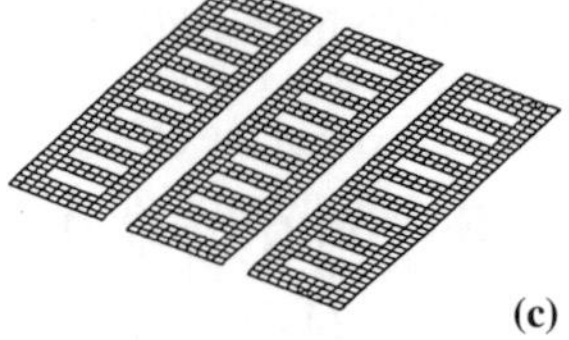

(c)

Fig. 8.13 FE model and selection of members for lightweight EV: (a) total structure model; (b) non-floor frame members; (c) floor members; (d) floor member cross-section; (e) members chosen for optimization; (f) optimization results.

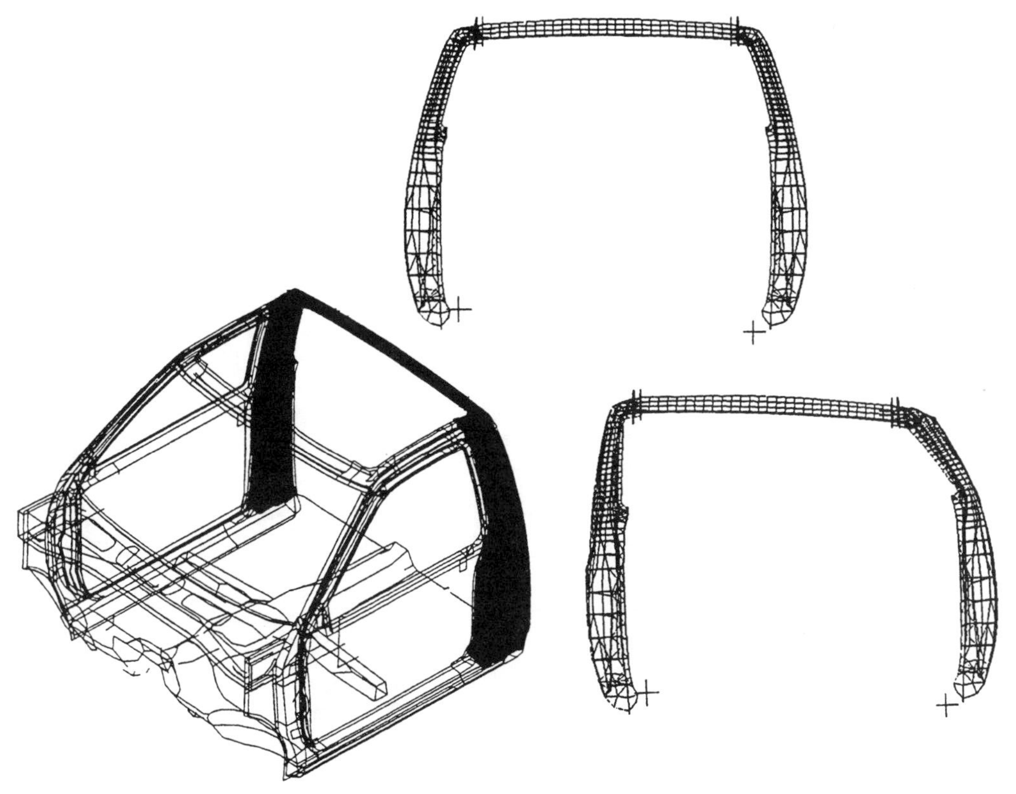

Fig. 8.14 FE model and B-pillar ring normal and deformed.

gear. This is particularly so in parallel hybrid-drive systems were a near-conventional automotive drivetrain exists in parallel to the electrical propulsion system. In the remaining part of this chapter, inertia, rolling and gradient resistances are examined with respect to accelerative performance; self-levelling suspensions that can tolerate large differences between vehicle laden and unladen weight are reviewed; handling and steering calculations are provided for exploiting the potential improvement in balanced weight distribution; and the electrical-powered/electronic-controlled chassis systems that are made more attractive by electric traction are considered concluding with the important area of low rolling-drag tyres.

8.8.2 DETERMINING WEIGHT DISTRIBUTION

As discussed in the Introduction, an electric vehicle should not be designed by a stylist and then engineered by an automotive technologist, in conventional fashion, but approached at the concept stage by an 'integrated' designer/engineer who is able to trade off the aesthetic and performance functions as he or she proceeds. A gravimetric analysis is important even at the concept stage to ensure that weight reduction, Fig. 8.15, is kept in focus throughout the design/development life of the vehicle and that the effect of weight distribution on overall functional performance is continually checked. Accurate prediction of weights for each element of the vehicle structure, powertrain and running units (a) is aided by graphical representation (b). For this purpose, alignment charts can provide a valuable guide to the use of different materials in sheet and rail form. This will lead to an early appreciation, for example, of front-to-rear weight distribution upon which ride and handling performance is dependent — as well as the selection of tyre size and construction. The useful

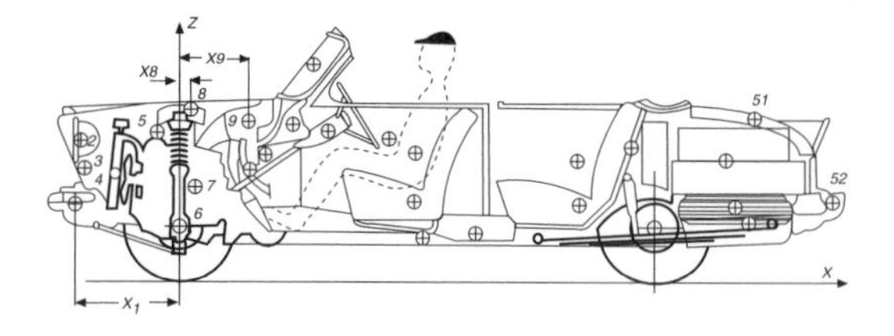

Lp.	Element	Q (kg)	X (m)	Δ *M* (kg.m)	
				−m	+m
1	Front bumper	4	−0.62	2.5	
2	Headlamps	6	−0.58	3.5	
3	Grille	5	−0.57	3.0	
4	Radiator with water	15	−0.50	7.5	
51	Boot lid	9	+3.10		27.8
52	Rear bimper	4	+3.50		14.0
			$\Sigma Q = Gc$	$\Sigma\Delta M = M$	

(a)

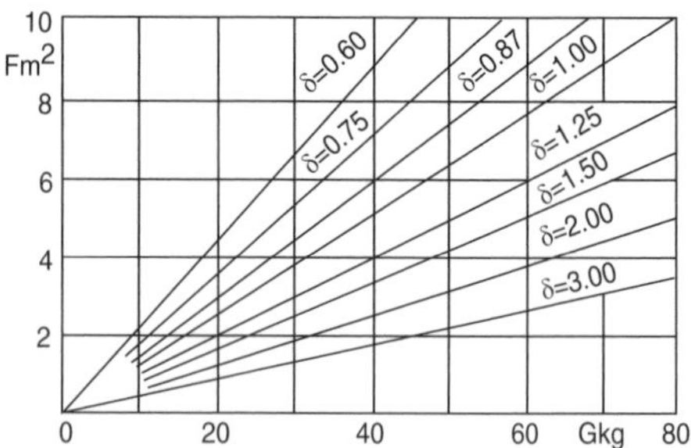

Alignment chart for the weight of steel sheets of different thicknesses. For small areas, 0-3 m², a larger scale graph is used

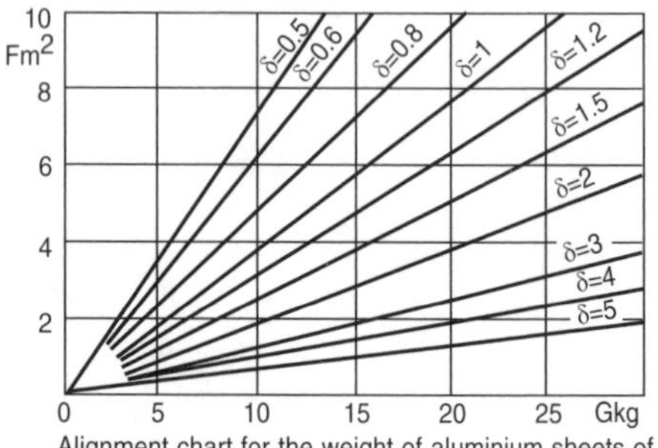

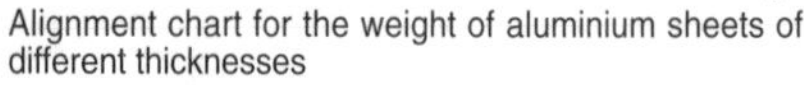

Alignment chart for the weight of aluminium sheets of different thicknesses

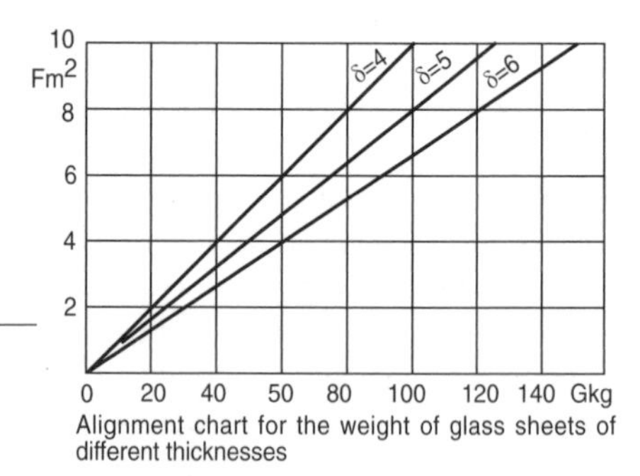

Alignment chart for the weight of glass sheets of different thicknesses

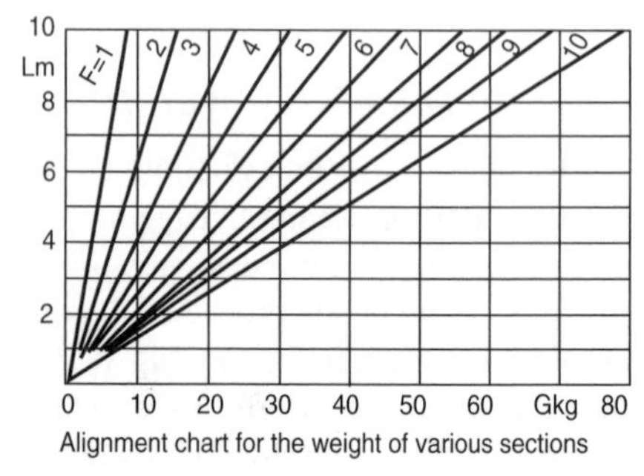

Alignment chart for the weight of various sections

(b)

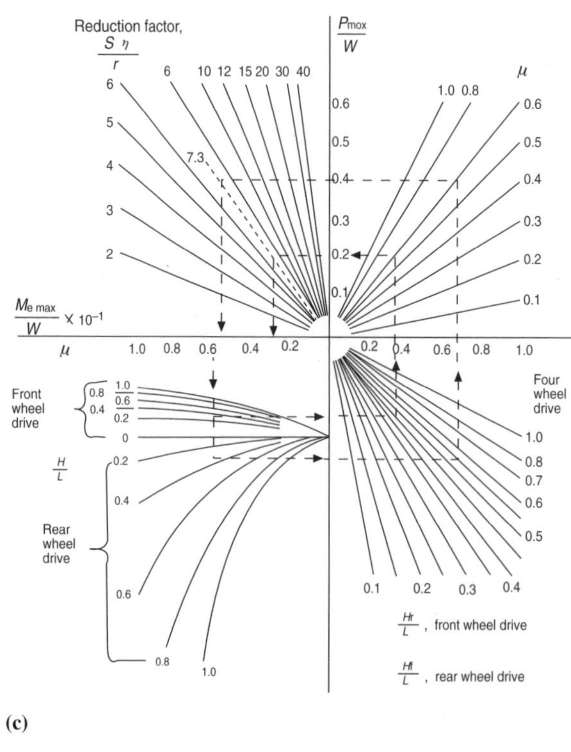

(c)

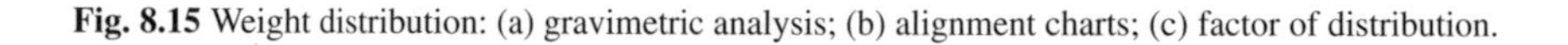

Fig. 8.15 Weight distribution: (a) gravimetric analysis; (b) alignment charts; (c) factor of distribution.

tabular approach to gravimetric analysis is combined with moment equations to determine axle loadings; the numbers locating the main assemblies on the vehicle correspond with the first column of the table and go on to include front wing (5), front suspension (6), power unit (7) and on to (51)/(52) as shown. The alignment charts are nomograms which represent the weight of a given material as a function of the surface area or length and help in the estimation of body structural weight, a surprisingly large percentage of the total on conventional cars. Successful prediction of body weight distribution is key to the early fixing of centre-of-gravity position upon which performance prediction depends.

The importance of vehicle weight is made clear by Taborek[9] who points out that a key relationship in performance considerations is that tractive force P is equal to load on the driving axle W times tyre-to-ground adhesion coefficient μ. But, while W for a static vehicle can easily be found by a moment calculation, the dynamic axle reactions of a moving vehicle are more difficult to work out. The dynamic weight transfer:

$$dW = (H/L)(T_d/r - fW) \qquad \text{and} \qquad P = (T_e RE)/r$$

where T_d is torque at drive axle, T_e torque at engine, R total reduction ratio, E transmission efficiency and r rolling radius. H is height of vehicle centre of gravity (CG) above ground, L the wheelbase and f the symbol for rolling resistance coefficient.

With front and rear-wheel drive, dynamic front axle load = $W(L_r + fH)/(L + \mu H)$ where L_r is the distance from CG to rear axle and μ is adhesion coefficient; this leads to the maximum transferable tractive force as

$$P_{f,max} = \mu W[(L_r + fH)/(L + \mu H)]$$

the square-bracketed term being the weight distribution factor d$W/W = w$ so that $W_{df} = W_{wf}\ P_{max} = \mu W_{wf}$ *and* $P_{r.max} = \mu Wdr$

With four-wheel-drive, $P_{4.max} = \mu W$ and $w_4 = 1$; otherwise the chart at (c) can be used to determine the relative distributions. The weight distributions derived above are dimensionless, being derived from ratios, and the chart also contains only dimensionless factors, with the weight of the vehicle being eliminated by dividing the tractive force equation by W, giving $P_{max}/W = \mu w$.

The dotted line shows an example for a vehicle with $h/L = 0.35$ and $L_f/L = 0.45$ defining its CG position, for which it is required to find maximum transferable tractive force and the maximum engine torque for a low gear reduction factor of $RE/r = 7.3$ – given also $\mu = 0.6$ for the adhesion coefficient. Starting in the lower left quadrant, the intersection of the $m = 0.6$ line cuts curves for front and rear wheel drive at $h/L = 0.35$. Horizontal projections from those points similarly lead to the P_{max}/W scale. Values thus obtained give the first solution while projecting into the top left quadrant gives the second. For a vehicle weight of 4000 kg, P_{max} = 1620 kg and $T_{e.max}$ = 224 kNm (front drive) or 840 kg and 112 kNm (rear drive).

In predicting performance, Fig. 8.16, this data can be used in a chart, such as at (a), to obtain available vehicle speeds on different gradients. The chart plots maximum values of tractive force at the wheel, as functions of speed, in different gears. Also on the chart, gradient slopes are plotted against vehicle weight. The 0% line is drawn at a slope designating a rolling resistance of 35.4 lb/ton (80 kg/tonne). Propelling force P lines are obtained by graphical subtraction of wind resistance from maximum tractive force, in this method due to Goldschmidt and Hadar of the Israel Institute of Metals. Dotted lines illustrate worked examples for vehicles weighing 1910 and 1875 lb respectively. The first is shown to be able to climb a slope of 10% at 36 mph reducing to 9% at the same speed if that speed is maintained. The second climbs an 8% slope at 43.5 and 38.5 mph in

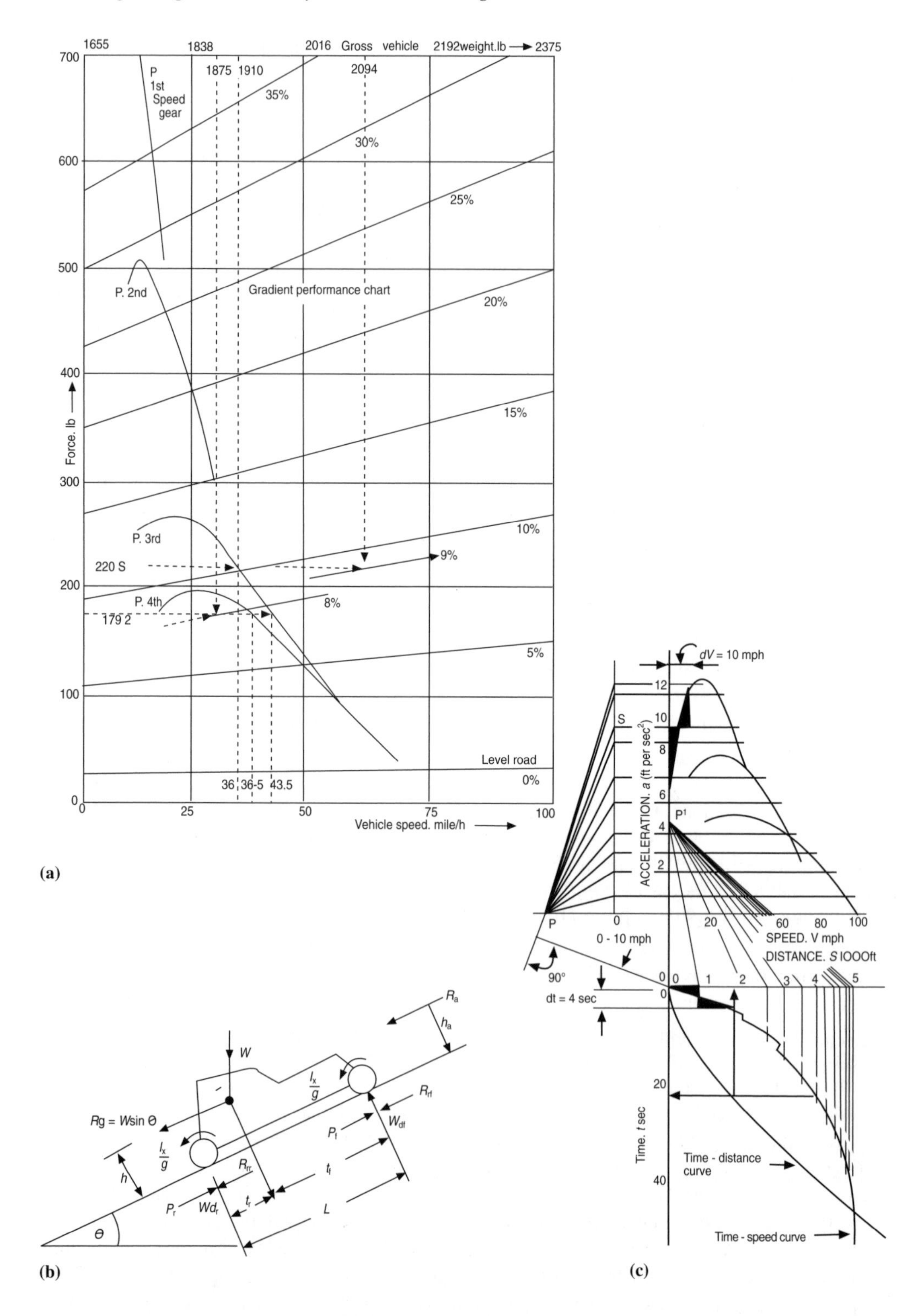

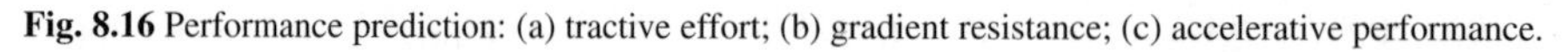
Fig. 8.16 Performance prediction: (a) tractive effort; (b) gradient resistance; (c) accelerative performance.

third and fourth gears respectively. Actual plots would of course depend on the torque/speed characteristics of electric drives.

8.8.3 ROTARY INERTIA, ROLLING AND GRADIENT RESISTANCES

While translatory-mass inertia calculation is relatively easy, that of rotary mass is more difficult. It is necessary to sum the rotating inertia of the separate transmission components by relating relative torque to the driving axle as $T_i = \Sigma IaR^2$ – for moment of inertia I and angular acceleration a. Since wheel circumferential speed is equal to vehicle translatory velocity, an 'equivalent mass' can be considered as concentrated at the rolling radius – having the same effect on the inertia of translatory motion as the summation of individual rotary inertias.

Thus effective inertia mass $M_i = M_e r^2 \alpha d = \Sigma IR^2$ where equivalent mass $M_e = \Sigma IR^2/r^2$, the effective mass being $M_i = M + M_e = MY$ where Y is termed the rotary mass factor, a valuable tool of the method. To find its value, the rotary masses are divided into parts rotating with wheels – and those rotating with power unit. The latter gain in importance in the lower ranges since M_e is proportional to R^2. Average values quoted in the literature for a fully laden commercial vehicle are fourth-gear, 1.09, then 1.2, 1.6 through to 2.5 in first. The value of Y is obtained from

$$Y = 1 + M_e/M = 1 + [(\Sigma I_w/Mr^2) + (\Sigma IR^2/Tr^2)]$$

where ΣI_w is wheel inertia and I_e is that for 'engine speed' parts and typically $Y = 1 + (0.4 + 0.0025R^2)$.

Rolling resistance R_r can be expressed as a dimensionless coefficient f (= 0.013 for radial tyres) while transmission resistance is obtained from a summation of the power consumed at various stages along the powertrain. Overall efficiency E depends on the transmission system and is conventionally assumed as $E = 0.90$ in direct drive and 0.85 in lower gears. Air resistance is obtained from

$$R_a = C_d A \rho V^2/2g$$

for vehicle speed V, gravity acceleration g, projected frontal area A and air density ρ (C_d rising to 0.8 for a bluff box van). Gradient resistance is the component of gravitational force parallel to vehicle attitude, (b). Since $W_d\mu = fW + W \tan \theta_{max}$, maximum gradient = $Wm - f$.

8.8.4 ACCELERATIVE PERFORMANCE

Selection of gear ratio will depend upon the chosen operation for the vehicle. The ratio for maximum acceleration can be estimated from

$$G_{max} = (Rr/T_e) \pm [(Rr^2/T_e) + (Wr^2/g + I_w/r)/I_e]^{1/2}$$

where motion resistance $R = R_r + R_a$ and I_e and I_w are engine and wheel inertias. Engine speed relates to vehicle speed as

$$V = 2pN_e\, r \,.\, 3600/12 \,.\, 60E_s G \,.\, 5280 = N_e r/168GE_s$$

and tyre slip efficiency E_s can be assumed equal to 0.965 for this calculation.

Thus N_e and T_e for each gear reduction may be tabulated against road speed by means of the engine torque/speed curve. Motion resistance forces are subtracted from P to give free tractive

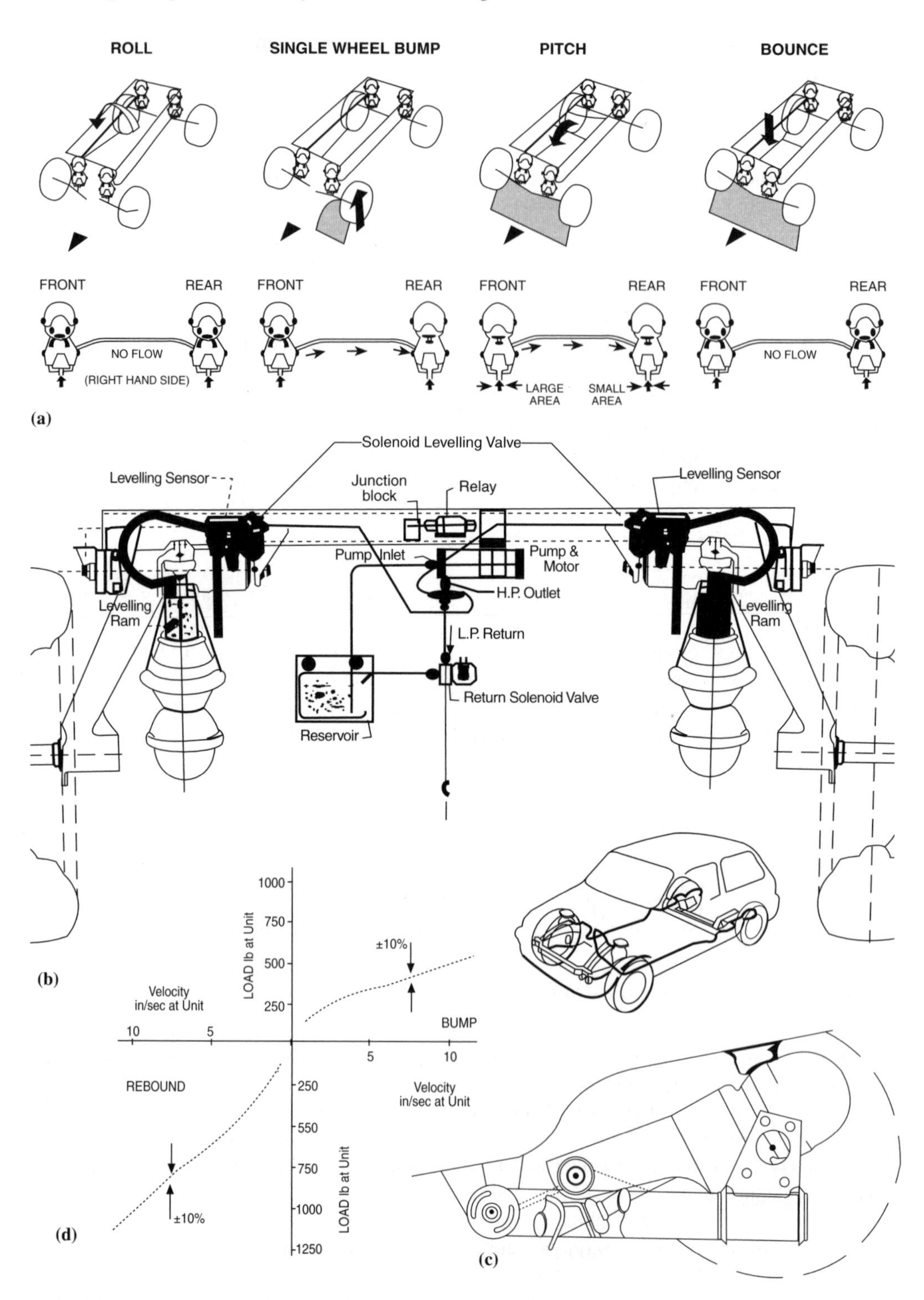

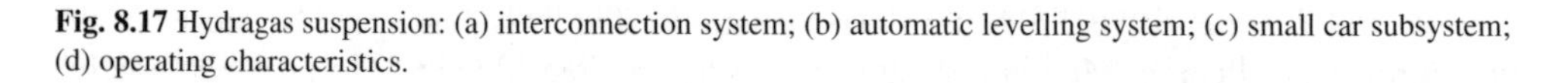
Fig. 8.17 Hydragas suspension: (a) interconnection system; (b) automatic levelling system; (c) small car subsystem; (d) operating characteristics.

force P_f then transferable tractive force is calculated. To obtain acceleration, $\alpha = T_f G/YW$ is used and values can be plotted in a curve such as at (c), a time/speed curve which yields times for accelerating up to (and between) given speeds.

8.9 Lightweight vehicle suspension

The ultra-lightweight vehicle presents a special challenge to suspension engineers, for in a 5-seater car the weight of four passengers and luggage at typically 700 lbs is significant compared with the maximum 1120 lb tare weight arbitrarily targeted for a lightweight electric car, with batteries on board. Substantial change therefore occurs in the suspended mass for laden and unladen vehicles. It is, however, an advantage of the near-symmetrical punt-type structure, proposed for the multipurpose electric vehicle in other chapters, and the large central battery tray gives the vehicle an inherently central centre gravity with little change in its position as passengers embark/disembark to/from the centrally disposed passenger compartment above that for the batteries. Suspension thus has to tolerate large changes in suspended mass though individual wheel units should be reasonably evenly weighted down.

8.9.1 SELF-LEVELLING SYSTEMS

An auto-levelling suspension is a useful advantage to help counteract ride-height change according to payload while air springing would be attractive proposition in terms of variable suspension rate and the ability to obtain soft ride without inordinately large vertical wheel displacements. In the interest of weight economy and minimizing servo-power requirement, Hydragas suspension, Fig. 8.17, is an attractive compromise. The ability to constrain the rubber spring element to provide a combination of direct-compressive and shear deformation provides high energy absorption at relatively low weight while the associated gas springs require no servo power. At the same time the Hydrolastic fluid connection between the springs provides auto-levelling. The system, specified on the latest MG*F*, has recently been updated and is now available with an electronic controller to sense vehicle acceleration. Below a preset speed the controller opens a valve between the front axle suspension units, reducing front roll stiffness. At the same time it closes a valve to maintain rear roll stiffness. Front-drive traction under accelerating conditions is thus improved.

Interconnection schemes for Hydragas are shown at (a) and by introducing gas springs, pitch frequencies were very substantially reduced and bounce frequencies reduced considerably over the earlier Hydralastic suspension. The spring units comprised two canisters welded together, nitrogen being retained in the upper half of the top canister by a butyl separator. Water-based fluid fills the space between separator and the diaphragm at the bottom of the lower canister. Gas pressure at static wheel load is 12.4 kN/m^2, rising to 54 at full bump. In the pitch mode (bump displacement at one end, rebound at the other) restoring force is created by change in diaphragm area, due to its taper, within the spring units. In level ride (single bump travel at each end) fluid can only be displaced through the damper valves further to compress the encapsulated gas above the rubber springs, in turn increasing fluid pressure. Combination of this and increased diaphragm area gives a higher rate in 'bounce' (and roll) than in pitch.

Make-up of rates involve combinations of so-called 'parasitic' and 'drop-angle' portions and are in turn referring to the rates due to rubber bushings within the suspension and that due to change of leverage with stroke. On a typical vehicle of 862 kg kerb weight, front/rear (combined) rates in kN/m for pitch, bounce and roll are 13.3/13.5 (26.78), 19.95/18.85 (38.5), 19.95/18.55 (38.5) compared with respective combined rates for a conventionally suspended vehicle of 32.55, 32.55 and 38.5 kN/m. Front and rear pitch rates must be chosen to optimize a balance between excessively soft rear pitch, leading to inadmissible attitude change with added load and maintaining front/rear pitch ratio to ensure flatness of ride.

Generally speaking the system needs neither anti-roll bars nor levelling systems. If, however, abnormally high load changes are experienced as with lightweight electric-drive vehicles then extending rams in series with the rear piston units can be provided. If necessary automatic levelling can be built in as illustrated at (b). An inherent advantage of interconnection is that height control at the rear is also reflected through to the front. The main feature of the latest Hydragas system (c), is the introduction of a preformed thermoplastic hose assembly to interconnect the front and rear Hydragas units, replacing the combination of steel piping and rubber pump hoses employed on earlier cars. The system is hermetically sealed and the interconnecting front-to-rear hose is filled with a methanol/water mixture, the weight of the car being carried by the pressure of this fluid acting on the displacers. Design criterion for the hose is to maintain a kerbside 27 bar operating pressure throughout its planned 10 year life with a 4:1 safety factor to cover a peak operating pressure of up to 67 bar. This is met by an extruded PA11 polyamide core reinforced with braided synthetic fibre and covered in a tough outer sheath. Operating characteristics are seen at (d).

8.10 Handling and steering

One of the important attributes for the electric vehicle, if it is to attract drivers from conventional petrol driven vehicles, is responsive handling. While tyres on EVs are required to have minimum rolling resistance there is likely to be some compromise in both ride rate and cornering characteristic, so attention to handling response in design is important. We have considered how drive systems, and the associated electrical equipment need to be arranged in the vehicle so as to balance front/

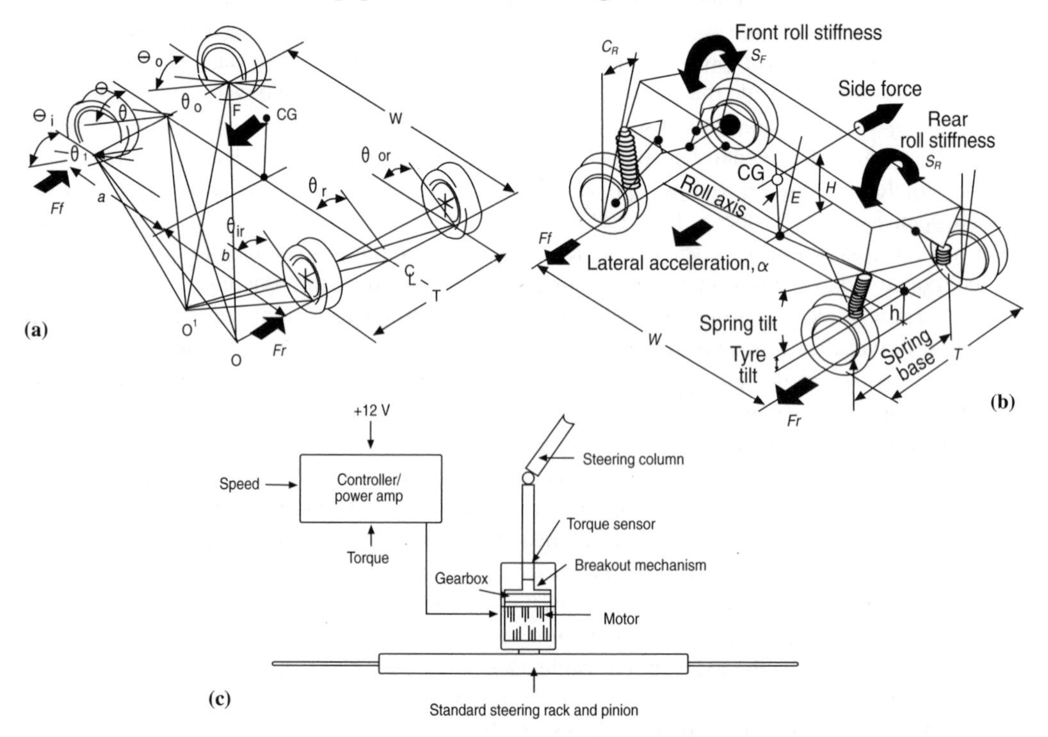

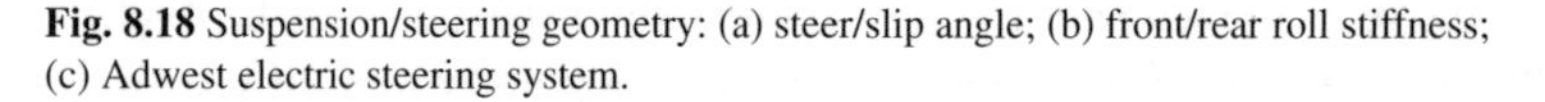

Fig. 8.18 Suspension/steering geometry: (a) steer/slip angle; (b) front/rear roll stiffness; (c) Adwest electric steering system.

rear weight distribution as closely as possible. This is inherently less difficult than on vehicles which have heavy IC-engine and transaxle units mounted at one end. Because both the suggested monocoque-tube and open-integral-punt structures are ideally symmetrical about the front and rear road-wheel axes, and in both cases the battery tray is centrally mounted between the wheel axes, an inherently low polar moment of inertia about the vertical (steer) axis of the vehicle results and design for near neutral steady-state handling response should not be difficult.

8.10.1 HANDLING RESPONSE

To fix some initial suspension/steering geometry values in design, Fig. 8.18, it is useful to consider the motions of the vehicle as though it moves in a horizontal plane only and, in cornering, that only fixed radius bends are encountered – at constant vehicle speed. This allows a 'steady-state' handling analysis to be carried out – with the simplifying assumptions yielding reasonably quick values for the geometric variables. The basic handling relationship for this condition is that the steer angle = heading angle + rear slip angle – front slip angle. There is a proportional (linear) relationship between cornering force and slip angle (between path and heading) until slip occurs (a) – then slip angle increases rapidly and aligning torque *AT* decreases. The effect of radial load *RL* increase is to increase *L* and therefore increase the cornering stiffness. In the figure, slip angle θ can be recognized from steer angle ϕ. When, at comparatively low speeds, the vehicle steers about a point O, then steer angles are related by

$$\text{cotan}\ \theta_o - \text{cotan}\ \theta_i = T/W. \qquad F_f = Fb/W \quad \text{and} \quad F_r = Fa/W$$

are drift forces which correspond to those required to resist centripetal force of the vehicle as vehicle speed increases and there is a build-up of slip angles. Cornering force (drift force) is also required to overcome thrust due to camber change in roll and differential induced drag caused by outward weight transfer. The latter leads to a torque in the ground plane which must be reacted by increased front- and decreased rear-side forces from the tyres. Also to be reacted are the combined self-centring torques due to tyre self-aligning, caster trail and kingpin inclination – also due to the product of tyre forces and kingpin offset as well as the effect of steered axle braking, in the opposite direction.

Roll modifies inter-axle weight distribution according to roll-axis position and the relative front to rear roll stiffness of the vehicle suspension systems. With the wishbone IFS and beam-axle IRS layout shown at (b), for lateral acceleration *a*, torque about the roll axis is $\alpha WH + WH \tan \varepsilon$ and for small roll angles respective axle torques are $(SR/SF + SR)(\alpha + \varepsilon)WH$ at the front and $(SR/SF + SR)(\alpha \quad)WH + FRh$ at the rear. Roll stiffness *S* relates spring and tyre tilt with spring base and track, and if the centre of gravity of the vehicle is positioned symmetrically on the vehicle longitudinal centre-line – at distances *a* from the front and *b* from the rear axle along the wheelbase *L* – then for cornering force *CF* on the whole vehicle

$$SR + CF/[1 + (b/L)] \text{ and } SR = CF/[1 + (a/b)]$$

also $CF = (W/g) - (V^2/r)$ relates vehicle speed with cornering radius *r* and the force can be expressed as a percentage *n* of vehicle weight *W*.

The degree of understeer, represented by the difference between front and rear slip angles, can be plotted against percentage of lateral acceleration *n* to give the response curve. It is then usual to determine the effect of cambering the tyres, say, to produce extra side thrust during roll (by adjusting roll-stiffness and suspension geometry), creating differential dynamic load transfer by varying roll-centre height through suspension geometry manipulation.

The logical procedure in constructing the curves would be to write down static weight distribution and CG position then determine by geometry the front and rear roll-centre positions for the suspensions. Then express the tilting moment in terms of x,n,g and equate it against the suspension restoring torques based on spring rate, spring base and roll angle. The resulting expression can be solved to obtain roll-angle and a table set up with equal increments of lateral acceleration corresponding to side loads derived from force = mass × acceleration. This table would show the minimum to maximum load-transfer condition. Finally, from side-force/slip-angle/radial-load curves available from tyre manufacturers, mean (between inner and outer) slip angles can be obtained to tabulate for front and rear of the vehicle in similar increments. These can be plotted against lateral acceleration as the response curve for the vehicle – yielding degree of understeer for percentage-g turn.

A particular aspect of urban electric vehicles is relatively short wheelbase and narrow track relative to their overall height. The experience of the first-production Mercedes-Benz A-class, which came to grief in the 'moose-test' applied on imports to the Scandiniavian market, is sufficient warning to designers to provide tolerance of severe chicane manoeuvres where build-up in lateral body oscillation can lead to rollover. In the case of the A-class the remedy was slightly lowered chassis, increase in anti-roll torque, extension of rear track, revised damper/tyre characteristics and redesign of the handling response to induce greater understeer. The company's electronic suspension control system was also extended to cover all models in the range.

8.10.2 ELECTRIC STEERING

Power steering mechanisms using electromagnetic actuation are now being introduced on petrol-driven cars and, of course, are ideally suited to the electric car. Because conventional hydraulic power-assisted steering reduces vehicle efficiency, by drawing power from the engine continuously, energy is wasted by driving the hydraulic pump even when assistance is not required, such as at high cruising speeds on motorways. A typical torque/speed curve for a conventional car hydraulic steer-assist pump shows that pressure in the idle hydraulic system is normally around 2 bar and to retain this back-pressure at a road speed of 70 mph (typically 3500 rpm engine speed) a 1.5 Nm torque is required, which can correspond to several hundred watts. In electrohydraulic systems (EHPAS) the engine-driven pump is replaced by an electric one. In the Adwest[10] system, an open-centred valve is used so that equal pressure exists either side of the rack-ram, in the straight-ahead (on-centre) position. Oil flow is thus fed back into the accumulator and the ram isolated. However, the system is inevitably more costly than traditional PAS.

With electric power-assisted steering (EPAS) the basic elements are an electric motor coupled to the steering column, or rack, a power amplifier and an ECU with appropriate sensors, the whole resulting in a smaller package than EHPAS but the sensing and control are more complex. Small cars with restricted underbonnet space are, of course, the best prospects for the system.

In the Adwest EPAS, (c), torque applied by the driver is sensed and the controller varies the assistance provided by an electric motor to minimize effort while retaining steering feel. The brushless motor is coupled to the column and provides up to 4 Nm torque at a 5:1 geared reduction resulting in maximum assistance of 20 Nm in addition to the torque of up to 5 Nm provided by the driver. The 25 Nm is adequate for static steering of a small car and straight-through mechanical connection is maintained, as well as a break-out mechanism, should either motor or gearbox jam. The torque transducer, mounted in line with the steering column, allows assistance to be provided above 1 Nm input by the driver. The motor requires approximately 30 A at full torque.

8.11 Traction and braking systems

A fundamental trade-off between longitudinal tractive force (or decelerating force in the braking situation) and lateral cornering force, developed at the tyre/ground contact patch is central to the balance between traction and stability, Fig. 8.19, of the wheeled vehicle. Road camber or lateral wind forces on the vehicle will rob the tyre frictional force available for traction (or braking) so a knowledge of the magnitude of the interacting forces is necessary in performance prediction calculations.

8.11.1 TRACTION VS STABILITY

Simplification of the vehicle to a rigid rectangular frame, carrying a wheel at each corner which cannot be steered, helps to visualize the force system involved. Lateral forces developed can be considered proportional to slip angle for the simulation and an axis system can be drawn in the ground plane as shown at (a). An analysis due to Rocard[11], dating from the 1950s, used this simplification to obtain an expression for the stability of the vehicle from which the oscillatory motion of the idealized vehicle in swerving from its straight-ahead path could be examined. These are described by Steeds[4] who rearranges them to define a critical speed for the vehicle at which the vehicle changes from a stable to an unstable state:

$$=\{[2K_1K_3(a+b)^2]/[M(K_1a - K_3b)]\}^{1/2}$$

where $K_{1,3}$ are cornering force coefficients for the different wheel stations and M is the vehicle mass acting through CG.

From this it can be seen that if K_3b is greater than K_1a the motion is stable at all speeds. Thus the choice of steering force characteristics for front and rear tyres in relation to the vehicle centre of gravity is fundamental to stability problems. As tractive forces (or braking ones) upset the balance of available side forces, these must be considered in the design.

When the wheels are steered there is also a difference between the lateral force developed in relation to the longitudinal axis of the vehicle (path) and that with respect to the plane of the tyre (heading), (b). The force along Y' is the lateral force comprising cornering force due to slip angle plus camber force due to camber angle. The force perpendicular to vehicle motion (path) is now referred to as the central force, acting with tractive force, and two axis systems are used to describe separately the wheel and the vehicle. The tractive force determines whether the vehicle rounds a corner at constant speed (steady-state handling) and is effective in developing tangential acceleration. The mechanism of friction development between tyre and ground is examined by Dixon[5] who has noted the dependence on sliding speed and temperature and the difficulty in quantifying the effects of temperature change. The speed dependence of friction is explained by there being one 'static' coefficient applicable to the non-sliding part of the tyre footprint and a lower 'dynamic' one applying to the sliding part. At high slip angles the sliding part will have reduced coefficient so that the cornering force peaks at moderate slip angle and then tails off – but its decline will be less dramatic than that of locked wheel braking as the relative sliding speed is lower.

A criterion of stability, handed on in earlier times by the aircraft industry, is the 'stability margin' between the centre of gravity of the vehicle and the 'neutral steer line'. This is such that, at any point along the line, lateral force can be applied without the tendency of the vehicle to rotate about a vertical axis – stability being achieved when the CG is ahead of the neutral steer line. There is also the concept of aerodynamic stability with lift and pitching moments transferring load from front to rear tyres. A simplified consideration of aerodynamic stability assumes the

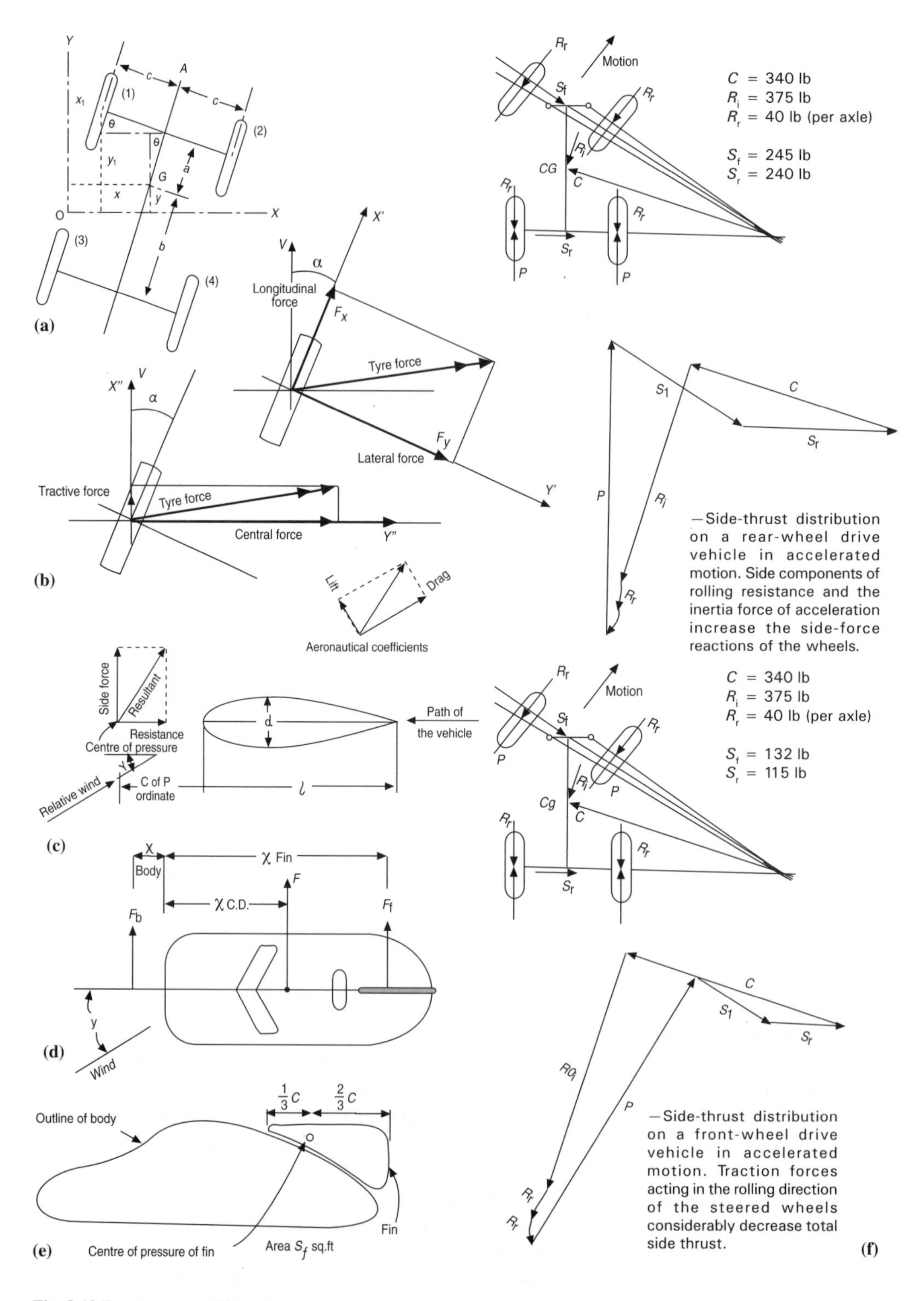

Fig. 8.19 Traction vs stability; (a) ground-plane axis system; (b) lateral and central forces; (c) resolved aerodynamic forces; (d) forces on streamline body; (e) effect of rear fins; (f) graphical method.

centre of pressure at the same height as the centre of gravity and side wind force produces only yawing moment. Longitudinal forces, side thrust and yawing moment may be combined vertically to give a single aerodynamic force acting horizontally through the centre of pressure as at (c). A side-thrust coefficient can thus be defined, $C_L \cos\gamma + C_D \sin\gamma$, where γ is the angle of attack. The view at (d) shows the results of measurements on a streamlined body with fineness ratio $l/d = 6$ such that the yawing couple = $(1/2)rV^2AlC_Y$.

For such a long streamlined body the centre of pressure lies well ahead of the nose and a small side force can produce considerable disturbance in direction. The dotted lines show that for a rectangular sectioned squat body the side force coefficient is considerably higher. C_S values limited to 1.0 apply to ground vehicles – involving considerable side force. So for ultra-light streamlined vehicles it is important to restrain the centre of pressure in a rearward position. This is achieved by rear fins, (e); that shown having its centre of pressure about one-third of its length back from the leading edge. The point at which the resultant side force takes effect can be found by taking moments. The crosswind response of vehicles is examined in detail by Howell[12].

Stability on a curve under vehicle inertia forces is also considered by Taborek[10] who suggests that traction and braking forces acting on the steered wheels should be balanced by other motion-affecting forces ideally acting through the vehicle centre of gravity. All these produce side force components which add to or subtract from the centrifugal force on the vehicle. Distribution of side forces to the wheels is the key factor in directional stability – safe manoeuvrability being determined by the wheel which first starts to slide. He commends vectorial analysis using graphical techniques to determine the distribution. In the graphical construction at (f), representing accelerated motion for a typical vehicle through a curve, example forces are centrifugal C = 340 lbf, total rolling resistance $R_r = W_f$ = 80 lb (356 N) and an acceleration a = 3 ft/s^2 (0.9 m/s^2). Thus the inertia force $R_i = (W/g)a$ = *375* lb (1.67 kN). Vectors are drawn in the figure for rear-wheel drive and front-wheel steering and the same with front-wheel drive. Compared with the case of the freewheeling vehicle without traction, the external forces all produce side components that increase the side force reactions on the rear-drive vehicle. On the front-drive vehicle, however, traction forces considerably decrease side thrust.

8.11.2 ELECTRONICALLY CONTROLLED BRAKING

The presence of electronic systems on EVs for drive motor control allows the possibility for integrating further control electronics, for electronic brake actuation and braking-by-wire, Fig. 8.20. Lucas Automotive have determined that the typical car driver is usually inhibited against applying optimum pressure to effect an emergency stop. The company are therefore advocating the use of an Electronic Actuation System (EAS) based on the electronic booster that has been replacing the fast-vacuum booster, fitted to many cars, and which was limited to providing assistance only at a fixed ratio and often reacted too slowly in emergency. A proportional control valve has been developed such that the output stroke is proportional to the input control current signalled through the brake pedal. A control loop is partially closed via an ECU which measures output/input force ratio and compares it with a vehicle-specific algorithm. An initial specification involved electronic booster control superimposed on a mechanical booster but a parallel development of the Long-Stroke (LS) booster meant that the benefits of electronic control could be applied to heavier cars and replace tandem booster arrangements which tend to be restricted to either high performance cars or heavier van derivatives. The company's LSC 115 T tandem power unit has an integrated electronically proportioning solenoid control valve; this valve is an electrical analogue of the servo control valve, but with input/output ratios compared electronically rather than by rubber reaction disk.

The company believe the full potential of such systems can only be realized when legal insistence on mechanical back-up is dropped and 100% reliability of the electronics can be guaranteed. If integrated with ABS and EMS, the ECU would be able to compare actual vehicle deceleration with the theoretical value based on a given booster output force, and thus forewarn of brake fade. The ECU could also relate engine torque to achieved acceleration for assessing loading on the vehicle so that the system is informed of prevailing road gradient. By integrating ABS, the booster could also provide the energy source and the combined system ECUs could help to create the next stage of automatic braking, in controlled traffic conditions, involving intelligent cruise control. Functions such as traction control, hill-hold and variable pressure boost for ABS could be incorporated.

The first stage of the project was seen on some Mercedes cars during 1996 as 'brake assistant'. For this application the vacuum chamber of a conventional vacuum booster is equipped with the Mercedes position sensor. This tells the ECU that the booster piston has travelled a given distance in less than a predefined time and switches on the solenoid valve. Atmospheric air then enters the booster working chamber to amplify driver effort with maximum available servo power. Without this device full decelerative advantage with ABS is lost. In an active stability control feature, the ECU helps to control side slipping out of a corner by selectively applying braking force to one of the outside wheels, generating braking force from the booster and modulating it by the ABS. In stage two, hill-hold will be added such that brake pressure will automatically be applied when a vehicle stops on a hill and automatically released when it pulls away. This will involve the proportional solenoid valve and the booster generating the necessary braking pressure by metering the air intake into its working chamber.

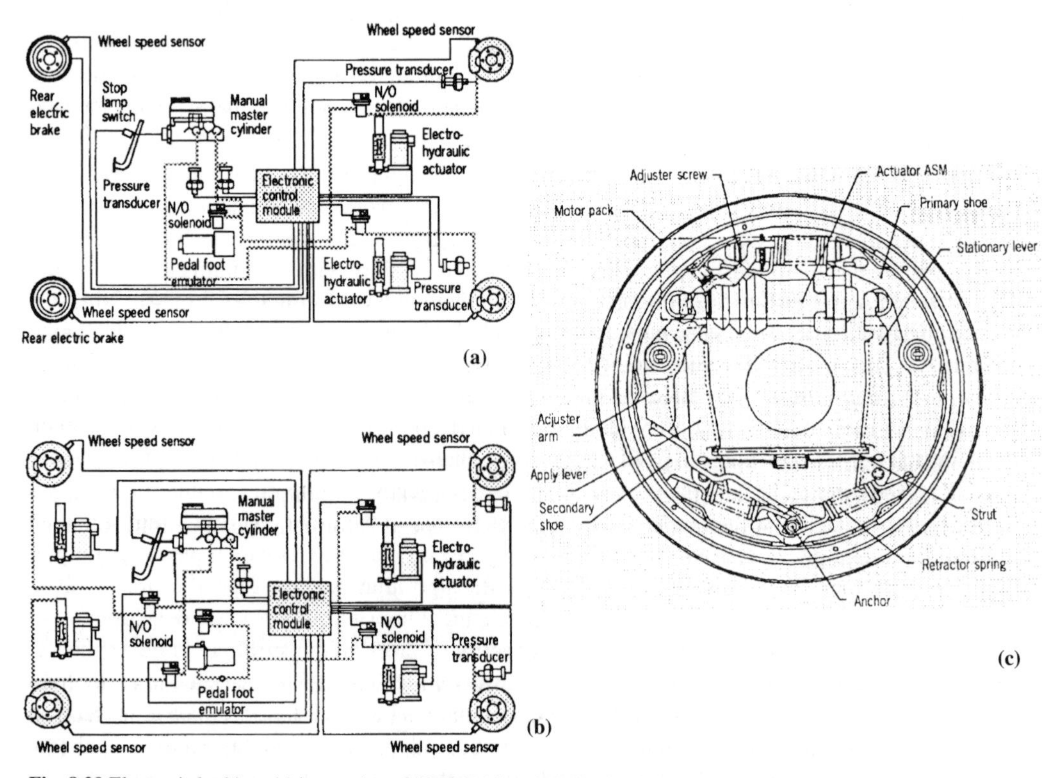

Fig. 8.20 Electronic braking: (a) interaction of ABS, EAS and EMS; (b) electrohydraulic apply and rear braking systems; (c) Electric wheel brake.

Such metering is also essential for automatic braking and Adaptive Cruise Control (ACC). Current cruise control, it is argued, can suffer a lack of sufficient engine deceleration on steep gradients such that the vehicle gains speed; EAS can supply precisely metered braking pressure to prevent this. Beyond this, EAS can generate pressure for ACC so that a vehicle is kept at a safe distance from the one ahead, without having to rely on engine braking which might prove inadequate. For stage three the problem of varying brake pressure requirement with degree of vehicle payload is tackled. Here, not only is brake booster pressure metered but also the pressure level boosted proportionally to the pedal force (measured by piston rod sensor), so that constant deceleration is obtained with constant pedal force irrespective of payload.

Brake-by-wire systems have now been developed which provide for not only a wired connection between pedal and brake actuator but also for electromagnetic actuation of the brake itself. The Delphi Chassis Galileo system allows dynamic brake effort apportioning, tunable pedal feel and variable boost ratio, without the need for hydraulic/vacuum assistance, as well as integration possibilities with stability enhancement systems. Potential exists for closed loop control of braking for front-to-rear and side-to-side braking balance; active independent wheel brake control allows ABS and ASR to interface with collision avoidance system, (a).

'Electrohydraulic apply' is one variant of the system, shown at (b), in which closure of the stop lamp switch, as braking is applied, causes the ECU to close normally open solenoids. Upstream of these brake fluid pressure caused by the application is sensed and transducers emit a proportional signal to actuate piston-displacement motors on the front brakes. Pressure is continuously modulated by the size of this signal and the rear brakes applied proportionately to provide a balance dictated by a complex slip control algorithm. The 'feel' on the brake pedal is achieved by displacing fluid from the master cylinder into an emulator device which provides a predefined force/displacement characteristic tunable by the brake algorithm. If a fault is detected during powered (motor-actuated) braking, the solenoids switch to the open position so that unassisted direct braking is available. ABS and ASR is achieved by overriding the driver input with a control command based on wheel speed.

The rear electric brake system (c) is intended for small weight-critical cars where removal of the park-brake cable assembly can save between 3 and 6 kg and works in conjunction with the brake system just described or a conventional hydraulic brake. The rear brake is a high-gain electromagnetic actuator designed to maximize torque capability with minimum electrical input. Closed-loop control ensures stability of the gain so that at an operating torque of 400 Nm, say, each wheel brake can respond at rates up to 4000 Nm/sec, to continually adjust dynamic brake output. The wheel brake incorporates a PM DC motor, gear train and ball-screw/nut mechanism, actuating the brake friction surfaces through a lever system. A 'backdrive' spring incorporates a park-brake latch mechanism. This is a bi-stable clutch device which locks the main motor shaft on demand. The park-brake holds until the switch deactivates, without electrical input. Closed loop control generally reduces the effect of component and operating condition variability on braking performance, algorithms based on differential wheel speed information compensate for load distribution variation.

8.11.3 ELECTRONICALLY CONTROLLED CONTINUOUSLY VARIABLE TRANSMISSION

While in series hybrid-drives batteries and IC engine typically power motor/generators, to drive the road wheels, in parallel configured, hybrid-electric/IC-engine drive vehicles there is mechanical drive between IC engine and road wheels, usually via continuously variable transmission, Fig. 8.21. A widely accepted CVT is the variable-pitch pulley and belt-drive type originating in the Van Doorne design; this transmission heralded the steel drive belt having separate tension and thrust members which considerably increased torque capacity. Axial force applied to the pulley sheaves tensioned the belt; then, the combination of lateral and radial force on the blocks was sufficient to transmit drive from pulley to belt. Maximum speed ratio was 6:1 – limited by the size of the pulleys involved – and input torque capacity 90 lb ft (122 Nm). To give a ratio spread of either 12:1 or 16:1 in the car, a fully automatic range-change was incorporated. Transmission efficiency of about 90% was claimed and gear ratios varied from 2.31 to 0.58:1. Unit weight complete with clutch was 60 kg, including variator mechanism.

VDT have developed 24 and 30 mm width-type belts. The 24 mm type is used in lower and medium torque applications, the 30 mm type in higher torque applications. In higher torque conditions, prestressing forces will be higher, and belt running radii shall be larger than in low torque applications. When the running radius of the belt increases, belt speed will increase as well. From this it may be clear that a 30 mm belt element, used for high torque application, cannot be seen as a scaled-up version of a 24 mm one. Its high belt speed capabilities must be better than those of 24 mm elements, in a comparable application (=engine speed). Transmission design affects mainly belt length, centre distance and element width. Because element width does no have to vary over a large range of specifications, the same transmission type can be used over a large range of engine type/vehicle combinations.

A CVT design concept (P884) extending applications into the 1.9–3.3 litres range has also been developed by the company, as illustrated. The transmission offers a choice of driving style – either comfort or sporting mode. In sporting mode there is lock up at low engine speed. Up to 17% better fuel economy is claimed, compared with an electronically controlled four step automatic. A hydraulic wet-plate clutch is used, with a torque converter for lock-up at start-off. The transmission uses a reduced number of belts by using wider, 30 mm rings. This leads to less internal friction and reduced production costs.

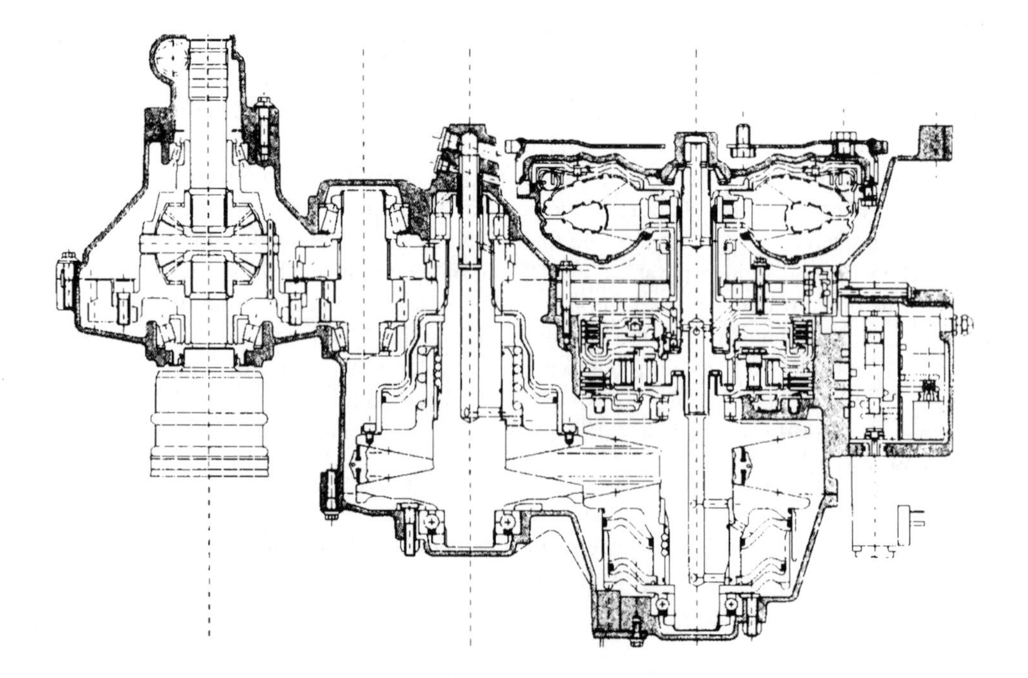

Fig. 8.21 Continuously variable transmission; the VDT P884 CVT.

8.12 Lightweight shafting, CV jointing and road wheels

Between the CVT and the tyre ground contact patch, there is considerable scope for weight saving and the reduction of rotational inertia in the drive-line. A development programme by GKN has led to substantial weight reductions in most elements of the drive-line for front-, rear- and all-wheel drive systems, Fig. 8.22. Reduction in weight has also had a knock-on effect in obtaining increased whirling speeds and therefore the ability to have prop-shaft UJs without the risk of vibration problems.

With the group's latest generation of half-shafts for both front- and rear-end application, there are 11 standardized sizes available. Lower weight is accompanied by higher resistance to strain, reduced vibration and easier assembly/repairability. In the case of prop-shafts, improved acoustics have been achieved by decoupling of torsional and axial vibration as well as less run-out resulting in reduced wear. Transmission efficiency of universal joints has also been increased and systems developed for providing controlled collapse of the shaft under front-end crash loading of the vehicle. A choice of steel, aluminium alloy and high strength composite systems allows tailoring shafts, and their couplings, for particular operational applications.

Steel shafts have been lightened by the use of high tensile materials along with the appropriate design modifications and revised joining methods. Aluminium alloy shafts can be up to 50% lighter than conventional steel shafts provided appropriate design modifications are made. Metal matrix/composite solutions, involving aluminium with ceramic inclusions, are also possible. Metal/composite combination shafts have also been successfully exploited on applications where three-shaft assemblies have been replaced by two-shaft systems. After a first generation of composite shafts has proven the effectiveness of resin systems reinforced by glass and carbon fibre in achieving up to 70% weight reductions over traditional shaft assemblies, the group has now announced a second generation. This involves end fittings, as well as the shafts, in polymer construction and has resulted in overall weight savings of up to 75%. In this second generation, one approach has been to replace the Hookes-type universal joint with a composite disk UJ where only small angularity is required. In such cases a typical steel shaft system weighing 10 kg can be replaced by a first-generation assembly weighing 5 kg and a second-generation one of only 2.7 kg, (a). Other advantages of second-generation shafts include increased torque capacity, (b), 40% increase in static break value and a new resin system which provides 15% increase in fracture value at temperatures of 120° C. Techniques of introducing damping into the fibre compound help acoustic performance.

In terms of assistance given by the prop-shaft to passenger protection in frontal impact, a key factor is the possible removal of the centre UJ/bearing where the shaft would potentially bend out of line on impact. By the shaft contributing to impact reaction, more protection can be provided to the footwell of the vehicle. By use of a drop-weight rig the group is able to produce crash-optimized cardan shafts in a variety of material combinations and geometrical configurations, (c). In the case of composite shafts, radially aligned reinforcing fibres can be used at the ends so that the end pieces are pushed into the main tube on initial impact such that they lie up against the joint shoulders. Thereafter the tube is split open on one or both sides with energy being reduced both by splitting and friction. With aluminium-alloy shafts, an axially weak point in the centre of the tube is the focal point; by using two different tube diameters either side of it, and carefully designing the transition area different impact characteristics can be obtained, by the smaller tube sliding inside the larger one. Special press fit connections of the end pieces can also be provided. The graphs in (d) and (e) show the critical shaft length and minimum required diameter plotted for different shaft materials, and fibre reinforcements, at a critical rotating speed of 7200 rpm. Fibre-reinforced types have strengths related to fibre orientation as shown in the second figure. As the shear strength

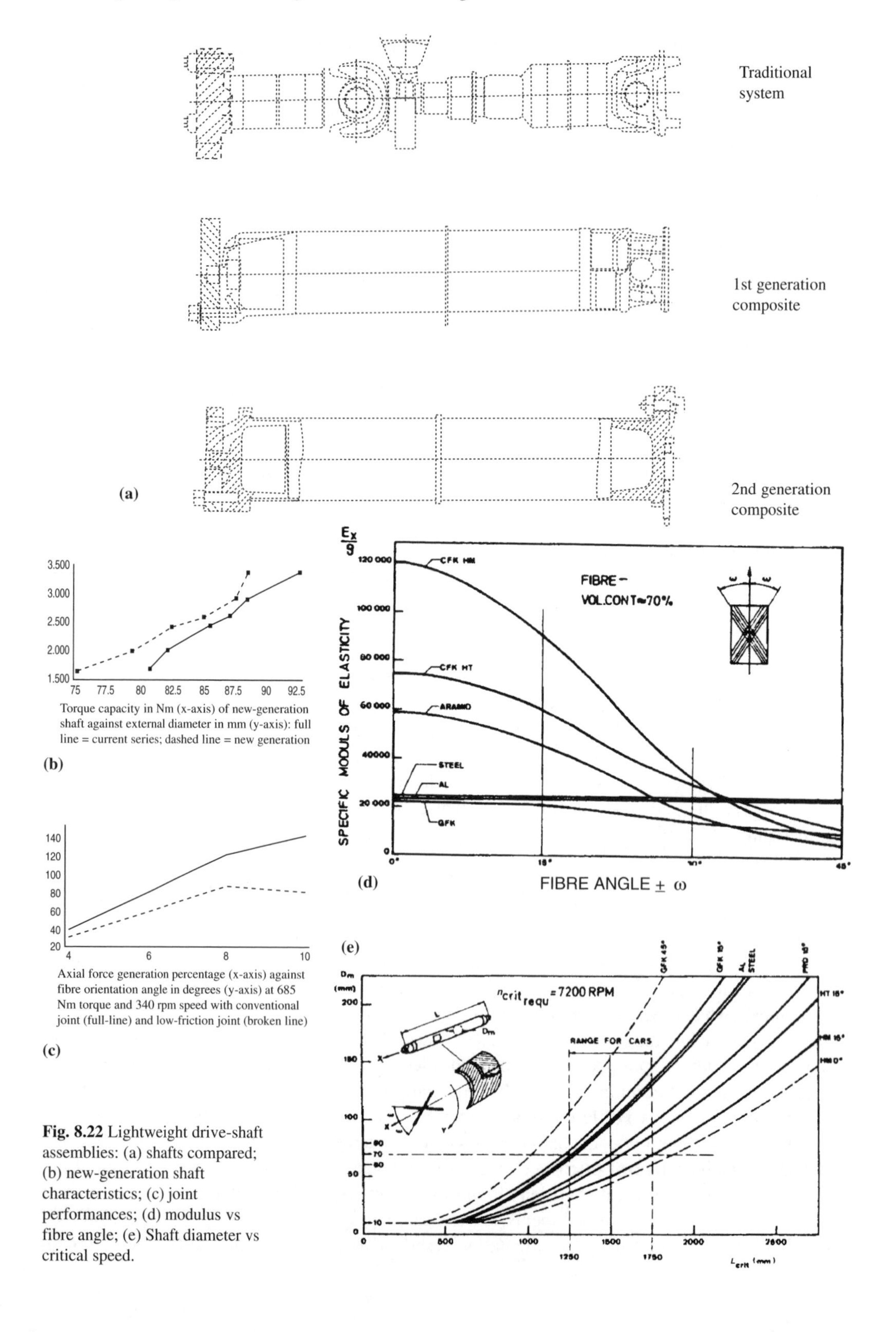

Fig. 8.22 Lightweight drive-shaft assemblies: (a) shafts compared; (b) new-generation shaft characteristics; (c) joint performances; (d) modulus vs fibre angle; (e) Shaft diameter vs critical speed.

of a shaft, with 0° fibre orientation, is comparatively low; (e) is based on a realistic orientation of 15°. Usually multi-layer composites are involved and the elastic behaviour of any defined combination of layers is generally derived from the properties of one unidirectional layer, defined by the modulus parallel and perpendicular to the fibres, one of the two Poisson's ratios and the in-plane shear modulus.

The company also recently released a series of lightweight shafts incorporating some of the technologies just discussed as well as considerably redeveloped universal joints. These assemblies involve constant velocity joints of the AC (ball type) and GI (tripode type). All shafts involve a boron steel with deep-case hardening. Within the downsized AC joint, point loadings of up to 3 GPa are involved for which a new grease has been developed to ensure the required performance. Tribological engineering was used in the analysis and has shown areas of subsurface 'contact' stresses beyond conventional Hertzian stress theory, surface, finish effects.

8.12.1 STRESS DISTRIBUTION IN ROAD WHEELS

While considerable weight savings have been achieved by constructing road wheels in aluminium and magnesium alloys there is also scope for lightening conventionally fabricated wheels by understanding the magnitude and direction of their internal loading. For those contemplating the design of special road wheels, Fig. 8.23(a), it is useful to consider the forces on an elemental length of rim (b) when trying to determine rim stresses. The stress derived from standard solid mechanics theory is given by

$$6\,P.l/h_1^2 + (P/h_1)$$

Work due to Svenson[13], (b), considered the imposition of stresses from both horizontal and vertical forces at the tyre/ground contact patch, leading to an expression for dynamic factoring of the statically derived stress values as

$$1 + 2.6C/P$$

where C is the spring rate of the tyre.

The complex shape of the wheel disc of the conventional pressed variety makes it difficult to calculate the stress from simple theory. Empirical work carried out at MIRA has shown that bending moments imposed on wheel discs are increased not only by lateral weight transfer but also by fore-and-aft weight transfer, body roll and tyre distortion, as illustrated at (c). Analysis shows torque on a stub axle to be:

$$M_A = -V_s e_s \quad \text{static and} \quad M_A = V_c e_c$$

in cornering while torque on wheel is

$$M_W = V_s cos\gamma_1 d + V_s sin\gamma_1 r_{s1} \quad \text{static and} \quad M_W = V_c \cos \gamma_2 d + V_c \sin \gamma_2 r_{c1} + H r_c$$

in cornering.

8.13 Rolling resistance

In a conventional passenger car travelling at 50 mph tyre rolling resistance typically absorbs 56% of the total power requirement at the driven wheels. In free rolling on a hard smooth surface energy losses are primarily due to tyre distortion, principally the bending of the tread as it rolls through the contact zone. Only about 5% of the energy loss is attributable to windage and surface friction. Resulting from the distortion of the tyre normal pressures between tyre and road are greater in the leading half of the contact zone than the trailing half, causing the virtual pressure centre to move forward in the direction of rolling, creating a torque about the wheel axis. In free rolling this torque is balanced by frictional forces in the contact zone, which are those which develop rolling drag, specific rolling resistance being the drag divided by the vertical load between tyre and road.

The tyre rolling resistance is effectively independent of size for tyres of the same type of construction but temperature, speed, deflection, inflation pressure and tyre materials/construction

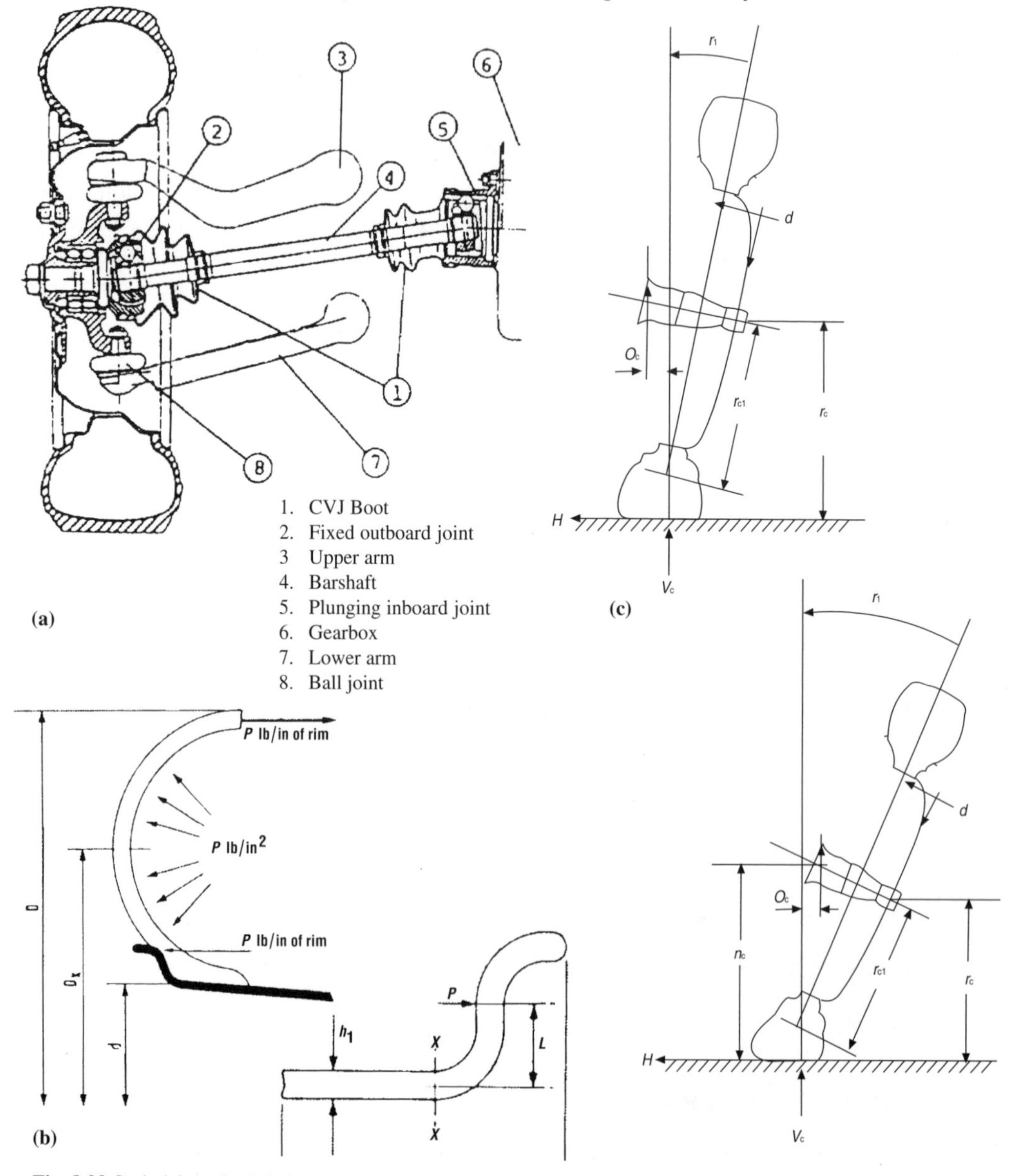

Fig. 8.23 Optimizing wheel design: (a) key elements to an independently suspended driven wheel; (b) element of rim section and tyre dimensions; (c) wheel disc analysis (static, top; cornering, bottom).

all have substantial effects, as do driving modes such as cornering and traction, Fig. 8.24. Temperature, and inflation pressure, effect is shown at (a), values of rolling resistance coefficient being given temperatures and speeds for constant loads and inflation pressures. Roll drag decreases with increase in temperature because hysteresis losses in the rubber compounds fall. The speed effect, seen at (b), is such that resistance increases slowly, with the square of speed, up to around 89 mph above which deformation of the casing in the form of a standing wave causes a marked increase, unless the tread is substantially braced as in the radial-ply tyre for which the standing-wave formation is shifted up the speed range by 40 mph. Pressure effect, see, at (c), is such that-despite the tyre being inflated to a certain pressure at ambient temperature, it suffers an increase in pressure due to temperature rise with speed, which in turn increases deflection and hence hysteresis loss, so drag too increases. An increase in inflation pressure of some 25% can reduce drag by 10%, however.

During traction and cornering, (d) and (e), rolling resistance increases with torque transmission due to induced slip, causing drag to rise by three times its free-rolling value. In cornering, as the tyre runs at a slip angle, its rolling drag is increased by the component of the side force given by the slip angle, at a rapidly rising rate. Construction effects centre around the increases in rolling resistance due to high-hysteresis tread compounds, thicker treads and a greater number of casing plies, the latter involving inter-ply shear effects. Radial-ply construction is now virtually the norm and involves a single radially disposed steel cord casing ply, or two textile ones, and three or more steel-cord bracing plies triangulated under the tread to form the reinforcing band.

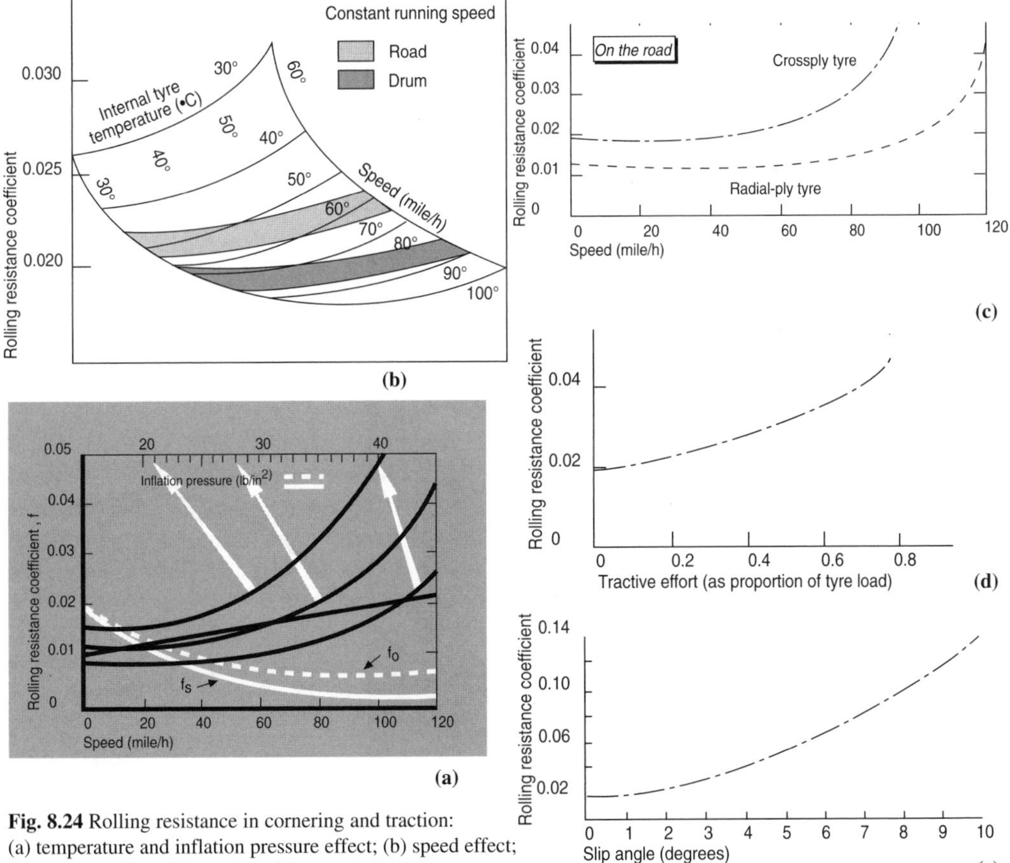

Fig. 8.24 Rolling resistance in cornering and traction: (a) temperature and inflation pressure effect; (b) speed effect; (c) pressure effect; (d) traction; (e) cornering.

Some researchers suggest that rolling resistance data for very small tyres, as may be used in lightweight urban electric cars, is not readily available from manufacturers. Work done by Margetts[14] using a towed vehicle fitted with test tyres, and load-cell towbar, has shown results as at (a) in Fig. 8.25 for such tyres. The results show a different relationship to that obtained by the classic formula due to Hoerner (dotted curve) for conventional-sized tyres.

Michelin's lower rolling-resistance 'Energy' tyres are said to compensate for the loss of wet-grip in moving to lower hysteresis tread compounds, for reducing drag, by new rubber formulations, carcass construction and tread patterns. By obtaining low hysteresis values at low frequency vibration, and the converse at high frequencies, the company have effectively reduced rolling resistance by 20% over conventional radial-ply construction, (b). Recently the company extended this technology to launch tyres particularly designed for electric-vehicle application, to obtain 35% rolling-resistance reduction against conventional tyres. The Proxima brand is designed for urban traffic conditions and claims unusually good grip on polished surfaces often encountered in city centres using a tread pattern with a high number of block edges. Fine cuts made in the centre of the tread blocks also help to achieve silent running, coupled with the plugging of tread drainage channels at their outer edges.

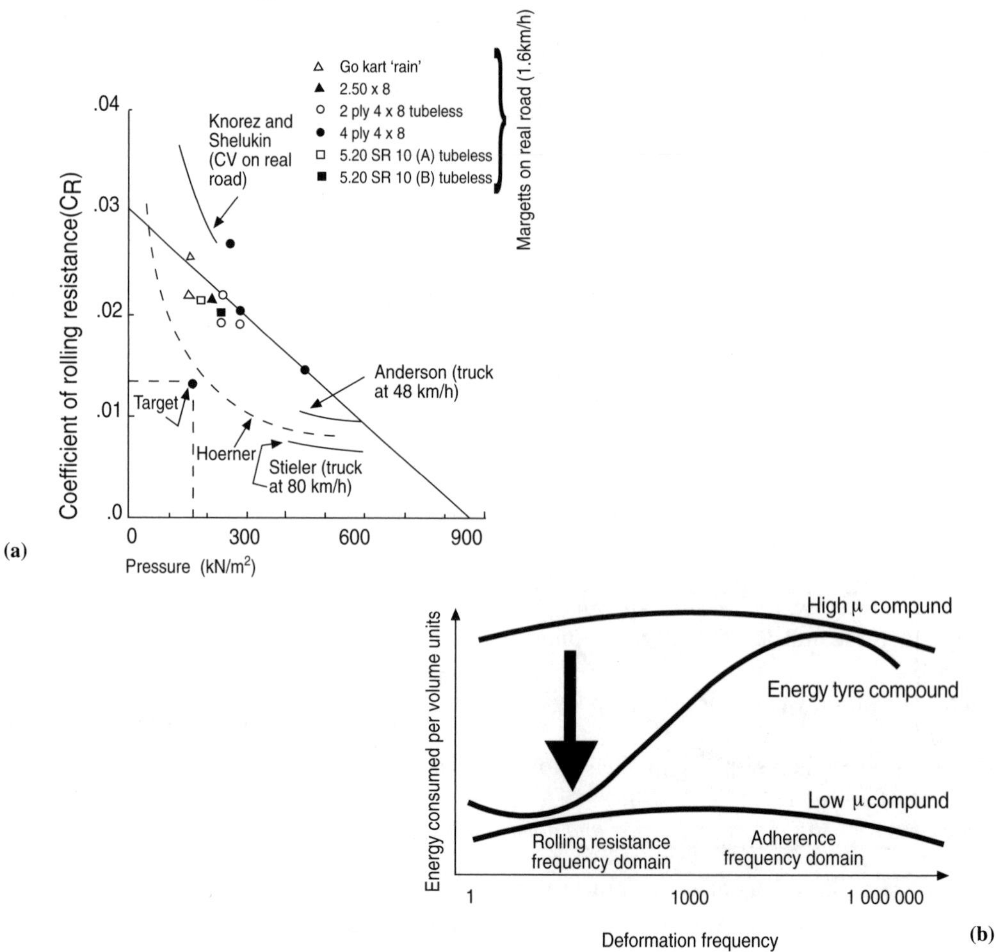

Fig. 8.25 Rolling resistance for small-wheeled and specialist tyres: (a) small tyres; (b) Michelin Energy tyres.

References

*Fenton, J., *Handbook of automotive body and systems design*, Professional Engineering Publishing, 1999

**Fenton, J., *Handbook of automotive powertrain and chassis design*, Professional Engineering Publishing, 1999

1. Beerman, H., *The analysis of commercial vehicle structures*, Mechanical Engineering Publications, 1989
2. Peery, D., *Aircraft structures,* New York, McGraw-Hill, 1950, 566 pp.
3. Tidbury, G., Stress analysis of vehicle structures, ASAE Note 1, College of Aeronautics, Cranfield, 1965
4. Fenton, J., Shell beams and the Argyris method, *Automotive Design Engineering*, March 1963
5. Pawlowski, J., *Vehicle Body engineering*, Business Books, 1969
6. Analytical method for chassisless vehicle design, *Automobile Engineer*, 1953, vol. 43, pp. 103–111; Structural design Part 2: The front end structure, *Automobile Engineer*, 1953, vol. 43, pp. 152–157.
7. Tae-Un *et al.*, Structural design of aluminium space frame with battery tray, SAE paper 960524
8. Lilley and Mani, Roof crush strength improvement using rigid polyurethane foam, SAE paper 960435
9. Taborek, Mechanics of vehicles, *Machine Design, Penton*, May to December 1957
10. Autotech paper C498/35/052, 1995
11. Rocard, Y., *L'instabilite en Mechanique*, Masson et Cie, Paris, 1954
12. Howell, Crosswind response of cars, Proc. SITEV, 1988
13. Svenson, The working stresses of wheel rims, *ATZ*, No. 8, 1967
14. Margetts, E., Influence of rolling resistance on very light cars, *Automotive Design Engineering*, April/May 1981

Further reading

Fenton, J., *Handbook of automotive body construction and design analysis*, PEP, 1998

Donald, I. (ed.), *Thin-walled structures*, Elsevier, 1990

Timoshenko and MacCulloch, *Elements of strength of materials*, Van Nostrand

Ali *et al.*, The application of finite element techniques to the analysis of an automobile structure, Proc IMechE, 1970–71

Fagan, M., *Finite element analysis*, Longman, 1992

Argyris and Kelsey, *Energy theorems and structural analysis*, Butterworth

Amos, R., Structural design analysis of commercial vehicle cabs, IMechE conference paper C134, 1984

Curle and Watson, Structural analysis, body design and development, IBCAM Boditek conference, 1985

Guignard, J., Human response to intense low-frequency noise and vibration, *Proc. I. Mech. E.*, Vol. 182, Part 1, No. 3, 1967/8

Gibbs and Richards, *Stress, vibration and noise analysis in vehicles*, Applied Science Publishers, 1975

Brughmans *et al.*, The application of FEM-EMA correlation and validation techniques on a body-in-white, paper 3 in IMechE Vehicle NVH and refinement conference report C487, 1994

Burton and Southall, Noise, vibration and harshness in the automotive industry, paper 3, session 3R, Design Engineering Show conference, 1982

Murakami, Y., *The rainflow method in fatigue*, Butterworth Heinemann, 1992
Jones and Williams, The fatigue properties of spot welded, adhesive bonded and weld-bonded joints in high-strength steels, SAE paper 860583, 1986
McRobert and Watson, Design optimization of forgings, Drop Forgings Congress, Cologne, 1983
Steeds, *Mechanics of road vehicles*, Iliffe, London, 1960
Dixon, *Tyres, Suspension and Handling*, Cambridge University Press, 1991
Short, M., Possibilities of aircraft structures in ground vehicles, SAE Transactions, 1945
Dr Hewitt, Paper C49X/3 1/240, Autotech, 1995
Vaschetto *et al.*, A significant weight saving application of magnesium to car body design, paper SIA9506B02, EAEC Congress, 1995
Basu, A., SAE Paper 860283, 1986
Doring, E., SAE Paper 860275, 1986

Notation

T	Torque
G	Shear modulus
a	Half section width
b	Half section depth
t	Wall thickness
τ	Shear stress
τ'	Complementry shear stress
y	Vertical distance
x	Horizontal distance
r	Radius
J	Polar moment of inertia
a''	Angle of twist
l	Length
b'	Breadth
d'	Depth
x,y,z	Distances along rectangular coordinates
$[A]$	Enclosed area
s	Arc length
$(\Sigma ds/t)$	Sum of elemental arc lengths per individual arc thickness
S	Shear force
q	Shear flow
h	Section height
M	Bending moment
P	Normal force
A_f	Boom section area
A_s	Stringer section area
dx,dz,dz	Elemental lengths
σ_v	Direct stress on vertical plate
[R]	Determinate force matrix
[X]	Redundant force matrix
[S]	Resultant force matrix
b_1	Member force due to virtual unit load
b_o	Member force due to original structure loading
f	Member flexibility
E	Elastic modulus (face$_f$ core$_c$)
G_c	Shear modulus of core
t_f,t_c	Thickness of face, core
K	Plate buckling stress coefficient
C	Honeycomb core coefficient
p	Uniformly distributed lateral pressure
w	Uniformly distributed load
σ_b	Buckling stress
Subs. m	Mean value
$X_{1\text{-}n}$	Plate forces in x-direction
$Z_{1\text{-}n}$	Plate forces in z-direction
$K_{1\text{-}n}$	Edge forces on plate
S,B	Bulkhead widths
$F_{x,y,z}$	Force vectors
I	Section second moment of area
R	Radius of curvature
θ	Slope of beam
v	Normal deflection of beam
Δ	Generalized deflection
F	Beam shear force
v	Poisson's ratio
N	Number of cycles
[r]	Nodal displacements matrix
[K]	Stiffnesses matrix
P	Tractive force
W	Axle load
μ	Tyre/ground adhesion coefficient
H	Centre of gravity (CG) height above ground
T_e	Motor torque
T_i	Inertia torque at drive axle
L	Wheelbase
f	Rolling resistance coefficient
$L_{r/f}$	CG to rear/front axle distance (a/b in pitch calculation)
w	Weight distribution factor
I	Moment of inertia
$I_{w/e}$	...at wheels /engine
R	Resistance
$R_{r/a}$	Rolling/air resistance
r	Radius
a	Angular acceleration
α	Translatory acceleration
M_e	Equivalent inertia mass
M_i	Effective inertia mass
Y	Rotary mass factor
E	Efficiency
E_s	Slip efficiency
C_d	Drag coefficient
A	Projected area
V	Velocity
g	Gravity acceleration
ρ	Air density
θ	Gradient slope
G	Overall gear ratio
N	Motor rotational speed
K	Pitch rate
f_n	Natural frequency
k	Spring stiffness
m	Suspended mass
v	Damping ratio
c	Damping coefficient
Z_d	Bump height
S	Roadholding force
σ	Roadholding/static force ratio

Handling nomenclature

AT	Aligning torque (tending to bring road wheel to its normal heading)
RL	Radial load (normal load at tyre/ground interface)
$F_{f/r}$	Front/rear drift forces (forces induced by tyre slip angle)
C	Camber angle (lateral tilt of wheel plane)
F_c	Camber thrust (induced lateral force)
ID	Induced drag (induced fore/aft force)
C	Caster angle (fore/aft tilt of steer axis)
CT	Caster trail (fore/aft displacement of contact patch)
k	Radius of gyration (offset of equivalent rotating inertial mass)
ε	Roll angle (lateral tilt of suspended mass)
CF	Cornering force (resultant lateral force induced by tyre slip angles)
C_L	Side force coefficient (as ratio of radial load)
C_Y	Yaw coefficient (yawing couple as ratio of radial load)
γ	Angle of attack (between vehicle path and heading)

Index